Roof Builder's Handbook

Roof Builder's Handbook

William C. McElroy

P T R Prentice Hall, *Englewood Cliffs, New Jersey 07632*

Library of Congress Cataloging-in-Publication Data

McElroy, William
    Roof builder's handbook / William C. McElroy, Jr.
      p. cm.
    Includes bibliographical references and index.
    ISBN 0–13–781816–5
    1. Roofs. 2. Roofing. I. Title.
TH2391.M315 1993
695—dc20

                                92–1090
                                  CIP

Editorial/production supervision: Mary P. Rottino
Cover design: Wanda Lubelska Designs
Prepress buyer: Mary E. McCartney
Manufacturing buyer: Susan Brunke
Acquisitions editor: Bernard Goodwin

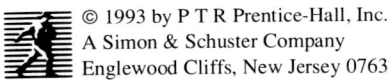

© 1993 by P T R Prentice-Hall, Inc.
A Simon & Schuster Company
Englewood Cliffs, New Jersey 07632

The publisher offers discounts on this book when ordered
in bulk quantities. For more information, write:

Special Sales/Professional Marketing
Prentice-Hall, Inc.
Professional Technical Reference Division
Englewood Cliffs, New Jersey 07632

Printed in the United States of America

10 9 8 7 6 5 4 3 2 1

ISBN 0-13-781816-5

Prentice-Hall International (UK) Limited, *London*
Prentice-Hall of Australia Pty. Limited, *Sydney*
Prentice-Hall Canada Inc., *Toronto*
Prentice-Hall Hispanoamericana, S.A., *Mexico*
Prentice-Hall of India Private Limited, *New Delhi*
Prentice-Hall of Japan, Inc., *Tokyo*
Simon & Schuster Asia Pte. Ltd., *Singapore*
Editora Prentice-Hall do Brasil, Ltda., *Rio de Janeiro*

Contents

6 NAILING INSTRUCTIONS   140

Introduction

I am an optimistic person. Optimistic in that I hope this manual will become the manual used by all persons interested in the field of roofing; the novice, the journeyman, the professional. Thus, I have written it with both the uninitiated and the professional in mind. There are sections on the very basics, which you may already understand. Skip these if you want, but I have looked at and studied hundreds upon hundreds of roofing jobs. Roofing jobs done by those who are in the know. Many, not all, of these professionals should review the basics, because some of you are making serious mistakes that are costly to the customer and contractor.

Do you want to be a roofer? It is a good trade to go into. You will find the work rewarding and possibly profitable. But, you must understand that roofing is one of those professions where a person not only has to learn his or her job skills but also those of others.

As a roofer you must understand the skills of a framer, carpenter, electrician, plumber, sheet metal worker, chemist, mason, architect, lawyer, and general manager. This is because people make mistakes and their mistakes can jeopardize your otherwise quality workmanship. Knowing their job will aid you in doing your job correctly. If you decide to own your roofing company, you also must understand the skills of a planner, designer, accountant, bookkeeper, drafter, and company owner.

You need to understand the construction techniques used in buildings. Poor construction will affect the quality of the finished roof. Also, you may be involved in building dormers, replacing rotted rafters and sheathing, and building roof overhangs over windows, doorways, and patios.

You should understand the safety aspects involved and you must understand how these devices can ruin your roofing job.

You will be roofing around plumbing vents and vent pipes, and you will be installing the flashing around vents. As a sheet metal worker you will be designing, cutting, fitting, and installing metal flashing in various locations around the roof.

You will be expected to specify the proper roofing materials for all sorts of mechanical and weather conditions. You must know how materials react chemically and physically with each other. Select the wrong materials and the job will fail.

You may be expected to repair chimneys and other brick work that is above the roof line and need to know about masonry. If you yourself are not to do the repair, then you must at least know when a repair is required.

You will be asked to suggest the best roof line and style for the building and neighborhood. The best roofing material for the job may be out of place architecturally for the job.

Are you a lawyer? No, but you may have to act as an attorney. You will want to draw up iron-clad contracts and make them stick. You may be dealing with a host of government agencies and their rules. You must protect yourself from those who are "suit" happy. One lost lawsuit and you can lose your livelihood.

As a manager, you must schedule your time, the arrival of materials and equipment. You must understand profit and loss. You must pay the bills, hire the employees, and do the hundred or more tasks that a business requires of you.

As a planner, you must plan your jobs completely. It is the little details that make a job profitable, or not.

You may not be the designer, but you may be asked to create or suggest a specific roof line for a building. Your roofing knowledge is a plus to an architect or designer and should be valued by them. They, though, are the responsible parties on the finished engineering and design. This is law in most states. You are not a drafter, but you must read and maybe even draw blueprints. You are not an engineer, but you must understand the engineering terms used in roofing and make proper selection of materials for the application.

Bookkeeping and accounting are the prime responsibilities of a business owner; you just cannot count on others to be as concerned about your livelihood as you should be. You may be just starting out as a helper, but someday you will want to be the boss—it happens to many tradespeople. Most fail due to lack of knowledge of the dollars and cents (sense) required.

A business manager does all the above, plus the following: Supervision of employees, interpreting labor laws, handling employee relations, doing the purchasing, scheduling, paying bills. I highly suggest that you study these subjects, not just the technical aspects of roofing.

Learning these skills may sound like a major task, but you can do it. Your starting point is the chapters to follow.

The reading is easy, up to date, and will show you how the professionals do the job. There are hundreds of pictures and drawings, dozens of reference charts, and a complete Glossary and Index.

So sit back and enjoy reading the *Roof Builder's Handbook*. The manual should become your roofing reference for years to come.

ACKNOWLEDGMENTS

The author would like to thank the following for their contributions and technical support in preparation of this manual.

CELOTEX CORPORATION
4010 Boy Scout Blvd.
Tampa, Florida 33607

CELOTEX CORPORATION
% PCI Communications Inc.
707 Franklin Street Mall
Tampa, Florida 33602

CRAFTSMAN BOOK COMPANY
6058 Corte del Cedro
Carlsbad, California 92008

DELEO CLAY TILE COMPANY
600 Chaney Street
Lake Elsinore, California 92330

DRILL-TEC DIVISION OF U.S. INTEC INC.
P.O. Box 2845
Port Arthur, Texas 77643

DURA TREND TILE CO.
2870 West Highland Ave.
Fontana, California 92336

GAF CORPORATION
1361 Alps Road
Wayne, New Jersey 07470

GERARD ROOFING TECHNOLOGIES, INC.
P.O. Box 9459
955 Columbia Street
Brea, California 92622-9459

GS ROOFING PRODUCTS COMPANY
5525 MacArthur Blvd. Suite 900
Irving, Texas 75038

HENRY COMPANY INC.
2911 Slauson Ave.
Huntington Park, California 90255

IP, INTERNATIONAL PERMALITE, INC.
300 North Haven Ave.
Ontario, California 91761

LUNDAY-THAGARD OIL CO.
Built Up Roofing Systems
9301 Garfield Ave.
P.O. Box 1519
Southgate, California 90280

MWELD/INC.
700 Highway #365
P.O. Drawer 1288
Nederland, Texas 77627

NEW PAECO INC.
One Executive Drive
P.O. Box 968
Toms River, New Jersey 08753

STANLEY BOSTITCH INC.
East Greenwich, Rhode Island 02818

U.S. INTEC, INC.
P.O. Box 2845
Port Authur, Texas 77643

WOLCOTTS FORMS INC.
15124 Downey Ave.
Paramount, California 90723

Special thanks to the following people who gave of their time and support:

Lawrence Jacobs, Craftsman Book Company.
Hannah T. McLaughlin, Pine Bush, New York.

1

Your Contracting Business

INTRODUCTION

You are considering a career in roofing and roof contracting. In which subfield of roofing are you interested? In this chapter we will explore some of these career fields and their requisite qualifications.

Here is a listing of the various roofing subjects we will discuss. I will expand on each later in this chapter.

- Residential roofing
- Commercial roof contracting
- Industrial roof contracting
- Rural farm roofing
- Preventive maintenance
- Repair maintenance
- Roof and truss design
- Architectural considerations
- Roofing sales with estimating

In addition, we will discuss the various roof coverings. Each type is a specialty of its own.

- Asphalt shingle
- Built-up roofing
- Concrete tile
- Clay tile
- Metal tile
- Metal sheet
- Plastic roofing panels
- Shake and wood shingles
- Slate tile
- Solar roofing

RESIDENTIAL ROOFING

Residential roofing is by far the easiest and least expensive roofing field to enter. The volume of available business is high. Every livable dwelling area in the world should be covered with a roof of some type. Many residences have additional out buildings used for storage and garages. Many have swimming pool enclosures, car ports, patio covers, and other assorted types of buildings, all of which require a roof of some sort. The customers for these roofs are building contractors, homeowners, landlords, and sometimes the government.

Start-up money is generally not a problem—you sell the job, obtain a deposit, and purchase the materials. At the completion of the job, you collect payment. The dollars involved are comparatively low, in the hundreds to a few thousand dollars. Turnaround is comparatively fast since most residential roofs can be completed within a few days to a few weeks.

Cost of tools and equipment required is low and ranges from under $100 for a hammer, a few hand tools and tape measure, to several thousand dollars for power tools. You need a vehicle, but you can use your car when first getting started.

Business setup can be as simple as word of mouth or as complex as obtaining full legal status. Full legal setup includes attaining a contractor's license, a salesman's license, a business license, a sales tax resale license, workman's compensation insurance, liability insurance, and compliance with IRS and OSHA safety rules.

I strongly recommend that as a beginner you do *not* invest money in obtaining legal status. Instead, install a roof or two for relatives, friends, or on your home. My reason for telling you this is simple. Roofing is not as easy as it looks. It is hard work and it can be dangerous. The roofing field requires knowledge of materials and installation procedures and very intense business management. Once you complete a job or two, you may find that it is not the career field in which you want to spend the rest of your life. Why waste hundreds or thousands of dollars setting up a legal business only to close it weeks later? (But be aware that if you do work for a relative or friend, workman's compensation and tax withholding laws may apply.)

If after a few jobs, you decide to pursue this career, then you must be fully prepared to invest several thousand dollars in your new business. Failure to set up your new business legally will get you in trouble. There are laws that you must comply with, state and local business laws, building department laws, OSHA safety laws, federal sales laws, tax laws, and contractor laws, to name a few.

COMMERCIAL ROOF CONTRACTING

Commercial roofs are abundant and there is ample work available to a good roofer. Commercial roofs are those of apartment buildings, strip malls, small office buildings, small storage facilities, and the like. The materials and installation procedures are not unlike those of residential roofing.

If you want to enter the commercial roofing field, then, I suggest that you work for a commercial roofing contractor for at least one to two years before you venture out on your own. Use this work experience to your advantage. Study everything, work at each level, and build your list of contacts.

The money required for commercial roof contracting can and will run into many tens of thousands of dollars. Commercial customers expect that you be fully legal, fully equipped, and properly insured. Commercial customers may expect you to finance the job in total. They may expect you to have financing available and offer payment terms over long time periods.

Most commercial customers work hard for their money and most will expect you to work hard for yours. I find this type of customer to be very demanding, quick to ask for a "freebee," and slow to pay.

Commercial construction has one additional problem of which you should be aware, that of you being a subcontractor. Most new buildings are built by a general contractor who subs out most if not all the work. These general contractors want to know that they are hiring qualified subcontractors who will not only do a quality job, but will be able to handle the expenses involved as well. No one wants a subcontractor going bankrupt halfway through a major construction project. Thus, most general contractors will only hire subcontractors with proven performance records.

INDUSTRIAL ROOF CONTRACTING

In both the commercial and industrial sectors, flat roofs are very common, and this is where your major market lies. It is also your major problem.

Industrial roof jobs are the most difficult types of jobs for a newcomer to obtain. Industrial roof jobs include factories, large shopping centers, large office buildings, and the like. Most of the roofs require built-up or membrane roofing. The roof surfaces cover many thousands of square feet and require special techniques to make them water- and weather-tight.

The industrial customer may be a corporation and you must sell your talents to a purchasing agent and to management. Selling to this type of customer may take from several days to several months. There will be written bids, and a lot of specifications, clarifications, and negotiation involved.

The industrial customer will expect you to be fully insured and bonded. They will expect full compliance with all aspects of the applicable laws and codes. They generally want terms or some sort of deferred billing. To you, the roofer, this means that you frequently "carry" them. To carry means that you extend short-term credit, waiting to get paid, while you have to pay your suppliers, workers, and taxes. This can easily run into thousands of dollars per month. The average industrial job can run into the millions of dollars. Large, well-established roofing companies can handle this situation, but you, as a beginner, probably cannot.

I consider the government an industrial customer. Dealing with any branch of any government involves paperwork and delays that are generally overwhelming. From experience, I have found that the higher up on the government ladder you sell to, the slower the payments will be to you. Therefore, government contracts generally are beyond the scope of many small companies.

RURAL FARM ROOFING

Rural farm roofing is much the same as residential roof contracting in that you are dealing one on one with the buyer. The major difference is that the job must be done inexpensively, as the available money is generally tightly budgeted. The materials you will be installing are of low cost compared to other roofing materials. Sheet metal roofing is prevalent. Looks and full weather tightness are not always critical requirements.

The rural farm is fast becoming a novelty. Urban spread, coupled with rising property taxes, is pushing the small farmer into extinction. The corporate farmer has taken over in many areas of the country. The point is that you, if specializing in rural farm roofing, will find less and less of a market for your trade. The existing market is more for replacement and maintenance of farm roofs than it is for new roofs. It is a shame, given that the United States is the worlds "bread basket." We were founded as and we do prevail in the world as the agricultural mecca, but this is becoming a thing of the past. Small farms are giving way to urban sprawl, crime, and parking lots. . . . That is my commentary for the day.

PREVENTIVE MAINTENANCE

Preventive maintenance is a very overlooked field. It can be a career field if you can put together a comprehensive maintenance package and convince people to buy it. Roofs, unfortunately, are in sight but out of mind for most people. It is only after a roof ceases to function that most people decide it is time for service, i.e., the roof is leaking.

There are dozens of minor roof problems that can crop up over the life expectancy of a roof. Each minor problem, if not taken care of regularly, will lead to a complete roof failure. I cover these minor problems in later chapters.

Your investment in equipment for preventive maintenance is small: you need hammer, ladder, hose, broom, and hand pump sprayers. You need insurance and a business license but not much more. License laws generally do not prevail for this type of work. The work itself consists of cleaning, preserving, and correcting minor problems such as loose nails, loose gutters, stains, mildew, and so on. You will need a large customer base as you only work a few hours per season on each roof. You probably will need four to eight customers per day to earn a reasonable living.

REPAIR MAINTENANCE

Repair maintenance is unlike preventive maintenance in that you make repairs on a per-call basis, whereas preventive maintenance is done according to a prescheduled contract. In preventive maintenance, you are looking for and fixing problems before they become major problems. Repair maintenance involves fixing a minor problem that has become a major problem.

I suggest that you consider selling preventive maintenance contracts whenever you make repair maintenance calls. It is good for customer relations and provides a more steady flow of income.

You will need insurance, a business license, and an assortment of small hand and power tools. Total cost should not exceed a few thousand dollars. You may, in some localities, need to be fully licensed as a roofing contractor. It is best to check with your local government agencies before getting into trouble.

ROOF AND TRUSS DESIGN

This is inside work and includes design engineering and drafting. Courses in drafting are recommended but not required. In some states you may be required to have a college degree and a Civil Engineering License before being allowed to design roof trusses. You should start this career working for someone else until you learn the trade.

Equipment needed is a pen and a pad. A computer with graphics will be a big plus, but is not a requirement. You will need power saws, layout jigs, and presses if you decide to build the trusses you design. This can run into several thousand dollars. One source of truss jig tables is Stanley-Bostitch, Briggs Drive, East Greenwich, Rhode Island, 02818.

You will find work with contractors, lumber yards, truss contractors, and owner builders. The average truss selling price range is as little as $30 to as much as $100. Most buildings will require 20 or more roof trusses.

ARCHITECTURAL DESIGN (Nondegree aid)

Again, this is an inside job that requires drafting. A study of building architecture is important. Engineering knowledge of structures and construction techniques is essential. A strong sense of visualization and style is important since you will be designing three-dimensional roofs in a one-dimensional plane. I recommend that you purchase a computer, equipped with the latest in three-dimensional, rotatable graphic arts. Cost of system with hard disc, color screen, keyboard, mouse, scanner, laser printer, and software will be upwards of ten thousand dollars.

Your customers will vary from an owner builder to the sophisticated and sometimes slightly offbeat architect. Your work will be used to sell buildings, to satisfy building inspection departments, to satisfy homeowner associations, and for contractor takeoffs. The term "takeoff" is the process used for determining materials and labor requirements by the contractor. You may have to prepare blueprints to scale and thus should know architectural drafting. You may have to prepare artist conception drawings using colored chalks or inks. You may have to prepare perspective drawings.

ROOFING SALES WITH ESTIMATING

Roofing sales are part of your business. Sales are essential for your survival and you must learn to be a top-notch salesperson.

Roofing sales are also separate contracting jobs. Many contractors, lumber yards, roofing manufacturers, and home improvement companies will hire salespeople dedicated to selling their product and services. They frequently hire salespeople as subcontractors rather than as employees. You act as your own boss and as if you own your business. You contract to sell roofing and receive a commission or fixed fee for each roof job you sell.

Salespeople must be licensed in many states, California and New York are but two. You must understand the principles behind good salesmanship and closing a deal. You must understand the federal and state laws regarding in-home sales of home improvements.

You will need a vehicle, a sample kit, ladder, pencil, paper, and a tape measure, as well as an inexpensive pocket calculator.

You will be required to make measurements on all sorts of roofs in all sorts of weather. You will be required to know every possible roof problem there is and make recommendations on how to fix each one. You must know how to calculate materials and labor with their added overheads and profit. You should know basic mathematics and geometry to the extent you can properly use formulas for determining square footage and volume of basic shapes. You should understand percentages, decimals, and metrics.

Is roof contracting a gender-oriented business? For the most part I would have to say yes. [The work is dangerous, difficult, and somewhat tedious. It involves heavy lifting, lots of climbing, the use of chemicals, and the use of sometimes bulky and dangerous equipment.] Since many females do not consider a career in roofing, the male gender dominates the field. There are opportunities for both genders in roof contracting that includes sales, estimating, designing, and business management.

ROOF CONTRACTING SUBCLASSIFICATIONS

This section pertains to the roof contracting subclassifications that revolve around the materials used. You will find that it is somewhat geographical in that many roofing materials are available in specific regions more than in others. Certain geographical regions and thus the lifestyles of the region have an effect on the roofing materials used.

Asphalt Shingles

The use of asphalt shingles is to the point of being almost nongeographical. You will find asphalt shingles and asphalt shingle contractors in almost every community of the United States. The shingles themselves may be different from one section of the country to another. Some examples: Most asphalt shingling sold in the southern gulf states contains built-in mildew preventives. Several metric sized asphalt shingles are sold in the northwestern states. In Dade County, Florida, the shingles must pass special wind resistance testing. In Los Angeles, California, they must pass L.A. City fire codes. Each of these tests requires special manufacturing, and therefore, more expense. So, though the shingling techniques are the same, the specifications and cost will be different.

To specialize in asphalt shingle roofing, you will need a basic tool kit. I'll call this kit a BTK-1 since we will be using the same kit for other types of roof contracting. As follows:

- Chalk line, 50 ft minimum
- Claw hammer, 16 to 28 oz
- Clipboard with supply of graph paper
- Common screwdriver or screwdriver kit
- Hack saw with spare metal cutting blades
- Keyhole saw
- Ladders: 6, 8, 10, 16, 20, and 30 ft
- Marking pencils, oval in shape
- Nail stripper box or other nail container
- Roof seat or roof jacks
- Roofer's hatchet with a built-in shingle gauge
- Rubber-headed mallet
- Safety belt and safety lines
- Safety glasses
- Tape measure, 50 ft minimum
- Tin snips, straight cut, left hand and right hand
- Tin snips, curved cut
- Trowels, assorted
- Utility knife with extra blades

To this BTK-1 kit we can add the following equipment, which will make roof loading and shingle installation easier and faster. I'll call this the BTK-2 kit.

- Air compressor suitable to drive power tools
- Air hose, minimum 100 ft long
- Ladder hoist, power or manual
- Scaffold sections, minimum of 3 each, 5 ft high by 8 ft long

To this we can add the following:

- Power brake for metal bending
- Power nailer or stapler or both
- Power shear for metal cutting
- Circular saw with carbide blade

You will require some office equipment if you decide to own your roofing business. I'll call this the OEK-1 kit.

- Calculator, with memory
- Desk and chair
- File cabinet, 2 or 4 drawers
- Pencil sharpener
- Straight edge with rule
- Table or desk lamp
- Telephone
- Trash basket
- Typewriter, with carbon ribbon

If you have the money, you can make life easier with the addition of the following, the OEK-2 kit.

- Answering machine
- Computer, minimum 20 megabyte hard disk and one floppy drive
- Dot matrix printer with 9 or 24 pin print head
- Full size keyboard with numeric pad (101 key)

You should have a SVGA or VGA or CGA card installed if you want color. For graphics, I recommend a minimum Dot Pitch of .038 or 640 × 400 pixels on the video monitor. This will enable you to create and draw fliers, sketches, catalogs, and proposals.

If you want to further enhance your equipment, you can add:

- Color monitor with expanded graphics card
- Hard drive, 40 or 80 megabyte
- Laser printer, HP-3™ or equal
- Modem, 2400 baud minimum
- Mouse with mouse pad
- Scanner, hand or table top

Minimum suggested programs are:

- BASIC language programming
- Database manager
- DOS™ (Disc Operating System) 3.01, 4.01, or 5.0 (avoid 4.0)
- Spreadsheet, Lotus 1-2-3™ or equal
- Word processor, 8-in-1™ or equal

Later, add the graphics:

- Pro Design or Design Cad™ CAD (Computer Aided Drafting)
- FormTool 3.0 or equal

Kit BTK-1 will cost about $800 at this writing. Kit BTK-2 will cost approximately $1,500. The office kit (OEK-1) costs approximately $600 and the OEK-2 office kit costs about $1,500. Minimum suggested is the BTK-1 and the OEK-1. Computer programs, if not included with your computer, will run from $50 to $500+ each. Normally included with your new computer are DOS and BASIC. Power tools cost from under $100 to well over $1,000 each depending on tool, manufacturer, and quality. I do suggest you purchase the best quality you can afford. The metal working tools I suggested are for flashing work. You will be installing 28 gage sheet metal flashing on various parts of most roofs. For cutting long narrow strips of metal from large sheets, you will need a shear. To form the metal into the flashing shapes required you will need a brake. You can use hand cutting and bending methods at first, but will find that the power tools do a cleaner, faster job. Some localities may have all the flashing pieces you will require for most jobs, but then again, many localities do not. You may farm out your metal work to a separate contractor or metal shop. This is fine if you can include the extra cost in your job bids and remain competitive.

Concrete Tile Roofing

Concrete tiles are fast becoming the roofing material of choice. The material is low in cost, fire resistant, rot resistant, and becoming available in most localities. Additionally, concrete tile can be obtained in various textures and colors.

You will require the BTK-1 and the OEK-1 kits for this, as well as the following:

- BTK-2 (optional)
- OEK-2 (optional)
- Circular saw with masonry blades
- Drill, 3/8 or 1/2 in. with masonry bits
- Mixing tub for mixing small batches of concrete
- Weight distribution pad, 2 ft by 8 ft piece of plywood with 2 in. of foam rubber on bottom. Used to distribute your weight over several tiles when walking or working on the tiles.

- Wire bender, suggest heavy duty needle nose pliers
- Wire cutters, suggest lineman pliers

You will find additional information on tools in the specific chapters throughout this manual.

Clay Tiles

In the southern and southwestern sections of the United States, clay tiles are used extensively. The clay materials used to make the tiles are abundant in these areas. You will require the same tools as used for the concrete tiles.

Your customer generally will be financially upscale. There is very little replacement work since the tiles do not wear out very quickly. Your business will be confined to about 98% new project or new construction work. Many commercial buildings will use clay tile for the front overhangs and built-up roofing for the actual roof covering. Thus, you must specialize in both clay tile and built-up roofing to obtain this business.

Slate Tiles

In the southeast and in the northeast, slate tiles are used on the finer buildings. Elsewhere in the United States very few slate tiles are used due to the unavailability of the slate. The quarries decide the size, color, quality, and shape of the tiles. Slate tiles are very upscale and generally the most expensive roofing to install. Your customers will generally live in classic older neighborhoods.

The number of available slate tile jobs is small but the profit potential per job is high. Most of the work will be repair work to existing slate roofs. New building projects generally use asphalt shingles or concrete tiles to keep construction costs affordable. I suggest that you specialize in clay or cement tile work with slate tile work as an option.

The tools and equipment required are the same as used for clay and cement tile contracting.

Shake and Wood Shingles

This is a dying career field in some areas of the country. The combustibility of wood has caused many localities to ban its use for roofing. Shake colored and grained asphalt, metal or concrete tiles are fast becoming the roofing of choice when the customer wants the shake shingle look.

Tools used are the same as for asphalt roofs; the BTK-1 and the OEK-1 kits will get you in business.

Metal Roofing

As stated before, metal roofing is more a rural roofing than an urban roofing material. The older cities, especially in the northern United States, do have an abundance of metal covered roofs, but there is not that much call for replacement or new residential construction using sheet metal as a roofing material. You will, however, find that there is a large commercial market for sheet metal roofing in all areas. You will find that there is a large replacement market for metal tile roofs, mainly to replace wood shakes and shingles in the higher class neighborhoods.

Tools required are the BTK-1, BTK-2, and the OEK-1 office kit. The metal working equipment is essential for profitability.

Plastic Roofing

Building codes place very strict limits on the use of plastic for roofing. Plastic burns and when melting, will aid in the spread of a fire. Plastic roofing may be used to cover carports, patios, sun porches, swimming pools, entryways, and other areas not considered living or occupied areas. Plastic sheet roofing may be used in strict compliance of the building codes for skylights and occupied areas. It also may be installed on barns and storage buildings, subject to code approval.

Tools are the BTK-1 and the OEK-1 plus an assortment of various hand saws. Treat the material as if it were wood.

Solar Roofing

Solar roofing is a specialty of its own and is not covered in depth in this manual. It may become the future of roofing but at this writing is more of a novelty than a commercial business.

The techniques tried so far are that of burying the building or flooding the roof of the building. In-ground homes have been with us since the first tribes of Indians settled the West. The Indians used earth as a roof because it was available, it kept out the elements, and in desert areas kept the living areas cool. You will find a scattering of newly constructed earth-covered homes in most states in the United States. Cost to construct is high due to the added structural support required to hold tons of earth in place. Heating and cooling cost of an earth-roofed home can be low since the earth covering keeps the building at around 68 degrees year round.

Flooded roofs work best in the southwestern states where there is a minimum chance of freezing. The building's roof is lined with a plastic membrane and then flooded with a water/antifreeze or water/salt mixture to a depth of 1 to 2 ft. The water collects sunlight and heat throughout the day. An automatic motor-driven cover rolls out over the water during the evening. This cover keeps some heat contained and the heated water is circulated throughout the building during the cool hours of the night. The cover rolls back at daybreak. By this time the water has cooled and the cool water is circulated throughout the building to cool its interior. The major disadvantages of the system are the added structural supports required to hold tons of water, and the routine maintenance. Water is a good environment for algae and other small types of plant and animal life, and these must be killed and filtered out of the system regularly. It can be done automatically, but at several thousand dollars of added equipment cost.

Built-up Roofing

On residences, commercial properties, and industrial properties one can use built-up roofing. There are vast markets for the skills of qualified built-up roofing contractors. The jobs can be as small as a 2 ft by 4 ft second story patio or as large as a quarter mile long shopping mall.

The equipment needed is the same as for asphalt roofing, with these additions.

- Assortment of roofers' mops and push brooms
- Carry buckets for moving hot tar to the roof
- Class BC, 5 lb fire extinguisher, required by OSHA
- Gas cylinders and regulators for heating the pot
- Heated tar pot, mobile kettle
- Roller, portable one man

In addition, if you are doing membrane work you will need the following equipment.

- Flame and roller bar with gas tanks
- Hoist, portable
- Specialty equipment recommended by roof membrane manufacturer

2

Job Safety

INTRODUCTION

Worker safety is of primary importance in any job. It is much more so in roofing. Roofers work high off the ground and with no or few safety lines. A scaffold system, shown in Fig. 2–1, can be a life saver.

The scaffold will break your fall if you should accidentally forget where you are. Scaffolding will also help prevent injury to those on the ground from falling items. The cost to rent the scaffolding shown in the figure is around $800 a week.

STEEPLEJACKING

Roofers are steeplejacks according to the insurance companies. Roofing related injuries rate among the top three for the most frequent and most harmful accidental injury insurance claims. Insurance premiums for roofers are very high. The insurance companies know that roofing is dangerous work. Do you?

Figure 2–1   Scaffolding makes the job easier

SAFETY CONSIDERATIONS

We should discuss some safety precautions other than avoiding falling off the roof. The following information is a review of the OSHA safety regulations and should be studied by anyone preparing for their contractor's license exam.

USING HOT POTS

Flat roof work normally requires the use of hot asphalt emulsions. The working temperature of most are several hundred degrees fahrenheit and is hot enough to give you a serious burn. Plus, the heated asphalt will stick to your skin and continue to cause damage.

To get this material hot, it is cooked in a hot pot or kettle. OSHA (Occupational Safety and Health Administration), has some very strict rules on the use of hot pots, hot materials, and the like. Here is a summary of those rules. Failure to comply can result in accidental injury, destruction of the building, and stiff fines.

OSHA RULES FOR HEATED KETTLES AND GAS CYLINDERS

• You may not carry a bucket of heated material up a ladder. Fill the buckets once you are on the roof.

• Carry buckets are to be in good condition and be constructed of 24 gauge metal or thicker. Maximum containment volume is 6 gal.

- Carry bucket may not be overfilled. The asphalt level is to be a minimum of 4 in. from the top rim.

- There must be at least one person watching a heated kettle. This person must remain within 100 ft of the kettle.

- A Class BC fire extinguisher is required within 25 ft of a kettle while the kettle is being heated. Extinguisher size varies with kettle size. Use a 20-lb extinguisher for kettles larger than 350 gal, a 16-lb extinguisher for kettles of 150 to 350 gal, and an 8-lb extinguisher for kettles under 150 gal.

- Gas cylinders must be far enough away from the kettle flame not to rise in temperature more than 10 degrees Fahrenheit. This is with the kettle going full blast for a minimum of 1 hr.

- Mobile tankers or kettles are not to be transported with a full load of heated asphalt. One must keep the asphalt level a minimum of 5 in. from the top of the container during transport.

- Gas cylinders must be properly supported always. A cylinder cart is highly recommended.

- When transporting gas cylinders, they are to be securely locked in a rack.

- When lifting gas cylinders and kettles to the roof, approved safety slings are to be used. The use of home-made chain or rope slings is not recommended.

- Oxygen cylinders should not be stored closer than 20 ft to LPG (liquefied petroleum gas), acetylene, or other flammable gases. Five ft is permitted provided a fire-resistant wall separates the tanks. The wall must be 5 or more ft in height.

- Gas cylinders and supply hoses are to be color coded. Oxygen is usually green, flammables usually red or orange. Check your local code for proper color coding.

OSHA RULES FOR WELDING EQUIPMENT

- Eye protection such as safety glasses, welding glasses, or welding screens must be used. Welding produces U/V (ultraviolet) radiation that will cause blindness.

- Passersby are to be shielded from direct viewing of the welding process by glasses or welding screens.

- All rules for proper handling of gas cylinders should be complied with.

- Hoses for gas welders and cables for electric arc welders are to be kept in serviceable condition. No cuts, frays, splices, or loose fittings permitted. Exception: Splices in arc welding cables are permitted provided the splice is more than 10 ft away from the electrode holders.

- Tips and holders are to be used with the proper welding rods.

- A 6-lb fire extinguisher, Class BC, must be kept at the welding area.

- Items that may catch on fire are to be removed to a minimum of 15 ft from the welding area. If not removed they must be protected from heat and sparks with a fire-resistant material.

- A fresh air supply or proper ventilation is required for all welding. The air must change at a minimum rate of 100 cu ft per min. This applies to any job site where there are fumes, vapors, dust, or other airborne contamination.

- A self-contained, air supplied respirator is required when welding heavy metals. Heavy metals include fluorides, chromium, mercury, lead, zinc, and beryllium. I believe that galvanized steel and Terne are included since they contain lead and zinc. Stainless steel and Monel are included as they contain chromium, nickel, and traces of other metals. Use a respirator.

OSHA RULES FOR SCAFFOLDS

- A licensed civil engineer must supervise the building of wood pole scaffolds over 60 ft high.

- All scaffolds, of any material, that are over 125 ft in height must be designed by a licensed civil engineer.

- Lumber used to build a scaffold must be structural grade, milled or dressed lumber, cleated or nailed together.

- Platforms are to be no wider than 36 in., no narrower than 14 in.

- Platform span for 2 × 10 lumber is 10 ft maximum. For 2 × 12 lumber it is 12-ft maximum.

- Scaffolds are not to be made from stacked piles of block, brick, wood, or other construction material.

OSHA RULES FOR LADDERS

- Ladders are not to exceed 44 ft in height.

- Two-section ladders up to 33 ft in height must have a minimum 3 ft overlap between sections.

- Two- or three-section ladders from 33 ft to 44 ft in height must have a minimum overlap of 4 ft for each section.

- You are not permitted to stand on the top two rungs of a ladder.

- You are not permitted to place scaffold planks on the top rungs of a ladder.

- You are not permitted to overextend yourself. Do not stretch or lean to reach the work; move the ladder to the work.

- You should not place a metal ladder within 4 ft of a power line.

- You are not permitted to block walkways, driveways, or paths on which others travel. Do not place the ladder where it may be bumped, hit, or knocked over.

- You are not permitted to tie, lash, bolt, screw, or nail two or more ladders together in an attempt to make an extension ladder.

- You are not permitted to paint over markings or the grain on a wood ladder so that the markings or grain do not show.

- Rungs on ladders must be 11 1/2 in. or more wide, spaced no more than 12 in. apart.

- Load capacity of a ladder must be complied with.

- The top of the ladder should extend a minimum of 3 ft above the spot where it contacts the roof, your loading and unloading area.

- Your loading and unloading platform on the roof must be unrestricted and be at least 4 sq ft with a minimum width of 20 in.

- Rungs are to be attached with approved fasteners or with a minimum of 3 eight-penny nails per side.

- Rung supports are to be used. Supports can be straps, blocking, or slots cut into the rails. If slots are to be used, then the slots are to be 1/2 in. deep and the rails are to be 2 × 4 in. lumber.

- Double-headed nails are not permitted for nailing rungs to the rails.

- Each rung, starting at the bottom rung, is to be 1/4 in. narrower than the last rung.

- Bottom rung widths are a minimum of: 16 in. for ladders up to 12 ft in height, and 18 in. for ladders of 12 to 20 ft in height. You must use 19 in. for ladders over 20 ft in height.

- Height is restricted to 26 ft maximum.

- Lumber used for rungs should be dressed structural 1 × 3 in. or better if ladder is under 20 ft in height, and 1 × 4 in. or better if over 20 ft.

- Lumber used for rails should be dressed structural 2 × 3 in. or better if ladder is under 12 ft in height. Use 2 × 4 in. or better if over 12 ft in height. Use 2 × 6 in. or better if over 20 ft.

- Lumber used for rails should be dressed structural 2×3 in. or better if ladder is under 12 ft in height. Use 2×4 in. or better if over 12 ft in height. Use 2×6 in. or better if over 20 ft.

OSHA RULES FOR ROPE

- Rope used as a lifting, pulling, or safety line must be a minimum of 5/8 in., 4,400-lb test, manila rope.

- Sisal rope may be used if downgraded by 20%. Sisal rope is white in color and is generally made in Mexico.

- All rope, if old or worn, has a maximum load capacity of 600 or less lb.

OSHA RULE FOR NAILERS AND STAPLERS

- Power driven nailers and staplers should be used a minimum of 10 ft away from other workers.

OSHA RULES FOR PORTABLE CIRCULAR SAWS

- Portable circular saws are to be used with the blade guard in place. The blade guard must automatically adjust to the thickness of the material sawed.

OSHA RULES FOR TABLE SAWS

- All operators are to be properly trained in use and safety requirements of use.

- Anti-kickback devices are to be used when rip cutting.

- Push sticks are to be used when it is dangerous to push material into the blade.

- Manually adjusted blade guards are to be set within 1/2 in. of the material being cut.

- Automatic blade guards should contact the material.

- Blade, under the cutting table, is to be protected or covered.

OSHA RULES FOR RADIAL ARM SAWS

- Arbors and the upper half of the blade must be shielded.

- Anti kickback device is required if rip cutting.

OSHA RULES FOR AIR COMPRESSORS

- Air compressors are to be equipped with a safety valve, a drain valve, a belt shield, a fan shield, and wheel locks.

- It is recommended that the unit be drained of condensate once per day.

OSHA RULES FOR AIRLESS SPRAY UNITS

* A safety is required on the gun to prevent unwanted discharge.

* A tip guard is required.

OSHA RULES FOR FIRE EXTINGUISHERS

* A class BC, 5-lb extinguisher must be kept within 75 ft of any construction area.

* A class BC, 5-lb extinguisher must be in or near all heavy construction equipment. Construction equipment includes cranes, power shovels, hoist, loaders, trucks, etc.

* Fire extinguishers are to be fully charged, maintained regularly, and properly tagged.

OSHA RULES FOR LIFTING OR HAULING EQUIPMENT

* Trucks capable of speeds over 15 mi per hour must have fenders.

* Trucks must have all applicable safety equipment installed and working.

* A horn, beeper, or other sounding device must be installed on all 2 1/2 ton and up haulers. The device must be heard for a minimum distance of 200 ft and must come on automatically when the vehicle is in reverse.

* A safety log is to be kept and updated once per shift. Condition of the lights, brakes, and safety equipment is to be entered in the log.

* Vehicles are not to be used until all brakes, lights, and safety equipment are in good working order.

* An operator cab shield is required on all hauling vehicles.

* Operators must be licensed, if required, to operate vehicles and heavy equipment.

* Operators of derricks, cranes, cableways, and equipment capable of lifting over 3 tons must be licensed by the U.S. Department of Labor.

* All equipment capable of lifting over 3 tons must be certified as safe by a civil engineer or the U.S. Department of Labor.

OSHA RULES FOR THE WORK SITE

* Water must be supplied at the work site. Containers are to have a faucet or spout and dipping is not allowed. Drinking water must be labeled as "Potable" water.

* Employees are to be provided with disposable drinking cups; shared cups are not permitted.

* Nondrinkable water is to be labeled "Nonpotable." Nondrinking water is water from a source other than an approved drinking water source.

- One toilet is required for each 20 workers. Not required if work crew is mobile and toilet facilities are nearby.

- Field toilets are to be Porta Johns or equal, or chemical, combustion, or recirculating.

- Smoking is not permitted within 50 ft of explosives or explosive gases.

- Explosives must be kept a minimum of 100 ft from any voltage line of 250 VAC or more, 25 ft if under 250 VAC.

- Excavations deeper than 5 ft must be sloped, covered, or shored.

- Excavations deeper than 4 ft must have a safe entry or exit every 25 lateral feet.

- On-site storage of materials may not exceed 16 ft in height, 20 ft high if movable by forklift.

- Traffic controls must be used if there is a danger to worker(s). Traffic controls are flagmen, flags, signs, road markings, and barriers.

- Fuel-driven equipment should have mufflers installed. Approved spark arrestors are required.

- Ear protection is required if sound levels exceed 90 decibels.

That is about it for the OSHA requirements. Most are just common sense items that everyone follows anyway. For you California readers, this entire safety discussion may be on the contractor's exam. Here are a few more safety items required by other organizations and common sense.

PSDNs

PSDNs are Product Safety Disclosure Notices. These are supplied by the manufacturer of the product to the customer, you, or your boss. A PSDN outlines the hazards of the material purchased and used. It explains the safe handling procedures and the consequences of not following the procedures.

You, as a boss or supervisor, are required by law to present a copy of every PSDN to your subordinates. They must read the PSDNs and initial the PSDNs as read and understood. If this procedure is not followed and someone gets harmed, you may be held legally responsible in a court of law.

TAILGATE MEETINGS

IT IS THE LAW. As a supervisor, you must conduct a safety meeting with your subordinates whenever starting a new job. The hazards of the job and the job site must be pointed out.

When you purchase a new piece of equipment, you must conduct a safety meeting with each operator of the new equipment. The safety meeting is termed a tailgate meeting since this type of meeting began on the tailgate of a truck at a job site.

Shop workers must attend a safety meeting at least once per month, which may be presided over by a supervisor or a company designated safety officer. Outside consultants may be used if required. People from the Red Cross and other organizations will be more than happy to help. Red Cross CPR training is also a must these days.

EMERGENCY KITS

Eventually you or a co-worker will be injured on the job. It may be as simple as a cut or as complex as a broken neck from a fall. "Oh, I work very safely, I don't need a first aid kit." It only takes one time, one injury to convince most contractors that they DO need a first aid kit on the job site.

Your first aid kit should contain the following:

- A first aid book
- Antiseptic
- Aspirin
- Bandages, assorted sizes
- Drinking glass
- Gauze and tape
- Scissors
- Snake bite kit
- Splint
- Sprain elastic tape
- Tourniquet
- Tweezers

You should consider including a neck brace and warm blanket. The blanket is for relief of shock. Include a local area map with the locations of medical facilities and their phone numbers. List numbers of ambulance corps and police as well.

Safety kits are to be housed in a waterproof container, not in a cardboard box. Everyone at the job site is to be informed of the presence of the kit and what it contains.

INFORM OTHERS OF YOUR LOCATION

Let someone know where you are if working alone. Tell them your plans for the day and when you will return. It is not uncommon for a roofer to slip and be knocked unconscious. If someone expects you at a certain time and you don't show up, they can come looking for you. Letting others know your plans may save your life.

SAFETY TOOLS YOU REQUIRE

Besides your normal tool kit and the first aid kit, you will require tools for safe working.

- Eye protection is essential. Hammering nails all day will eventually lead to a nail head splitting and hitting you.
- A dust mask or approved respirator is essential if you will be cutting insulation, masonry, or treated lumber.
- A fire extinguisher is a requirement of law whenever you are heating asphalt in a hot kettle.
- Ladder jacks or roof jacks are necessary on all roof slopes over 4:12.
- Safety lines should be worn whenever working on a roof of over 6:12 slope.
- A hard hat may be required at many job sites. Wear a hard hat if you are working on the ground and there are other workers on the roof. A misswung hammer can put a pretty large dent in your head.
- Sunburn lotion or sun screen should be used whenever you are working on a roof. The sun does not have to be shining for you to receive a severe sunburn.
- Clothing should be loose-fitting, comfortable, and full length. Tee shirts and shorts are not very good protection from splashing tar, bugs, abrasion, and sunburn.
- Shoes should be designed to give proper ankle support and they should provide a firm grip on the roofing being installed. Be careful, some soles are great on dry surfaces, extremely poor on wet surfaces.

SAFETY TIPS

- Keep all cutting tools sharp. Dull tools will lead to accidents.
- Keep all swinging tools in good shape. A hatchet or hammer with a loose head or a cracked handle is an invitation to injury.
- Watch where you are stepping always. It only takes one nail through your foot to send you to the hospital.
- Pick up after yourself. If you drop a nail on the roof, pick it up. You may just prevent a harmful slip. If you drop a nail off the roof, make a mental picture of its location. Pick it up when you return to the ground.
- When working in the sun or on hot days, drink plenty of water. Do NOT drink ice water, it will cause stomach pains. If working in 90 degree plus heat, take salt tablets. Your body sweats out salt very quickly. Better yet, do not work in 90 degrees plus heat. Find something to do in the shade.
- If you start feeling too hot or dizzy, sit, there is a good possibility that you are about to faint. Put your head between your legs for a minute or two. When feeling better, get off the roof. Stay grounded until you know you are better. Take a break, a cup of coffee, a sandwich, a walk, whatever.

- If you start to get a cramp, stop working and massage it. If it is a leg cramp, get off the roof. Go for a walk and work out the cramp.
- If the roof is wet or icy, get off it. You cannot install shingles or other roofing on a wet roof anyway. Why chance a fall?
- Never work on a roof when hung over or when taking medicine that cautions you about drowsiness. Roofing work is dangerous and you need to be fully conscious always.

TREATED LUMBER PRECAUTIONS

- Wash hands after working with treated lumber.
- Do not burn treated lumber.
- Do not use treated lumber where it will be in contact with food, pets, or people.
- Wear a dust mask when cutting, shaving, milling, or sanding treated lumber.
- Dispose of treated lumber according to your local sanitation department recommendations.

3

Common Roofing Problems

INTRODUCTION

While writing this manual, I traveled the United States from coast to coast gathering information and observing various roofs. I took hundreds of notes and pictures. The average roof, for the most part, is installed correctly and does its job well. I found that there are many very talented roofers out there, that many are creative, and most do care. As always, there will be the "bad apples," the persons who do shoddy work. They may just not know what they are doing, although some do and just do not care.

I found in my travels roofs of ages from new to hundreds of years old. I found defects and poor workmanship on roofs in every locality. This chapter details the most common problems and the reasons and cures for these problems. Once you understand what can happen to the various components of a roof system, I trust you will not make the mistakes of others.

I detail the problems associated with water damage in another chapter. I highly suggest that you read both chapters a few times. Knowing what problems can develop is as important as knowing how to install a roof. I will begin with a more visual portion of a roof, the fascia.

FASCIA

Fascia does as the name implies, "faces off" the rafter tails. Fascia is an appearance board that hides the rough cut lumber it covers. Fascia provides a nailing surface for gutter systems. It holds the rafter tails in alignment, preventing warping and twisting of the rafters. Fascia provides a mounting surface for many types of soffit and soffit venting.

Unprotected fascia board will eventually need repair (see Fig. 3–1). Shown is the butt joint of two such boards. The protective paint has long since evaporated or washed away and the bare wood is now splitting. This could be prevented with regular preventive maintenance, that of painting. It could be delayed by proper installation.

What was done wrong? First, untreated wood was used, and insects and severe weather will destroy untreated wood. Also, there is a tremendous amount of water running off the roof that will destroy the untreated wood fascia. Plus, there may be rain gutters nailed to the fascia, which create spaces for water, dirt, and bacteria to collect. Fascia board should be painted, treated lumber for long-lasting results.

Second, the butted cut ends of the lumber should be sealed before being installed. Water entering the joint soaks into the bare wood and causes it to swell. This swelling splits the wood fibers and upon drying, the wood splits. Water entry is a good starting place for dry rot. Always prepaint the cut edges of weather exposed lumber before installing lumber.

The third problem is the nailing. Nails driven within 3 in. of the cut edge of a piece of lumber will eventually split that lumber. The nail, upon entry into the lumber, cuts and spreads the wood fibers. This weakens the lumber and starts the splitting process. Predrilling of these nail holes will greatly aid in preventing this splitting.

The last problem shown is that of the drip edge overhang. Water exiting the roof drips

Figure 3–1 Fascia problems

directly on to the fascia. Water exiting the drip edge molding gets sucked against and under the drip edge. Keep the shingle and drip edging overhangs from ½ to ¾ in. away from the fascia. What is the cure for this example?

- First, inspect the soffit and the rafters to which the fascia boards are nailed. They may be rotted from water soaking through the butt joint. If needed, make the necessary repairs. Replace or repair all damaged lumber with treated lumber.
- Second, drill new holes through the fascia and install plated screws or nails.
- Third, fill the cracks with exterior spackle or exterior wood putty. The new epoxy fillers do an excellent job. Finish with an exterior stain or use a coat of exterior primer with a top coat of exterior latex paint.
- The fourth step is to remove the drip edge from the fascia. If doing a reroofing then this isn't much of a problem. Tear off the old roof and its drip edging and correctly replace it with new material. If you are reroofing without doing a tear off, then you must bend this drip edging away from the fascia. This can be accomplished with wedges. Wedges ¼ in. wide by ⅛ in. thick of aluminum, wood, or plastic will do the job (see Fig. 3–2).

You will need sufficient length of strip to fit under all drip edging. The wood and aluminum strips should be prepainted with the proper type of paint to match the fascia color. Do not paint the plastic, as paint does not adhere well to plastics. Coat one side of the strips with construction glue, available in caulking tubes at most builders' supply houses. Insert these strips, glue side toward the fascia, between the fascia and the drip edging. Use

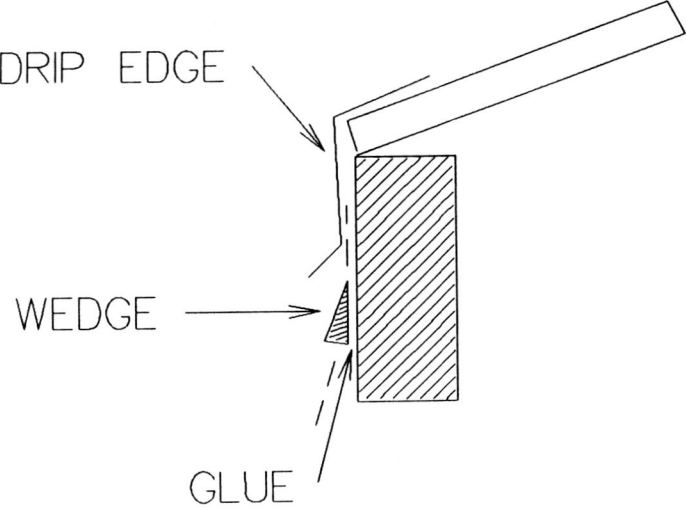

Figure 3–2 Moving drip edge away from fascia

a stick to force the strips up, thus forcing the front of the drip edging outward. Using a proper solvent, remove excess glue, let dry a day or two, and paint.

• Finally, reinstall your rain gutters.

DRIP EDGE

The photo in Fig. 3–3 is not crooked. The roof slope is from right to left. See the problem? The upper slope drip edging was installed before the lower slope edging, hence the overlap is incorrect. Water flowing down the upper drip edge will enter the roofing at this point. You must remove one of the pieces and reinstall it properly to fix this problem.

Figure 3–3 Incorrect overlap of drip edging

Figure 3–4 depicts the results of not having enough shingle and drip edge overhang. The water ran behind the gutter rather than into it. Dirt, leaves, and other roof accumulations collected between the gutter and the fascia. This blockage then became a home for insects, mold, and dry rot. As shown, it also became a home for a plant or two. Major repairs are required because some roofers didn't do their job, that of assuring proper roof water runoff.

The roofing must be removed and the roof sheathing must be replaced. The gutters must be removed so that the fascia, soffit, and rafter(s) can be replaced. The entire section must be repainted.

Figure 3–4   Not enough shingle overhang

SIDING

Siding is not a roofer's problem? Yes it is. You must remember the purpose of a roof, that of protecting the building under it. If the roof design and installation are not proper, the roof will not do its intended job. As a roofer, you must consider the effects of your roof on all elements of the building. You must consider the wind, sun, and rain that hits the building. You must consider the architectural style and physiological effects of the people using the building.

Both Fig. 3–5 and Fig. 3–6 show roof problems. Shown are the lower (Fig. 3–5) and the upper (Fig. 3–6) portions of this building's siding. The building is in southern California and receives much sunlight and little rain. Paint is flaking and peeling from the siding in both pictures. The reasons, though, are different. The upper wall is not getting enough rain to wash off the chemical smog prevalent in Los Angeles. The lower wall is getting too much direct sunlight and is reaching temperatures of 110 degrees or more. Both conditions are not good for paint adhesion.

The problem is not one of painting, although a good stain would have been a better choice of coating. The problem is the roof overhang, or I should say the lack of an adequate roof overhang. The overhang is about 12 in. wide. Just enough to prevent rain from cleaning the upper wall, too narrow to prevent direct sunlight on the lower wall. There probably is no solution, since changing the overhang would be a major cost. Possible options might be a sun screen patio or wall to break up the sun's direct rays, or a line of shade trees blocking the sun.

The actual solution used involved days of scraping and water blasting followed by a good wood stain. A schedule for hosing off and restaining was presented to the homeowner, every 3 months on the hosing, every 3 years on the staining. A few shade trees were planted.

Figure 3–5   Unprotected siding

Figure 3–6   Overhang too small

ANTENNAS

Figure 3–7 shows an antenna mast sitting on shingles. This will eventually grind its way through the shingles and a leak will result. The small pieces of roofing were the results of old dry shingles and a T.V. repairman.

The solution here was to lift the antenna mast and install a 6 in. by 6 in. by 2 in. block under it. Prepaint the block and cement it to the roof shingles with roofing cement. Now, the antenna mast cannot dig through the shingles. What about the old dry shingles? Well it's almost time to reroof, isn't it?

Figure 3–7   Improper antenna installation

SHINGLES

Sunlight is a long-term killer of roofs (see Fig. 3–8). The sun produces heat, sometimes very intense heat. At 90 degrees air temperature your roofing material can approach upwards of 140 degrees on its surface, 200+ degrees in the roofing material layers. This heat dries out the oils and binders that give roofing a protective coating. As the material dries out, it loses its hold on the mineral surfaced granules that are supposed to reflect the sun's rays. A good wind or rain will cause the granules to blow off, exposing more of the asphalt layer. Eventually, the roof crumbles, cracks, or evaporates and allows water entry to the

OILS, SOLVENTS, BINDERS

SUN

Figure 3–8 Effect of sun on roof

building. The process is a long one, 15, 20 years or more. The weight of the shingles, 235 lb, 250 lb, etc., is the weight of the asphalt and protective granules. The heavier the weight, the longer it will take the sun to destroy the roof.

Sunlight not only produces heat, it also produces U/V radiation. U/V is the unseen waves of light that give you a sunburn. U/V will fade colors and destroy the microscopic bonding layers between the asphalt and mineral granules, U/V destroys shingles, paint, metals, wood, etc. Many shingles sold in desert sun areas are now impregnated with U/V inhibitors.

In the past, roofs in high sunlight areas were built-up roofs covered with very white pebbles and rocks. The idea was that the irregular surface would scatter the sun's rays and, thus, make it easier for the roof to absorb the heat and U/V radiation without self-destructing. For the most part it worked. The down side was the weight involved, as the weight of the stones required low sloped roofs and heavy T&G sheathing. Then there is the problem of keeping gutters clean, since the pebbles would wash or blow into them. Most of these roofs are slowly being replaced with U/V impregnated asphalt shingles or clay/cement tiles.

Asphalt shingles that display either curling or buckling require investigation (see Fig. 3–9). The first question is, how old are the shingles? If over 15 years old, then this is normal deterioration. If under 15 years old, there is generally a problem. Is there a chimney or vent further up the roof? If there is, then the chances are that it is not extended high enough into the air stream. The fumes are attacking the shingles.

No chimney or vent? Are all the shingles like this? If so, is there a smoke or fume producing factory nearby? The smoke or fumes may be deteriorating the shingles. There isn't a factory nearby? Are you in an acid rain area of the country, primarily the northeast? No acid rain? Then are you in a very hot desert area?

The vent problem can be solved simply by adding extensions to the existing vent. Eighteen-in. extensions will usually get the fumes into the air stream. As for the age, acid rain, and sunlight problems, reroof using tile or a better grade of shingle, one formulated to

ASPHALT SHINGLES

HEAT CURLING

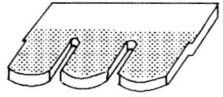

FUME BUCKLING

Figure 3–9　Shingle problems from heat and fumes

combat the problem. Talk to your local manufacturer's representative. He or she probably has the solution.

Spiderweb shingles are the results of constant sunlight and heat (see Fig. 3–10). The asphalt binder is dry and brittle and the granules are blowing off. They are not leaking yet, but given enough time, they will.

Figure 3–10　Shingles have deteriorated from sun and age

Figure 3–11   Poorly installed roofing

See Fig. 3–11. I should make this one the final test. How many roof problems can you spot? I count nine.

1. The shingle overhang is well over 1/2 in.
2. Felt is showing over slow sign.
3. The shingle cutouts end in valley.
4. Bottom shingle of the valley is not properly installed.
5. Two horizontal lines where sheathing is sagging.
6. There is a vertical line where the shingles are joined.
7. Vertical bump under twig where building addition starts.
8. No flashing between roofing and siding, above sign area.
9. The peak's extension is starting to sag.

The last five problems stem from poor roof sheathing and frame out construction. The first four problems stem from poor workmanship by the roofers.

SECOND ROOF

Figure 3–12 depicts the result of an improperly installed second roof. The first roof was either the standard three tab or shake shingles and should have been removed before locking shingles were installed. The heat of the sun has allowed the new shingles to sag into the

Figure 3–12 Second roof sags from heat

voids. The alternate solution would have been to use standard tab shingles in a butt-up procedure. As it is now, the voids under these shingles are subject to damage from people walking on the roof.

ROOFING FELT

Roofing felt sometimes buckles under the shingles and creates a bump in the roofing. Why does this problem manifest itself? What probably happens is that the felt gets wet with overnight dew or rain. The roofer then installs the shingles over the flat but wet felt. Felt is not waterproof, it will soak up water and in doing so, it expands. As it dries out, it shrinks and if firmly attached to the roof sheathing, will buckle. The solution is to tear off the roofing and start over.

SHINGLE MISALIGNMENT

A single vertical line of visually misplaced shingles or vertical dip or bump suggests a room addition. The tie in from the old roof to the new roof was not smooth. Asphalt shingles, when cold, are stiff. When hot, as heated by the sun, they can become limp as rags. When hot, they will soften and then mold themselves to most surfaces. If the sheathing is not perfectly smooth, free of voids, knots, bumps, dirt, etc., the nonsmoothness will transfer to the shingles. Sight down the roof from one rake to the other at roof level. This visual check will usually show you whether the sheathing is flat and smooth. If items like this tie in are not correct, then get the carpenter back to correct it properly. If not corrected, your roofing job will appear to be defective.

Figure 3–13 took a little investigation. A horizontal line of shingles was visually out of place, not laying flat to the roof. The line was several feet from the eave and on both sides of the building.

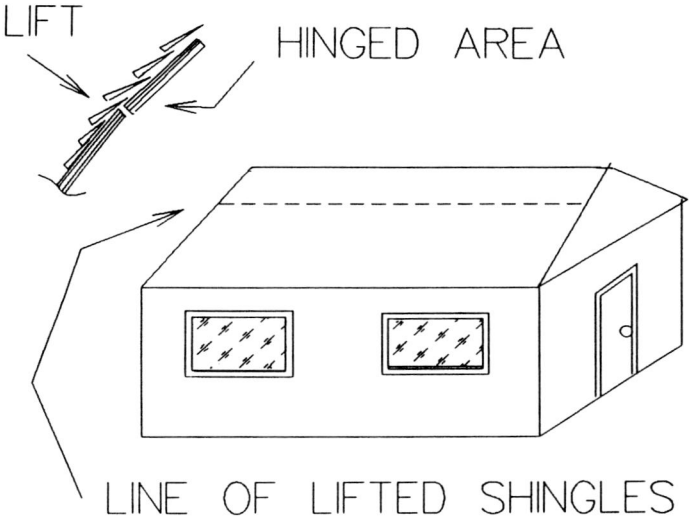

Figure 3–13   Roofing problem on prefabricated home

As it turned out, this was a factory-built house. Factory-assembled houses are made in sections, trucked to the site, and slid in place. The height of each section required that the truck driver take roads that had sufficient vertical clearance for the roof peak to clear. This often added hundreds of miles to route the truck driver had to follow. To solve this problem, some manufacturers of factory-built homes have hinged their roofs. The top section hinges downward during transport allowing shorter routes to be taken by the driver, due to the lower clearance requirements. The installers raised the roof into place at the building site. As a roofer, you cannot do much about the problem.

The problem in Fig. 3–14 is similar to the one in Fig. 3–13 except that the line of displaced shingles is only a few feet from the eaves. The problem shows up mostly on buildings designed to omit soffit. The roof sheathing from the eave to the wall line is thicker than the sheathing from the wall line to the ridge. This is done so that the roofer can use longer nails when installing shingles at the eaves. Longer nails help prevent wind lift blow-off of the shingles at the eaves.

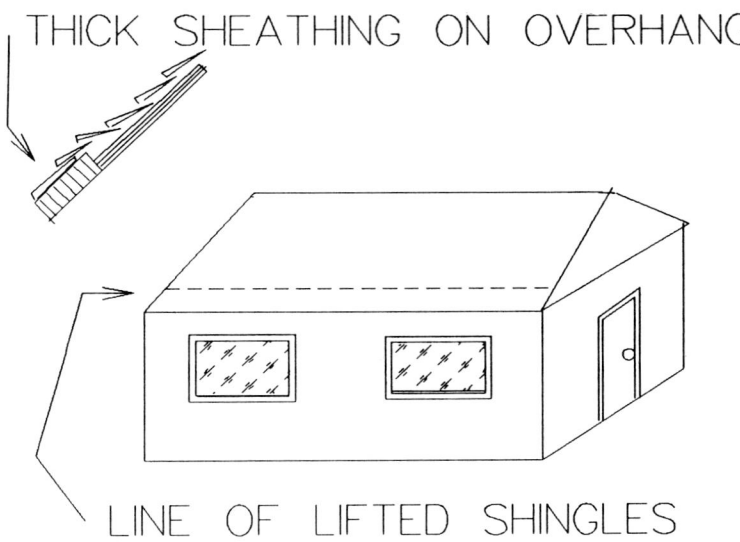

THICK SHEATHING ON OVERHANG

LINE OF LIFTED SHINGLES

Figure 3–14 Lack of shims raises shingle's edges

This horizontal shingle displacement is caused by the shingles relaxing and sagging into the small void between the thick and thin sheathing. This causes the leading edge of the shingles to tilt upward leaving an entry space for wind-blown water or ice. To prevent the problem, the void at the intersection of the two different thicknesses of decking should be filled before the shingles are installed. This can be accomplished by installing low-cost wood shims in this void. These shims are available at most lumber yards. If less than 5 in. of the shingle falls on the lumber sheathing to plywood sheathing interface, I recommend you add the filler.

Every course of shingles being laid should be examined for proper spacing. Any spacing not to specifications can be adjusted in either that course or the next. If you wait too long, you may find that the adjustment is too large to make. If the problem is not immediately caught and if the repair cannot be made in the next course, then make the fix with several small adjustments from course to course. The overall appearance and roof integrity will be better.

RIDGE CAP SHINGLES

Ridge cap shingles are subjected to various types of wind conditions and, therefore, must be secured tightly. Wind hitting the windward side of a building travels up the roof and under the roof peak shingles. As wind travels over the roof peak, it creates a low pressure vacuum that tries to suck the shingles off the roof. The combination of uplift on the windward side and the low pressure sucking motion on the wayward side will quickly tear off any ridge cap shingles that are not properly secured. In canyon or mountainous areas, the problem is multiplied by the strong wind turbulence that is present.

Missing ridge cap shingles are the direct result of the roofer not using the proper length nails. The solution is twofold. First, use the proper length nails. The nails used to secure ridge shingles must penetrate three layers of shingle and three quarters of the way into the sheathing. Second, start your ridge shingles from the rake that receives the most wind. This way, the overlap is away from the prevailing wind.

Figure 3–15 shows another improper installation of ridge cap shingles. There is not enough overlap and the nails are showing. The solution is to start over, this time, overlapping each shingle by one half and not nailing in the weather exposure surface of the shingles.

Figure 3–15 Poorly installed ridge cap shingles

BROKEN OR MISSING SHINGLES

Broken or missing shingles in the roof field are a direct result of using nails that are too short. They are also the direct result of not removing the plastic protection strip from the glued-down strip on self-sealing shingles. Sometimes, this can be the result of kids or vandals.

One must first figure out if the protective covering over the self-sealer strips was removed? If so, then were the correct length of nails or staples used? If the protective strips were not removed, can they now be removed? If so, then do so. If not, then apply dabs of mastic under the tabs of the shingles. If the nails are too short, lift the shingles and renail using longer nails or staples. If this is not possible, then reroof. If the correct length nails were used, and the protective sealer strips were removed, then the problem is most likely vandalism.

Missing shingles on a corner are usually the result of using nails that are too short. On buildings that do not have soffit, it can be the result of not using heavier sheathing on the overhang. The shingles could not be nailed to the thin sheathing with the longer nails required at the starter edge, the eaves.

Wind hitting the side of a building must go somewhere. It may just follow along the wall and continue its journey. More often than not, it tries to go up, over the roof. In doing so, it generates tremendous uplift currents that catch the shingles and blow them off. Building corners are subject to uplift and turbulence, the air, in trying to escape the walls, swirls and forms mini-tornadoes. This violent turbulence can tear off well-anchored shingles, sometimes even the sheathing.

If the problem continues, even after proper construction and shingling, then planting some trees, set back 10 to 15 ft from the corner, may help. Trees act as a windbreak and help break up the speed and turbulence of the wind.

Figure 3–16 shows shingles that have puffed up rather than curled. Puffy shingles are caused by heat entering under the shingles and drying out the asphalt under the shingle's reinforcement mat. Puffy shingles are a good indication that the building is under-insulated or hasn't any insulation. Figure 3–17 (top) shows a normal shingle cross section.

The shingle lays flat when it is new. Outside heat from the sun dries the exposed asphalt oils, and the asphalt shrinks (see Fig. 3–17, middle). As the shingle shrinks, it warps upward at its edges and curls. In Fig. 3–17 (bottom) the heat source is from the building. The lower side of the shingles shrunk and bowed or as I call it, puffing.

Puffy shingles show up on the field rather than the eaves and rakes of a building. The eaves and rake areas are not over the heated interior, they stay cooler and do not lose their natural oils as readily as the field shingles.

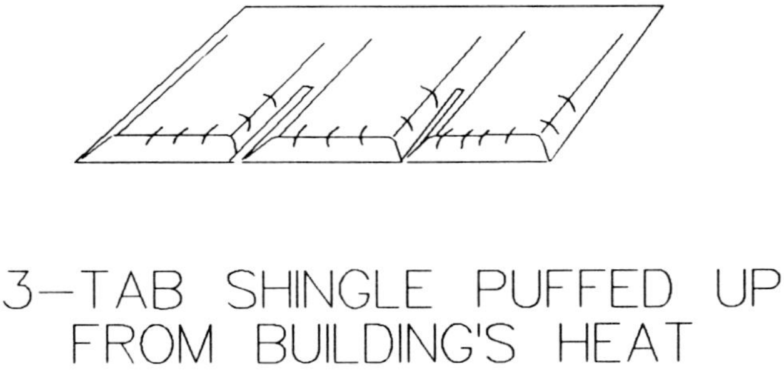

3–TAB SHINGLE PUFFED UP FROM BUILDING'S HEAT

Figure 3–16 Puffed shingles indicate problems

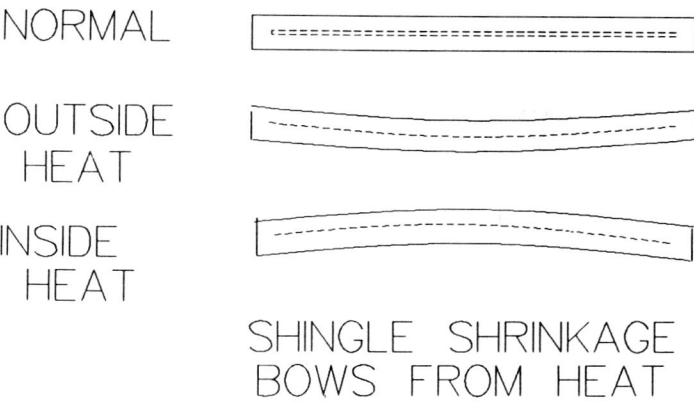

Figure 3–17   Bow indicates heat source

SHEATHING

The problem shown in Fig. 3–18 occurs at times of the day when the sunlight hits the roof at an angle that cast shadows. The dips in the roof are across the field and at 16, 24, 30, 36, or 48 in. intervals. The interval depends on the rafter spacing. What has happened is that the sheathing has relaxed and sagged. Three-ply sheathing, too thin a sheathing, too large of a sheathing span from rafter to rafter, or wet sheathing can cause the problem.

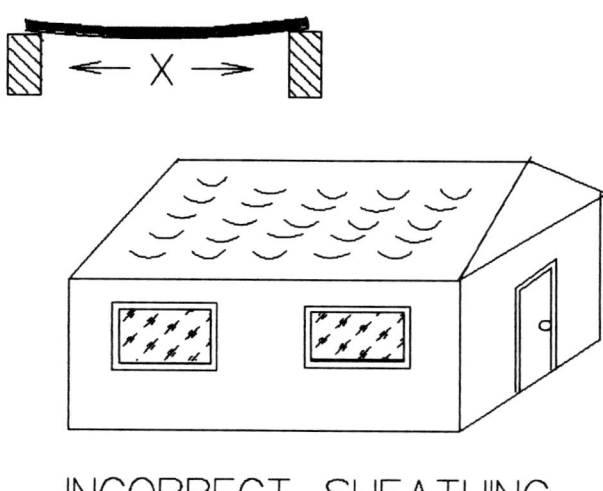

Figure 3–18   Result of using improper sheathing thickness or type

The remedy is found during the construction stage. Use proper sheathing that is dry and of sufficient strength to prevent sagging of not more than ¼ in. (see Tables 4 and 6, Appendix B). Use properly installed vapor barriers and attic ventilation. After the roof is installed, the only solution is to strip the roof of shingles and then resheath and reshingle. This can be a very expensive proposition.

Properly installed roof sheathing is the key to a successful roof (see Figure 3–19). Sheathing should run with the long side parallel to the eaves. All sheathing should be installed with the grain in the same direction. All sheathing edges should be supported, either by the rafters or by T&G grooving or with clips or blocking. Each course of sheathing should be staggered.

The edges of the plywood sheathing lift upward due to the weight of the sheathing over the unsupported rafter span. These edges deform from moisture absorption, being nailed too tight, or from aging and delamination. When installed in a random staggered pattern, the maximum bump generated is only 4 ft in length at any one roof section. This is generally not noticeable. When installed with all sheathing in line, the bump can show itself from eave to roof peak, and is usually very noticeable. Who do you think will get the blame for a poor shingling job, the roofer who laid in the shingles or the carpenter who installed the sheathing? If you see a bad situation like this, then stop, do not shingle over it. Get the person responsible to take care of the problem before proceeding.

PROPER SHEATHING INSTALLATION

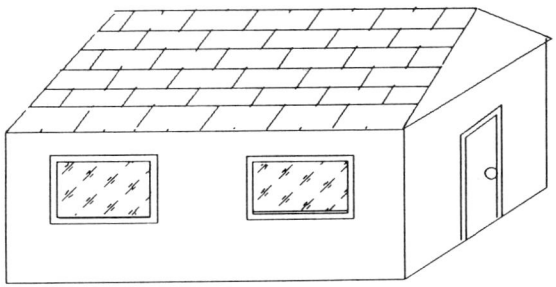

Figure 3–19   Proper sheathing installation

BEAM AND RAFTER CONSTRUCTION

Figure 3–20 shows the saddle back roof line. The primary cause is lack of sufficient bracing under the ridge beam. All materials, wood, concrete, steel, etc., when suspended between two points, will eventually relax and sag. It may take years—this saddle back roof problem may not show up for 10 years or more—but it will happen.

All buildings are subject to wind, rain, sunlight, settling, expansion, contraction, drying,

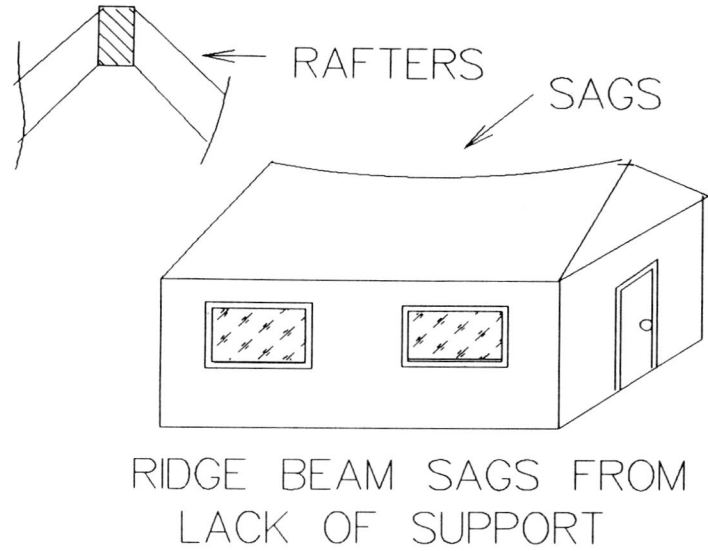

Figure 3–20   Roof beam not properly supported

water absorption, pollution, and relaxing (sagging). Building design and construction must take each of these factors into consideration. In our example, Fig. 3–20, are you really going to install a 30 ft long 2 × 12 ridge board without any vertical support under it? As a professional roofer, are you really going to shingle over a roof you know is designed and constructed improperly?

Figure 3–21 illustrates an unintended Japanese arch roof. The cause is the same as in

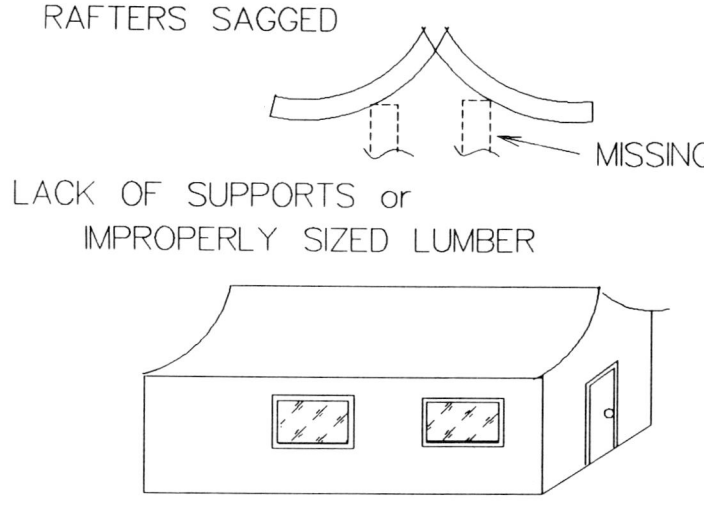

Figure 3–21   Rafters improperly supported or too small

Fig. 3–20, only this time the rafters lack support. This can occur on supported rafters if the rafter boards are too small. Example: Using 2 × 3 lumber as top cords of a truss when the truss is intended for use in a heavy snow area. Trusses made from 2 × 3 lumber are acceptable in southern and desert areas.

IMPROPER OVERHANG SUPPORT

Figure 3–22 shows an improperly braced or leveled gable overhang, which will result in the roof line drooping at the gable ends. Gable overhangs are not integral to the gable truss wall. The overhangs are added during construction and are generally of 2 × 4 construction. A good many builders rely on the roof sheathing, usually 1/2 in. plywood, to hold the gable overhang in place. In other words, the 2 × 4's are suspended from the sheathing rather than the sheathing being held up by the 2 × 4's. This is poor construction and will result in the gable overhangs sagging. It may cause the roof sheathing to be permanently damaged and cause water leakage under the shingles. The overhang's ridge board should be part of the ridge beam in a site-built roof. The overhang's ridge board should have a cornice supporting if the overhang is not an integral part of the building.

Many builders think that since they used either 2 × 6 T&G roof decking or fly rafters, extended from the roof to the gable overhang, this problem will not occur. After all, 2 × 6 T&G will support considerable weight. Trust me, it will happen. All material suspended horizontally and not properly supported will relax and sag. The solution is

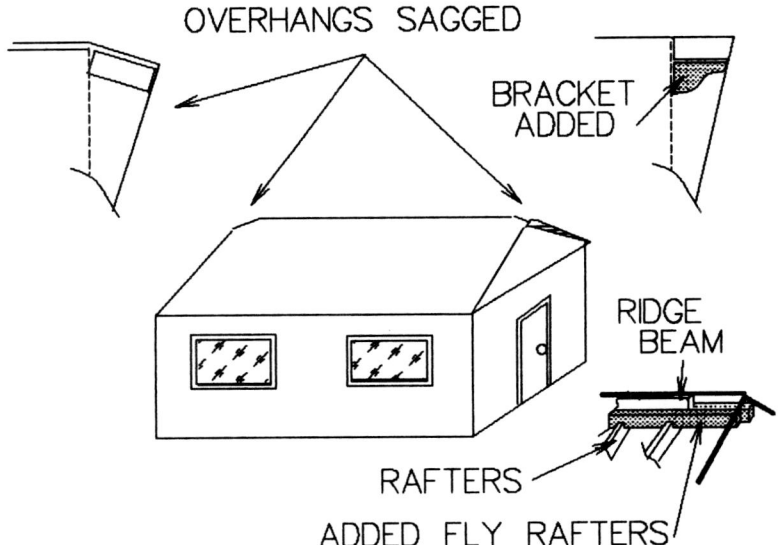

Figure 3–22   Two methods of fixing sagging roof overhangs

removal of the roof right to the rafters and then properly rebuilding it using brackets or gussets.

CROOKED RIDGE LINE

See Fig. 3–23. Sighting down the ridge from rake to rake, we see a very crooked ridge line. This is a common problem when trusses are installed. The builder did not line the trusses up from eave to eave. If very far out of alignment, it will be difficult for the roofer to install the ridge cap shingles. The misalignments will be visually noticeable from the ground in that the roof field will appear wavy. Prevention entails using a string line from gable to gable and aligning the intermediate trusses with that string line. After the sheathing is installed, there is little or nothing that can be done other than a complete, very expensive rebuild.

A crooked ridge line may make the installation of ridge vents very difficult. Ridge vents require a straight, smooth surface for their support and proper operation. See the chapter on flashing and venting.

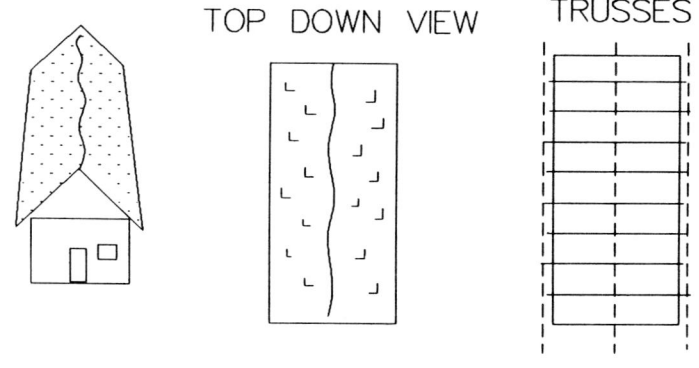

TOP DOWN VIEW   TRUSSES

STRINGLINE WAS NOT USED FROM RAKE TO RAKE WHILE SETTING TRUSSES IN PLACE

Figure 3–23 Ridge line crooked

ROOF STAINS

Roof stains come in a few different varieties. The most common is rust staining caused by metal flashing. This can be prevented by proper periodic maintenance, that being the painting of the flashing. The second type of staining is mineral stain from mortar and brick used to fabricate chimneys. All mortar and brick contain minerals that can dissolve in water. The third type of stain is oxidation stain caused by using exterior self-cleaning paint. Oxidation stain may be found along and below dormers or walls above the roof line.

SELF-CLEANING PAINTS

Most exterior paints are chalking paints, self-cleaning. They are designed to oxidize and wash away. This action is controlled and takes many years before a new paint job is required. As each outer layer of paint washes off, it removes harmful chemicals and dirt. It exposes new, nonoxidized paint and gives the building a fresh painted look. The building's protected areas, under the eaves, do not get rained on. The soffit and the upper wall surface of most buildings are subject to this lack of self-cleaning and should be rinsed off periodically.

I do not recommend the use of self-cleaning exterior paint on surfaces above the roof line. When water hits the siding, the oxides wash free. Roofing, being rough in texture, will trap the oxides washed onto it, and these oxides will appear as a stain. If allowed to remain on the roof they will eventually be embedded into the roof materials and become a permanent stain.

The solution for all the above stains is paint. Exterior nonchalking latex enamel will do an excellent job. Chemically clean all wood, paint, and metal using a liquid sanding fluid. Then, prime with an acceptable primer and paint. The brick chimney can be painted or clear coated. The source of entry of water into the bricks should be found and fixed for best results.

Removal of the roof stain is a little more difficult. First wash the area with soap and water followed with a clear water rinse. Always wash and rinse from the top down to prevent water from entering under the shingles. If the stain persists, try a second wash using one cup of household bleach to a gallon of water. TEST this mixture on a small spot first. Wait a few days, if the stain is gone and the roof did not discolor, then do the entire area.

If this does not work, try diluted phosphoric acid, navel jelly™, for rust stains. Use methyl chloride, a paint stripper or acetone or liquid sander for removing paint. Again, test a small area first and wait a few days. These chemicals do not normally harm asphalt, but they may harm the granules protecting the asphalt. Do not use mineral spirits or hydrocarbon base cleaners, as these will dissolve the asphalt. Finish any chemical cleaning operation with soap and water and a clear water rinse.

For removing stains from wood shingles or shakes, you can try the items listed above. If they do not work, then try a commercially available wood bleach. You can also try CWF™, Clear Wood Finish. And there is always the final option—reroof.

For stain removal from metal shingles or metal sheet, try the above items or sanding paper. You can use a shellac base primer over the stain to seal it, followed by a top coat of latex nonchalking paint.

For removing stains from clay or cement tile, try the above items. If they do not work, then try a hydrochloric base cleaner. Be careful that you do not over clean the tile with the acid. You may remove the glaze coating of the tile. If you do, the tile will no longer be fully water resistant and must be replaced.

SMALL DARK AREAS ON LIGHT ASPHALT ROOF

Dark areas on an otherwise light colored asphalt roof usually suggests the absence of mineral granules. This can be caused by sunlight, U/V radiation, rain, or wind. The U/V radiation and the sun's heat have dried out the asphalt and the resulting loose granules wash or blow away.

The second cause is physical damage from workers. In hot weather, over 90 degrees fahrenheit, the surface temperature of the roof can approach upwards of 140 degrees fahrenheit. This melts the asphalt to near its liquid state. Workers walking on the roof then displace the asphalt and the protective granules, thus, exposing unprotected asphalt.

Unprotected asphalt will deteriorate very fast and roof failure will result. Most shingle manufacturers can and will, upon request, supply you with loose granules. These can be spread on the unprotected asphalt areas. Use a heat gun, waxed paper, and a rolling pin to embed the new granules. Heat the bare spot to just below the melting point, spread on the granules, cover with the waxed paper, and apply pressure with the rolling pin. When cool, remove the waxed paper and sweep off any loose granules.

ELECTRICAL LINES

Many buildings have their electrical power brought into them with overhead lines. The lines terminate in a power head. To prevent the lines from snapping, both by their weight and by high winds, there is a guy wire installed. This guy wire is attached from the pole to a solid portion of the building. The guy wire carries no electricity, it is there for the live wire(s) support and strain relief. Very often, the guy wire is installed directly into the roof sheathing and presents the roofer with problems.

To properly reroof around the guy wire, one must remove it. To reinstall it after the roofing is on, one must retension the guy wire. Personally, I prefer calling the power company. They have the tools and expertise to do the job without electrocuting themselves. They know how much tension to apply to the guy wire and they become responsible for any current or future line-in problems.

It is advisable to install a 2 × 10 in. × 1 ft long board on the roof where the guy wire is to be attached. This board provides a solid foundation for the guy wire to roof connection. It provides a built-up area that you can shingle around without removing the guy wire. It may prevent the guy wire eye screw from penetrating through the roof sheathing into the attic, depending on screw length used. Glue and nail this block in place and flash with metal. Some that I have seen are nailed and flashed with mastic, a potential leak source.

If there is a roof antenna or two, then I suggest 2 × 6 × 6 in. blocks be installed where the antenna guy wires connect to the roof. Put one under the mast as well. These blocks can be painted, or you can cement a shingle to them, to make them blend in with the roof shingle color(s).

POWER FOR YOUR EQUIPMENT

Figure 3–24 shows two power head examples. Three wires in the head suggests that the home has 115 volt AC (Alternating Current) service. This is usually an older home with 60 to 100 ampere service. Four wires in the power head suggests a newer home with 115/230 volt AC, 100 to 200 ampere service.

The reason I'm pointing this out is for your protection. You may be using an air compressor to power some of your tools and nailers. Most quality air compressors draw 12 to 15 amperes of power while running, 60 to 80 amperes during start-up. The outlet you plug into may not be able to carry the current load of your compressor. This can result in your compressor burning out. It can also result in a wall fire in the home.

Plug your compressor into an outlet that is close to the meter or panel box. This location gives the most voltage and current. Why do I tell you this? Well, many years ago, I was working on an older home and plugged my compressor into an outlet that was 60 ft from the panel box. The amount of current drawn by the compressor burned up the wire somewhere between the panel box and the outlet. It cost me two days lost work, a customer, and several hundred dollars for an electrician. It could have caught the house on fire.

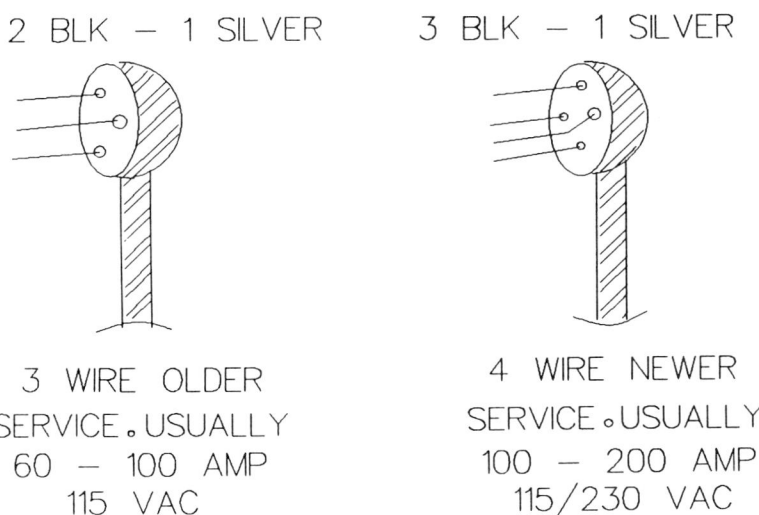

Figure 3–24   Power head type indicates available voltage

LACK OF GUTTERS

This is not really a roofer's problem, but since you are there putting on a new roof, why not earn a few extra bucks. Check the union and licensing rules before doing work other than the work you are authorized to do.

Figure 3–25 is a pool equipment storage building with an extended roof to provide cover over the pathway. The roofing job looked complete. Do you see the potential problem? The problem is trees. The trees in the background drop leaves all over this roof. During rain storms, the runoff water with leaves and dirt drains into the swimming pool. The trick is how to install rain gutters on the eave. The eave overhangs the pool by several feet and one must work upside down from the roof.

Figure 3–25   Lack of gutter causes problems

DORMER OR "L" INTERFERENCE

Can you spot the problem in Figure 3–26? What caused the problem? The roofer shingled from the eaves upward on both sides of the gable dormer. The shingles meet horizontally as they should. The mistake is with the vertical alignment. To prevent this, one must shingle up one side of the gable, and then back down the other side to establish a proper starting point (see Fig. 3–27).

On a dormer, there are two problems that generally result from poor planning by the shingle installer. The first is that he/she ends at the roof with an odd amount of exposed shingle. One starts from the end of the dormer and works toward the existing roof. The

Figure 3–26   Poor shingle installation due to lack of planning

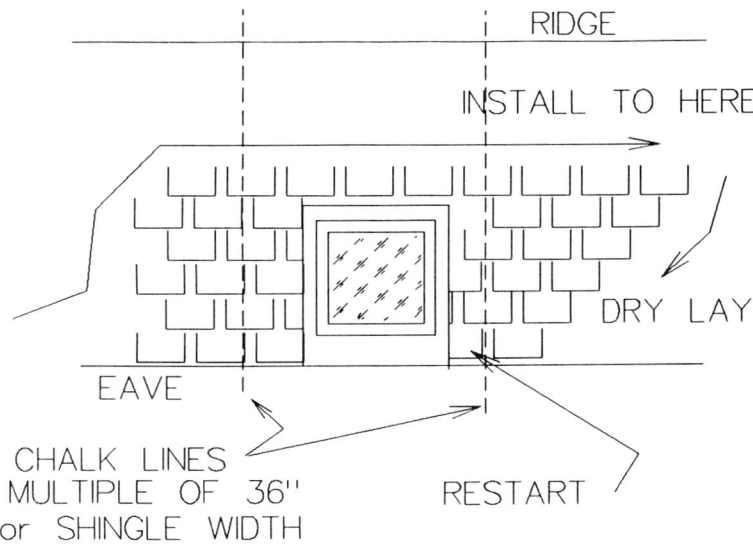

Figure 3–27   Proper method of shingling around a dormer

second problem is the tie-in. Many put the dormer ridge shingle over rather than under the roof shingles. This allows water to flow off the roof, under the dormer ridge shingle, and into the building. See the chapter on flashing for proper method of making this valley.

PARAPET WALLS

See Fig. 3–28. I recently had an opportunity to exam a parapet wall at a large shopping center. The contractor correctly brought the roofing material up the parapet walls in an attempt to seal the walls. The roofing on the sidewalls buckled in one corner and the contractor was accused of shoddy workmanship.

The mall management filed several lawsuits against the roofing contractor in their attempt to get the problem fixed. And yes, the roofer did make several installation mistakes. This was not one of them. What was happening was that the building was settling into the swamp on which it was built. The watertable was within 2 ft of ground level. The roof problem was a foundation problem in that the building's foundation did not have piers sunk down to bedrock. The building was slowly sinking out of sight.

Management's solution was to sue the building contractor. His solution, in turn, was to sue the foundation and roofing subcontractors. They then sued the design engineer, architects, and the building contractor. Meanwhile, a year later, the roof leaks, the walls have cracked, and the management of the stores under the leaky roof have become very hostile.

The correct solution is to dig out that corner down to bedrock, jack up the building, and pour concrete piers. Then, repair the walls and roof.

Here is another built-up roof problem, this time on a residence.

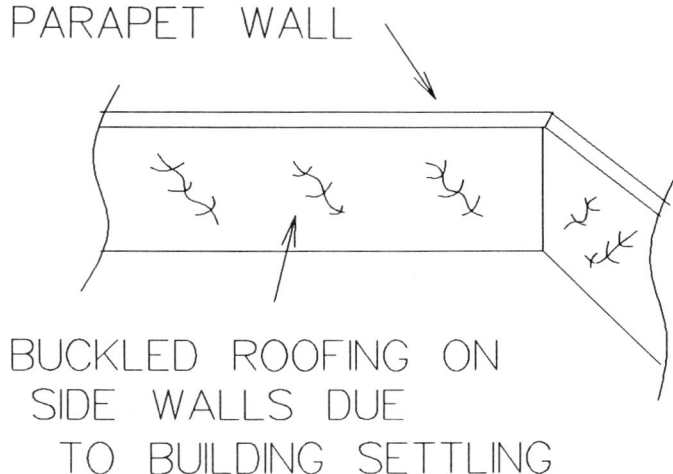

PARAPET WALL

BUCKLED ROOFING ON
SIDE WALLS DUE
TO BUILDING SETTLING

Figure 3–28 You must investigate the cause; a roof problem may be caused by other problems in the building

Figure 3–29 shows a residential built-up roof on a lean-to type building. The home is nestled under several trees and the droppage is constant. The owner has to rake and sweep the roof surface several times a year in order to keep it clean. Figure 3–30 provides a close-up view. Too much asphalt was applied during roofing and it has squeezed out over the years creating a mess. This may be a potential leak location in the future. It is interesting that the asphalt, being exposed to the weather, has dried and cracked. It is no longer a sealer or a protector. The solution is to remove the old asphalt and apply new material. This roof is getting old and is nearing its expected life span.

ROOF AGING

See Fig. 3–31. The roll roofing material has aged considerably at this corner of the roof, which is in bright sunlight from dawn to dusk. The solvents in the asphalt have long since evaporated and the asphalt is turning to dust and blowing away. The drip edge undoubtedly contributed to the process by conducting the sun's intense heat to the under surface of the roofing.

As shown, there is damage to the drip edge—it is rusted through. The sheathing rotted due to dirt and moisture becoming trapped under the rusted drip edge. The entire corner can be easily lifted off the roof by hand. There isn't a water leak inside the building since this is a 2-ft wide overhang and the building is in an area of little rain.

The remedy is to remove all deteriorated wood, metal, and roofing and rebuild, this

Figure 3–29   Too much asphalt used

Figure 3–30   Close-up view of Fig. 3–29

Figure 3–31   Roofing dried and turned to dust

time applying a finished roofing coat of aluminum fibered, reflective material. If you cannot purchase reflective roof coating at your roof material supplier, try a R.V. dealer. Recreational vehicles are frequently coated with reflective roofing.

JOB COMPLETION

See Fig. 3–32. The roofers did their job, now where is that painter? The roofers replaced the fascia, the felt, and the shingles. They didn't paint the new fascia. They didn't trim the last hip shingle. They used 1 in. by 6 in. lumber for the fascia and it bent and warped. They spliced two lengths of wood together instead of using one continuous length of fascia. They didn't include gutters and drains in the job. Did they do a good job? You be the judge.

It is your reputation on the line. It is your business and your profits on the line. Complete the job you started to the best of your ability. You will sometimes lose money on an individual job that you underestimated. But why leave the job unfinished and open to criticism by all who view it?

Many problems can develop in roofing a building. Knowing what to look for is half the job in roofing. Knowing what to do about the problems found is the other half.

Figure 3–32   Incomplete job

4

Water-Related Problems

INTRODUCTION

Before discussing the construction of a roof or the application of shingles to the same, we should consider the most important aspect of a roof: its uses.

A roof provides security to the people and items contained under it, security from intrusion by people, animals, insects, objects, and weather. For the most part, the average roof does its job very well. This chapter is about those times when a roof fails to function up to par.

In the United States there are about one hundred million roofs. These are on private homes, apartment houses, office buildings, factories, and a multitude of other buildings. On average, I would say that eight out of ten of these roofs require some type of repair. A typical roof has a life expectancy of about 12 to 20 years. If we use 15 years as an average roof life, it translates to about 6,700,000 reroofing or roof repairing jobs a year. This does not include roofs on new construction, which averages about 800,000 buildings per year. When you add in the roofs on buildings in all the other countries of the world, you find that you have a sizable market for your trade.

The largest single complaint about a roof is that it leaks. Insurance companies pay out billions of dollars each year to building owners who have had property damaged by this water intrusion. Walls and ceilings are the usual victims, but sometimes, foundations are ruined or weakened. Often, furniture and equipment are ruined or stained. Work sometimes has to be stopped while the rain pours in and again during the repairs. The strange part about all this is that most of it could be prevented. Proper building design, proper application of the roofing, proper maintenance of the roof, and just a little detective work can prevent most leaks, or at least prevent a minor leak from becoming a major problem.

YOU ARE A DETECTIVE

You, as a roofing contractor, should be involved in all the above causes of action. You are the expert, the detective. As such, you will be called in to find the leak and fix the leak forever. Good thought, except you are called during a rain storm when it is too wet to do anything on a roof. Or, you will be called in days or weeks or months later when there is not a drop of water to be found. People tend to delay spending the big bucks until the first sprinkle of the season reminds them that their roof leaked in past years.

You have to locate the exact spot of the water leak(s) and effect repairs. To do so, you should first understand a little about water.

Water is a solvent. Anything left in water long enough will eventually dissolve and wash away, including metals, plastics, wood, paints. Some items take a lot longer than others to dissolve, but they all do. That is a "clue" for you. The items that dissolve on a roof are metals, the chemicals in the wood, in the shingles, and in cements or concretes. The metals are the nails and flashings, the traceable item is rust. Metal, in contact with water and oxygen, oxidizes into rust and this rust is carried by running water to other areas. You find the source of the rust and chances are you will find the source of the water leak. Paints, woods, and shingles are all products that contain chemicals, salts, and minerals. These dissolve in the water and flow with the water to other locations. Again, find the source and you have probably found the water leak. Mortar, cement, and concrete are porous materials and will soak in water if not properly finished and sealed. Most will exhibit what is known as *efflorescence*. Efflorescence is the conversion of certain minerals into salts via water and shows up as a white powder on the sides of buildings, walls, and bricks. It is a clue to where the water is coming into a building.

CLUES TO WATER STAINS

How do you use these three clues? Well, one way is by noting the color. We are looking at water stains made by these leaks. We find that one water stain is reddish in color. Rust is red in color so it is likely that there is rusted metal on the roof. Looking around the area, you may find a roof vent that is improperly sealed and is now rusting.

The second water stain is brown in color, which suggests that the leak is from or near some exposed wood. You find a missing shingle that has allowed the water to penetrate the roof sheathing.

The third stain is white in color, suggesting that it came from an item made of cement. There is a chimney on the building and close inspection reveals that the mortar is loose in several places.

There are a few more colors that can provide clues for you. A green stain indicates the presence of vanadium. It comes from bricks such as those used on a chimney. A black stain is usually associated with redwood lumber. Redwood lumber contains tannic acid, which when combined with iron or steel and water, forms the stain. Always nail redwood with aluminum or brass nails. Any flashing used should be aluminum or brass or insulated from contact with the redwood by butyl rubber.

Water stains usually lead back to the source of the water entry into and under the shingles. Do not assume that the location of the stain is the location of the leak, as it generally is not. Water can and does do strange things. The force of gravity exerts a downward pull on all items, water included. Thus, one would expect to find the water leak upward, above the water stain or water dripping, but this may or may not be true. Water under pressure can run uphill, downhill, and sideways. Also, water clings to items by surface tension, thus it can turn corners and travel many feet along beams, rafters, shingles, wires, pipes, etc. Water can be absorbed into porous materials via capillary action. It can enter an exposed rafter at the cut end, travel upward inside the rafter, and exit at a wall surface. Many lower wall leaks are attributed to water leaking down inside the wall from a roof leak. The fact is that the water came up the wall studs from the foundation.

WATER CHARACTERISTICS

Water exhibits a few other characteristics not found in many other materials. One is that heated water expands and eventually forms steam. If you have ever looked at a roof on a hot day after a rain, you see steam evaporating off the roof. Water trapped under shingles or paint can expand to the point that its volume can increase a thousandfold. If it lacks a place to go, it can blister both paint and shingles. When water gets cold enough, it turns to ice, a solid. As it does, it expands. This is a major roofing problem in the northern portions of the country. Steam usually can escape through small openings, ice cannot. Ice buildup on a roof can destroy a roof in only a few hours. How much force can ice generate? I've seen boulders 20 feet across split in two just from water getting into a small crack and then freezing. Most roofs are not as strong as that boulder.

See Fig. 4–1. A drop of water at room temperature is a drop of water. At 212 degrees Fahrenheit, it is steam, steam that now occupies 1,400 times the volume of the drop from which it originated.

See Fig. 4–2. At 32 degrees Fahrenheit, or lower, our drop of water becomes ice. Ice will occupy a volume 14% greater than liquid water. If water is trapped under roofing material, you will be replacing the roofing material.

Water Has Weight

Water has weight, about 8.4 lb per gallon. During a heavy rain the average 1,400 sq ft home can have upwards of 5 tons of water dumped on it. If that water is not allowed to drain off, there is a high probability of the roof collapsing. Thus, most roofs are built with a slope. Even flat roofs have a curvature or slope built in to help drain off this water.

Since water does have weight, and since water is a liquid, it can do something else

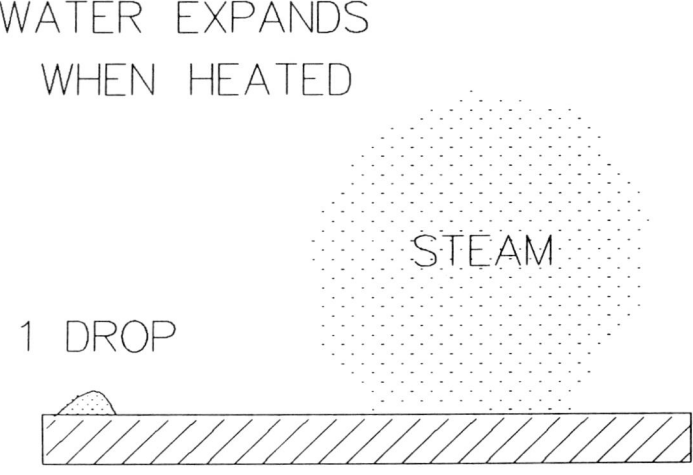

WATER EXPANDS
WHEN HEATED

STEAM

1 DROP

Figure 4–1 Steam expands and is destructive

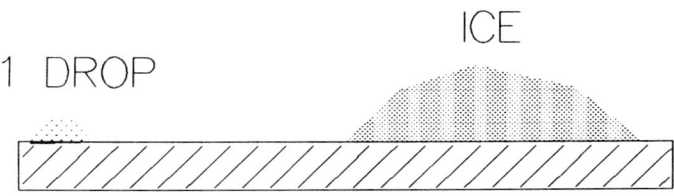

WATER EXPANDS
WHEN FROZEN

ICE

1 DROP

Figure 4–2 Ice is equally destructive

that most substances cannot do. Water exerts force in all directions, down, up, and sideways. A cubic foot of water weighs 62.4 lb. If you have a container that is 12 in. long, 12 in. wide, and 12 in. deep, you have a cubic foot. Fill with water and it will weigh just slightly more than 62.4 lb. If you measure the lateral or side push of the water starting at the top 1 in. of the container, you will find that this force is 5.2 lb. As you measure down inch by inch, you will find an additive effect. The bottom 1 in. will have 62.4 lb of pressure on it. What does this mean to you as a roofer? Well, it is unlikely that you will ever get a puddle 12 in. high on a roof, but it can and does happen to a lesser extent in ice-prone areas. The ice buildup along the drip edge of a roof prevents the water above it from exiting the roof surface. This water wants to go somewhere and it generally does, right back under your shingles and into the building.

Water Is Abrasive

Water is also an abrasive and can wear away a surface just as a piece of sandpaper can. Entire mountains have worn away over time by the abrasive action of water. If you point a garden hose at the ground, the water's force quickly wears away the earth and a hole forms. Now consider a roof design where one roof overhangs another lower roof surface and drains to the lower surface. At the microscopic level, we will find that during a rain storm the falling water is chipping away at the surface of the lower roof. Over a period of years, it will have worn holes or grooves in the lower roof's roofing material, thus becoming a leak source.

WHERE TO LOOK AND WHAT TO LOOK FOR

Roof Age

You now have an idea of some water leak problems. In looking for other problems, you should first determine the age of the roof. A new roof, under 5 years of age, shouldn't have too many problems. All roofs, new and old, have some problems, usually a small leak or two, but these little leaks generally do not show up as an interior problem for several years. When inspecting a new roof, the most probable areas for a leak are in the workmanship or via people.

Workmanship

Look for obvious things first—a missing shingle or tile, a cracked or broken tile or shingle, nail heads popping through the shingles, bulges or lumps in the roof surface, wrong type of flashing used or the right type installed incorrectly. Check chimneys and walls above the roof line for the lack of counterflashing.

A big workmanship problem is improper nailing of tile and shingles to the sheathing. Using the wrong nail or incorrect size nail can cause several problems depending on the roofing material used. As stated before, do not use steel or even galvanized nails when applying redwood or cedar shake to a roof. The tannic acid in the lumber will react with the steel and cause staining. Eventually, the metal will dissolve and the roof will fall apart. Use aluminum or brass nails and flashing. If you must use steel or galvanized steel flashing, then insulate the metal from the shingles with butyl rubber or roofing felt. See the chapter on nailing for more information.

Holes in the Roof: Some Common Offenders

Roofers sometimes inadvertently put holes in a roof, but they are not the only ones who do. Think of the T.V. repair person while you are doing your inspection. Where and to what did he or she attach the antenna? Most T.V. antenna installers are trained to fasten the antenna to the fireplace chimney. They then run three guy wires to the roof and secure them with eye bolts. These wires keep the antenna stabilized during high winds. The eye bolts are screwed through the roofing and into the sheathing, creating a source of leaks. The installation may have been correct, but with years of wind-created vibration, the bolts may come loose or enlarge their holes.

The bottom of the antenna mast is also a potential spot for leaks. Antenna masts should be set in their anchor cup about 2 in. above the roof surface, but most are not. Twisting and vibration of a mast touching the roofing can eventually eat a hole through the roofing. Antenna masts are usually anchored to a chimney if there is one, so check for damage to the chimney. When a chimney is not available, some antenna installers will use a vent pipe as an anchor. As the antenna twist and vibrates in the wind, so does the vent pipe, thus breaking the seal between it and its flashing, another leak spot.

Then you have the decorator, the person who installs Santa and his reindeer on the roof once a year and now 6 months later wonders why the roof is leaking. Never mind that Santa was installed with 12D nails, now removed.

The people from the power company who brought their feed line from the pole to the roof may also be the guilty party. A big screw-in insulator or two will hold these wires in place. The span from the pole to the roof is better than 50 ft and the wire weighs about 50 lb. That is 50 lb of vibrating side load on the insulator screw and the roof, which is sure to leak eventually. Good roofers and contractors provide an anchor plate for this purpose.

Many years ago, a customer called me to his home in Malibu, California to inspect a Spanish clay tile roof that was leaking. What I found was a roof clear of power lines, antennas, etc. I also found about two dozen broken roof tiles starting from one corner of the building and going all the way up to the ridge. Months before, a big brush fire had occurred. The firemen, to find a vantage point, had climbed on the roof. Clay tile does not support much weight, especially the weight of a 185-lb firefighter in full gear, hence, the broken tiles. When working on clay tile roofs, you must distribute your weight evenly over several tiles. A 2 ft by 4 ft piece of plywood with 2 to 4 in. of foam rubber glued to its bottom works well as a weight distribution surface.

Some people like to wash down their roof occasionally, washing from the bottom up, right under the shingles. Always wash from the top down, the same way the water normally flows.

People, sometimes kids retrieving toys, walk on their roofs for a variety of reasons—to see the fireworks better during the Fourth of July, to clean it off, to inspect it, etc. They often do more damage than good. Roofing is somewhat fragile, and more so the older it gets. Look for damaged shingles, tiles, and shakes. Look for broken or abraded edges on the shingles. Look for shingles that no longer have any surface granules on them (asphalt shingles).

People often use vent pipes for hand grips. Check the tightness of the vent pipes.

Check the chimney for tightness and flashing damage. After all, the chimney sweeps may have been banging around up there.

Chimney Leaks

Most chimneys are constructed very well and should outlast the building. But, as always, there are sloppy workers in that field also. New chimneys will leak water into a building if the mortar has not been properly "struck-off." Often the newly mixed mortar is getting old and dry just about the time the top row of bricks is laid. The bricks, set in this dry mortar, did not adhere properly and are now loose. Sometimes the capping mortar is applied too wet and "mud" cracks, allowing water seepage. There may not be a rain deflector over the flue opening, which allows water to run down inside the chimney and into the home.

Here are some other chimney problems. The mortar has gotten old and is falling out from between the bricks. Several earthquakes have vibrated the mortar bonds and cracks have formed. There was a chimney fire and the mortar melted out from between the bricks. The chimney is built after the roof is put on and no counter flashing is used, sometimes no flashing at all. Often the cricket is omitted from the design. A cricket is a small dormer type area on the up slope from the chimney. It deflects the water out and away from the chimney. Remember our discussion on water exerting force in all directions? If you allow water to build up behind a chimney it will back up under the shingles. A cricket prevents this from happening. These are all things to be concerned about, for they will show up as a roof leaks.

Leaks Around Vents

Next, check for leaks around the vents. Most roofs have dozens of pipes exiting through them. Often it is an unsightly mess, with eight or ten vent pipes within a foot or so of each other. To me, this is not only a potential source of water leaks but also an unnecessary one. There is a process called *reventing*. Reventing is covered in the chapter on flashing and vents.

Years ago, the sewer vents were made of metal or clay pipe and were sturdy. Today most are made of PVC, a flexible and sometimes brittle plastic. PVC is polyvinyl chloride. It is strong and when new does not break easily. It does, though, bend and flex. Metal, clay, and PVC will expand and contract with changes of temperature. This expansion and contraction will sometimes break the seal between the pipe and the flashing and create a roof leak.

Plumbers can sometimes create leaks in the roofing. They are working in the building on a plumbing leak and move pipes around. The vent pipe gets moved or jiggled, thus creating a roof leak. You will find that plumbers are not carpenters and when installing the roof vents they sometimes cut too large a hole through the roof sheathing, which must be covered up and sealed with vent pipe flashing.

You may see other vents that are covered with a hood. They are about 4 to 8 in. in diameter. These are either cooking range vents or gas furnace or gas water heater vents. They may be warm or hot to the touch. Again, beware of recent work done on the appliances, as the vents may have moved or vibrated loose.

Check for prior workmanship at all vents. Are there any broken or loose shingles near them? Do two or more shingles butt together directly above the vent? Is there at least 1 in. of clear space between the vent and the side shingles? Has someone piled mounds of sealer or roofing cement around the vent in attempting to fix a leak? Is the vent rusted? Is anything tied to the vent pipe, such as a rope to hoist some item up, a flag pole, or a T.V. antenna? Does the metal flashing cover and run out onto the lower shingles or does it run under them? It is supposed to be on top of the lower shingles, under the upper shingles. See the chapter on flashing for proper installation of vent flashing.

Lack of Proper Air Venting

Lack of sufficient vents can create problems as well. I learned this through personal experience. The first building my father and I put up some 20 years ago was not very large, about 24 ft by 35 ft. We poured the foundation, framed it out, sided it, and we roofed it. We sat on things for several months and had no roof leaks. When it was time to finish the interior, we put up the drywall and the painting. We were drowned out during the first good rain. The ceiling drywall got soaked and began to sag and warp, but not because of roof leaks. Moisture entered the building through the concrete floor and evaporated. It then traveled upward and through the ceiling drywall. The space between the ceiling and the roof became saturated with moisture which in turn became rain. We did not know about vapor barriers, or air venting of attic spaces. It is necessary to vent this space, as the air flow through this space carries off the moisture and keeps things dry.

What else can happen if the attic area is not vented? Well, not only does the moisture collected there ruin the ceiling of the room below the attic, but the moisture tries to escape up through the roof. This rots out the rafters, the sheathing, and the shingles. This moisture collects under the metal flashing and rust forms. On the roof surface everything will look fine, while inside the roof is rotting away. Advise your customer if you see this problem. It is a lot cheaper to put in a few vents than to replace an entire roof.

One would expect to see this air vent problem in northern areas of the country, and on older homes you do. But I lived in Los Angeles and saw it all the time. Builders don't see the need for much air venting since Los Angeles doesn't get that much rain. Plus, they do not over insulate the homes in California, if they insulate at all. No insulation used, thus no vapor barriers exist, right? Where does all the internal moisture come from? Many homes, especially west coast homes, are build on cement slabs in direct contact with the ground, so the moisture comes from the ground up.

You will also find moisture problems directly above kitchens, laundry rooms, and baths. These are the three areas in a home that can generate considerable moisture, which must be removed by proper venting.

Moisture-Collecting Items

Some other potential problem areas may include trees directly above the roof, branches and leaves falling and remaining on the roof, and branches that are touching the roof. These may present a problem because not only is there a possibility of physical damage from a branch falling through the roof, but there is a moisture problem. Branches and leaves will hold moisture on the roof and not allow the roof to dry after a rain. Leaves and branches can form dams that will block the flow of water off the roof and the water will run back under the shingles. All branches and leaves should be removed. You may need a chain saw and some branch cutters in your tool kit. Also, often the branches will get in your way while working on the roof, a dangerous situation that must be eliminated before starting.

Gutter Blockage

Leaves have a habit of collecting in the rain gutter systems and blocking the flow of water to the down spouts. In areas such as the southwest, there are many roofs coated with pebbles and crushed rock. These pebbles blow or wash into the gutters. One such home I worked on had 4 in. of this material in the gutter system and flowers were growing. Gutter blockages can and usually will back water up under the shingles.

Ice Dams

In cold areas of the country, freeze areas, the water trapped in the gutters forms massive ice dams. These ice dams prevent the melting roof snow from exiting the roof and water backs up under the shingles. These ice dams not only cause leaks into the building but frequently destroy the gutters, the shingles, and the fascia.

There is another reason for an ice dam. As snow melts off the roof surface it runs down to the drip edge and off the roof into the gutters. Often, it is going from the warm roof surface into very cold air at the drip edge and overhang area. This refreezes the snow into ice and ice dams form. To help prevent this, one wants to warm the snow to a temperature high enough where it will not instantly freeze as it leaves the roof. If it does start to freeze, then one wants to prevent it from freezing into an ice dam. Two methods of doing this are as follows.

The first is with heating wires installed in a zigzag fashion from the eaves upward about 2 to 3 ft and across the entire length of the overhang. The heater is controlled by a thermostat and turns on any time the air or overhang temperature drops to or near freezing, 32 degrees Fahrenheit.

The second method used to prevent freezing at the overhang is to roof that portion with galvanized sheet metal, aluminum, or copper. The metal roofing extends from the eaves and upward onto the roof by 2 to 4 ft. Metal, being a good conductor of heat, picks up some heat from the roof itself, but the primary heat is from the sun. The additional benefit is that metal is slippery when wet and most buildups of snow will break loose and slide off. This system works best with roofs of 8:12 slope or more.

Plants

I mentioned flowers growing in the gutter, but plants may also grow on the shingles themselves. The sun seldom shines on the north side of a building. This coupled with a buildup of dirt, leaves, and sometimes decomposing shingles forms a planter bed. Seeds blown by the wind lodge in these beds and begin to grow. The plant's roots eventually grow under the shingles and loosen them, creating a leak. You will need a pump sprayer and some commercial plant killer to dispose of this problem.

Another plant problem is ivy. One roof I worked on was in a project of homes built on steps down a hillside. The neighbors used ivy as a divider wall. The ivy grew and extended itself over a 6-ft wide path opening and onto the lower home. It anchored itself into the shingles of the lower home and in doing so tore off many shingles. The shingles that were not torn off were loosened and caused water leaks.

CONSTRUCTION-CAUSED LEAKS

Often the leaks are built into the roof either during construction or when someone reroofs the building. The following are some signs to look for.

Improper Material Selection

Sometimes the roofing material used was not the proper material for the job. You should never shingle a low slope or a flat-roofed building. Any roof with a slope of 3:12 or less should be roofed with roll roofing or with built-up layers of tar, felt, and gravel. Using shingles invites leaks since shingles do not seal out water, as they are designed to direct water off the roof.

One place that shingles are frequently misused is on a room or patio room addition. The original roof has a slope of 8:12, but the addition only has a slope of 2:12. The homeowner or roofer tries to keep the roof surface looking the same and roofs the addition with shingles of the same type as the main roof. This practice invites water leaks.

Leaking Valleys

The term *lace* in roofing is where one interweaves the shingles into a valley from both opposing roof surfaces. Any time you have a valley area you should inspect for possible leak-prone shingling. Lift the edge of a few shingles and see if the corners of the covered shingles are dubbed. Dubbed means clipped off at an angle. Dubbing keeps water from traveling along the top edge of the shingles rather than into the valley. Check for the presence of dirt and loose materials, which can hold water and rot the shingles. Check for rusted metal flashing. Check to see that the proper type of flashing was used. Check to see if the water runoff is in the center of the valley, and check for a splash diverter.

Incorrect Sheathing

If you feel any soft spots or lump spots, these are an indication of a leak and rotted roof sheathing. Lumpy spots are a sign of blisters or dirt or other objects under the shingles.

Often the roof sheathing is not completely cleaned off before the shingles are installed. Sometimes the building designer incorrectly calls out for more than one layer of felt underlay. Some builders use flakeboard for sheathing rather than plywood, and its chips might have come loose. Asphalt shingles should lie flat to the roof to protect the roof, lumps under the shingles form voids that can trap water, dirt, etc. Walking on a lumpy roof can tear holes in the shingles. Hailstones can puncture the shingles at the void areas. Lumps near the edge of the shingles can allow water entry under the shingles. If there are lumps, then the roof shingles should be removed before you reroof.

Chipboard is another name for flakeboard. It is made from pressed chips of wood and makes a fine underlay, providing it does not get too wet. Water, especially freezing water, that gets into the board will pop the individual chips.

With regard to the double layer of roofing felt, it is difficult enough to get one layer smooth, much less two layers. Chances are that you will get lumps in one or the other.

If there are any dormers or other building levels above the roof area you are on, you should inspect these areas carefully. Often these are additions to a smaller single-story house. A hole is cut in the original roof and the new area is built on. The problem is in how the connection, the tie-in from the old to the new, was made. Inspect this area carefully.

Dormers

If water leaks at the dormers, check to see that the window is closed or tight. Check the window glass and the framing while you're at it. Then, check the flashing: is it the proper type and is it installed properly? I find that many gable dormers are improperly flashed. See the chapter on flashing.

Room Additions

The area of concern is where the addition's siding meets the old roof. Many builders use a continuous piece of metal flashing at this junction. This is wrong. A continuous flashing will carry water under the shingles and create major leaks. The flashing should be *step* flashing installed on the edge of every shingle. The exception to this is when there is an upper roof overhang protecting the junction. See the chapter on flashing and vents.

Some builders use no flashing at all and figure that asphalt sealer cement will keep the wall to roof intersection sealed. Wrong again! Sealer by itself will work for awhile but will eventually pull loose, crack, or melt away. This is especially true if one is trying to seal wood shingles to a wood wall. Wood shingles absorb water and that water will break the bond between the shingle and the sealer cement. This bond breakage will cause leaks.

Installed Improperly Siding

Check the siding to see if all joints between each piece of siding are tight. If they are not, a water leak will develop. This is not a roofer's problem, but you probably will be blamed for creating a roof leak.

Drip Edging

Drip edging is metal flashing used at the edge of a roof. It is used to guide water off the roof and into the gutters and away from the fascia boards. Water sticks to surfaces like glue. A sharp edge, like that of the drip flashing, provides a place of minimum adhesion where the water will break off and fall away from the building. Many roofers make the mistake of not installing a drip edge thinking that the shingle overhang of 3/4 in., at the eaves, will allow proper drainage. What happens is that the water travels around the edge of the shingle and sticks to the underside of the shingle. It soaks into the roof sheathing or the fascia and causes rot. Check for the presence of the drip edge flashing and if it is there, check for its condition. At the building's corners, pull the flashing upward gently. If it doesn't move, it is probably doing its job. If it is loose, then suspect a leak at the corner and plan to replace a board or two.

Wind Currents

Some roofs are prone to weird wind currents, U-shaped buildings being one such type. The wind sweeps down into the inside of the U and forms mini-tornados. These wind currents can blow falling rain up the roof surface and under shingles that are not sound. Any shingles that are edge damaged or curled will need replacing.

Venting

As an added thought, when you were on the ground did you smell any sewer gases? Remember those sewer vents on the roof? Well, air often sticks to, and follows a surface the same as water will. A vent pipe that is not high enough above a roof can have its exhaust sucked into the air stream and carried to the ground level. A U-shaped building is very susceptible to this, as I found out after building one. The code states that vent pipes are to be a minimum of 6 in. above the roof surface to be out of this downward air flow. I had built to code but still had to add over a foot of pipe to eliminate the odor problem, so it's a good idea to carry a few sections of pipe and some couplings with you.

Overhang Areas

Now you should climb off the roof and do some inspecting from the ground. Walk around the building and check for any obvious wood rot in the overhang areas. Are they from water damage or insect damage? If from insects, you will need to have the customer obtain the services of an exterminator. If from water, where is it coming from? Is it a roof leak or is it condensation or dew? A simple test is to tape a sheet of aluminum foil to the area and wait a few days. If the foil is wet between the foil and lumber, suspect a water leak. If the foil is wet on its exposed surface, then the problem is condensation.

Building Design

Visually, check for tightness of the roof sheathing to the building's walls. Sounds strange? I had a house built for me several years ago. The framers used 1/2-in. plywood for the roof sheathing, which is fine. What was not too great was that they bent the plywood from roof surface to roof surface to form valleys. The roofers roofed right over it. A few days later the plywood decided to straighten itself out and lifted the roof off at each valley area. The contractor wasn't too smart but the roofer was even less smart. It cost them plenty to tear off the roofing, a portion of the roof, and then rebuild it.

WATER PROBLEM EXAMPLES

As a roofer, you must check the work of other contractors (see Fig. 4–3). You must think ahead to what happens during the rest of the year and in the years to come. Shown in the figure is a downspout tailpiece. It is mounted only inches above the ground. During the winter months there is snow 2 ft deep. This snow blocks the flow of water out of the drain. The water remaining in the drain freezes and forms an ice dam. This in turn prevents water in the gutters from draining and it too turns to ice and forms an ice dam. The sun comes out and heats the snow on the roof, it turns to water and flows to the now blocked up gutter. A pond forms and the water backs up under your new shingles. Night comes and this water, under the shingles, freezes. The expanding ice tears up your shingles, your roofing felt, your drip edge. As it thaws out, during the next sunny day, it seeps through the roof sheathing into the building destroying insulation, lumber, drywall, paint, and possibly furniture and carpets.

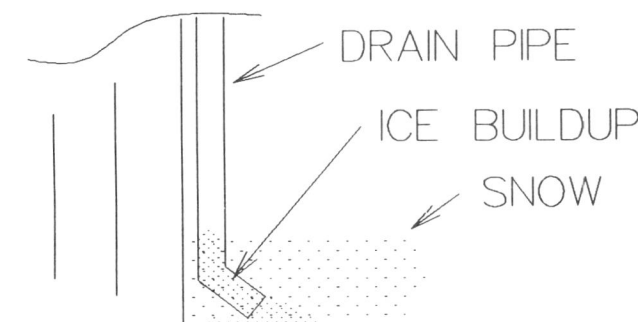

Figure 4–3 Poor drain location can ruin a roof

Who gets the call from the irate customer, the gutter installer or you? Who ends up with an expensive call back, you! Whose fault is it? Yours, you should have been watching and corrected the situation, that of the low drain, before it became a major problem.

If you think it can't happen, see Fig. 4–4. This not only took out a section of roof but also some of the siding. The only saving grace was that it also tore up the gutters, thus requiring the gutter installer to return for a callback.

The Drywell

In heavy rain and snow-prone areas, you should consider having the gutter system drain into a drywell (see Fig. 4–5). This helps prevent ice formation in the gutter and downspout. It also helps prevent water from saturating the ground near the building's foundation. This

Figure 4–4   Ice dam caused by poor gutter system

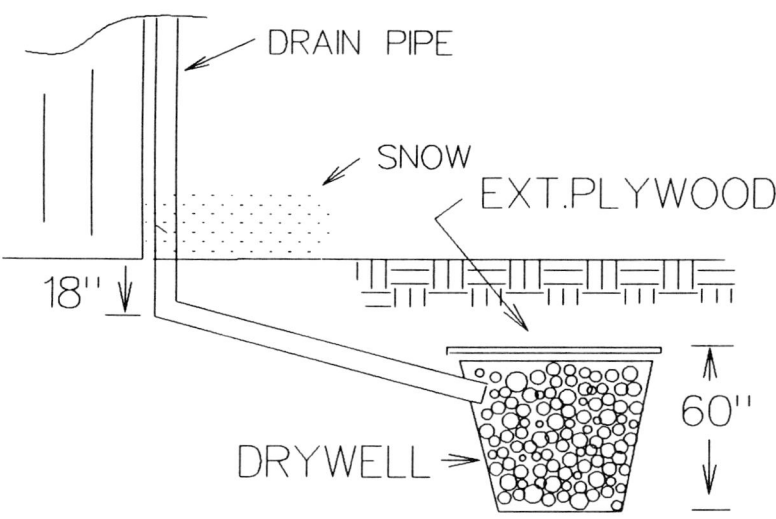

Figure 4–5   Drywell construction

in turn minimizes water entry into slabs, foundation walls, and basements, and thus slows the building's settling process.

A drywell is nothing more than a hole in the ground filled with 3- to 5-in. diameter rocks. Water draining from the gutter system flows to the drywell, collects, and eventually seeps into the ground. Build the drywell 8 or more ft away from the building. The drain line should be below the frost line and should enter the drywell at its top sidewall. Most localities recommend a minimum depth of 18 in. for the drain. In other words, below 18 in. the ground, and thus the water in the drain pipe, will not freeze as it is below the frost line. You should contact your local building department for the frost line depth in your area. The drywell should be about 3 to 5 ft in diameter and a minimum of 3 ft in depth. A sheet of CDX plywood is used to cover the drywell and prevent entry of earth into the drywell. The lateral, horizontal, drain from the downspout should slope at a minimum of 1/4 in. per 10 ft of length, more if possible. Why doesn't the water freeze in the downspout above the frost line? Running water, for the most part, does not freeze. The water will keep running if the drywell is not full.

Lack of a Gutter

Figure 4–6 shows a long-term roofing problem caused by someone trying to save a few dollars. The building has a roof above another roof. To save a few dollars, no gutter system was installed on the upper roof. Over the years the water runoff from the upper to the lower roof eats away the lower roof and starts a leak. Water is an abrasive. What happens here is that over the years the drip of the water abrades the lower shingles to the point that the protective granules loosen. The loosened granules wash or blow away and expose the shingle's

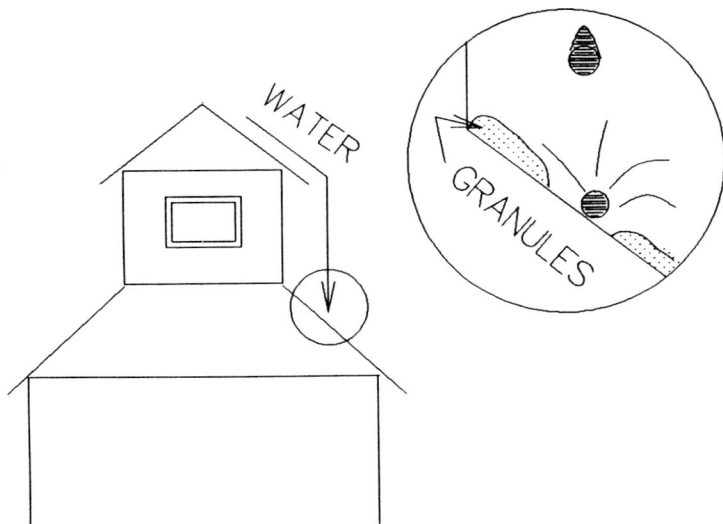

Figure 4–6   Lower roof ruined by lack of upper roof gutters

asphalt layer. The sun then gets to the exposed asphalt and decomposes it, and the roof cracks, crumbles, and leaks. The solution is to install gutters on that overhanging roof.

Gutters

Figure 4–7 shows the beginning of dry rot. But on an aluminum gutter? Metal is not subject to dry rot. The paint on the metal is subject to mold and mildew attachment and growth. The problem shown, the little black dots on the underside of this aluminum gutter, is to figure out what the dots are. Are they dirt or is it mildew?

Figure 4–7   Investigate this; it is potential problem

The solution to the problem is twofold. The first solution is soap and water. Wash the area with 1 cup of Soilex™ or other strong soap dissolved in 1 gal of water. If the spots wash off, then they were dirt. If they do not wash off, then wash with a solution of 1 cup household bleach in 1 gal of water. If the spots are mold or mildew, they should disappear.

If neither solution works, then what you have is a chemical reaction stain. Fumes and acid rain have collected and reacted with the paint. If this is the case, wash the surface with a solution of 5% muriatic acid or 3% hydrogen peroxide. Rinse with clear water and dry. Repaint with a Teflon™ base latex trim paint that includes a mildewcide.

Figure 4–8 shows the same gutter and same problem as in Fig. 4–7, but with one addition: the stains on the fascia and soffit aluminum. Cleanup and repainting is the same as that in Fig. 4–7. These stains are from water entering behind the gutter. Check for proper drip edge and roofing overhang. Check for gutter blockage and drainage slope. Make all repairs before repainting.

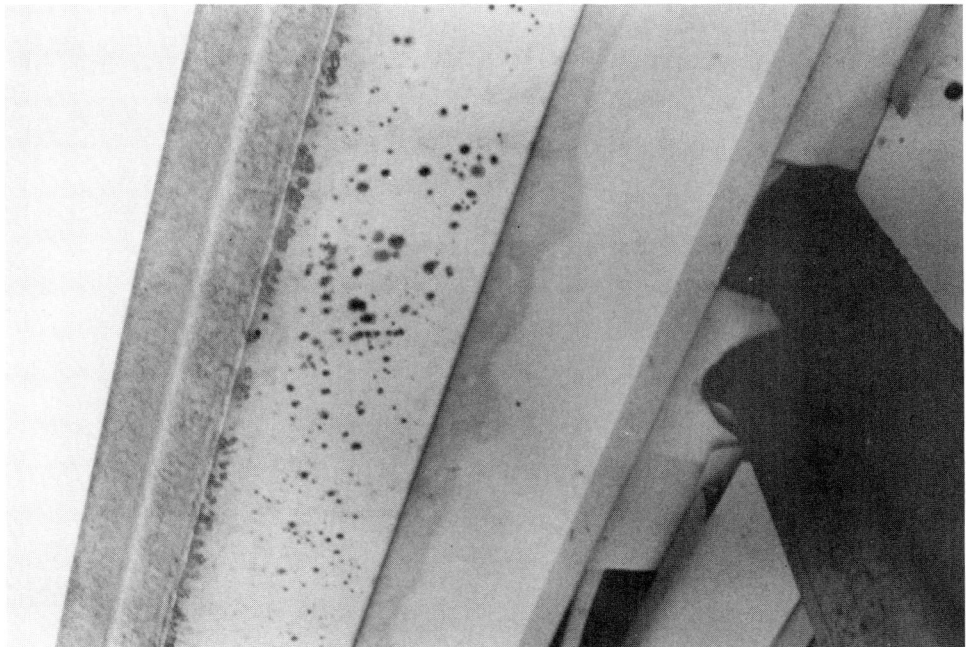

Figure 4–8 Possible lack of shingle overhang

Rotted roof shingles and sheathing at the center of the eave are a problem in snow-prone areas of the country. The destruction is the result of an ice dam. The gutter is not draining to the downspout and the backed up water has frozen. The gutter has sagged in the middle and water has collected from both ends. This water then freezes and forms an ice dam. When doing a roofing job, sight down the gutter from end to end.

Do this at the gutter plane. Many gutters look fine from the ground but are not in reality (see Fig. 4–9). This optical illusion is caused by the eye following the roof and fascia lines rather than the gutter line. By sighting down the gutter, one sees the dip (Fig. 4–10).

Snow and Ice

Snow and ice take their toll on a building. Then the upper roof overhang does not have a drain gutter, and the melting ice will eventually ruin the lower roof shingles. Sometimes the lower roof slopes are not sufficient to shed the ice and snow collected on them. The collection of ice and snow against the siding destroys it, and we have a water leak.

In Fig. 4–4, a home with a very high slope roof, one would expect very little snow collection. Sometimes, on the north side of a home, the high slope is what causes snow to collect. The roof blocks the sun's warmth from melting this snow and the roof remains cold long after the other roof sections have warmed up. The absence of direct sunlight not only creates snow buildup but will contribute to mold and mildew formation on a roof.

Figure 4–9   An optical illusion

Figure 4–10   Different angle of gutter in Fig. 4–9

Note the ice dam at the corner intersection. Ice is the result of water or snow melting and then freezing. If the roof doesn't get direct sunlight, then how did the snow melt? It melted from reflected light, that is, sunlight and heat reflected from the white siding. Unfortunately, when this melted water entered the shadows, it froze and formed an ice dam. This ice dam is ruining the shingles, the sheathing, the fascia, and the siding. The recommendation for this home is to install a thermostat-controlled roof heater cable, as in Fig. 4–11.

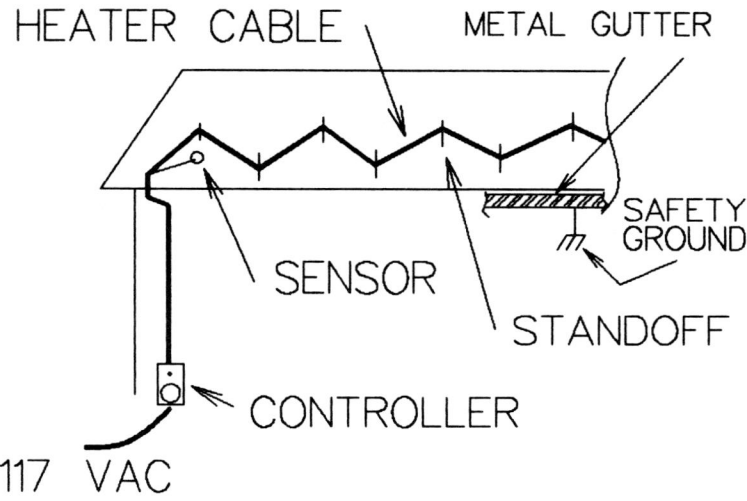

Figure 4–11   Roof heater cable used to prevent ice buildup

A nonenergy consuming alternative to the heater cable is shown in Fig. 4–12. The upper roof overhang is shingled with sheet metal. Normally one would expect the metal to be very cold and act as a collector of ice. What happens in practice is that the metal collects heat. The heat comes from the sun, the air, and the building. During a snowstorm, the first snowflakes will melt on contact with the metal. This creates a very slippery surface and further snow accumulation will slide off. After the snowstorm it will be the metal that warms up first. This again provides a slippery surface that prevents snow accumulation. The snow melting off the shingle section of the roof will continue to fall over the edge to the gutter, roof, or ground below. The main disadvantage to this metal trim system is appearance.

Chemical Fog and Mist

Figure 4–13 shows a roof leak that is not a roof leak. The peeling and flaking paint on the soffit would suggest that water has entered the back side of the soffit and loosened the paint. This could very well happen in a leaky roof situation. Water dissolves minerals and chem-

Figure 4–12   Metal roofing used to prevent ice buildup

Figure 4–13   Appearances may be deceiving

icals from most substances it passes through. Water aids in the oxidation process of most materials and destroys them. Most minerals and some chemical compounds will oxidize into salts, salts that are deposited by the water on or under the paint. Paint will not bond to a salt and will crack and flake.

We do have a water, chemical salt, problem in Fig. 4–13. It is not from a leaky roof, it is from the lack of water. Morning dew and fog have carried dissolved salts and chemicals to the soffit. These chemicals have attacked the paint and destroyed its protective properties. The chemicals have broken the bond between the paint and the soffit material and flaking paint is the result. The prevention of this problem is to wash this area with water once or twice a year. Clean water removes these chemicals and salts before they can do much damage.

Bath Shower Pans

Figure 4–14 shows a water leak in the outside corner of a room. The wallpapered ceiling is rotting and falling down. This is the result of a plumbing leak. This was the lower floor of a two-story house. The bath was over this section. The bath has a shower installed with a ceramic tile floor. Water had entered through the unsealed tile grouting and into the steel shower pan. The pan, used to prevent this problem, rusted and eventually leaked. The job was for a plumber and a tile man, not a roofer. Again, you as a roofer must be a detective. A customer's first reaction to inside water damage is "The roof leaks." As shown, this is not always true.

Figure 4–14   Is this a roof leak problem?

Freshly Painted Plaster

Figure 4–15 illustrates another inside ceiling showing results of water damage. Again, this is not a roof leak. What happened here is that this is a plaster ceiling. The painter painted the plaster while it was wet and uncured and this lead to the condition shown. Plaster as applied is wet and must dry for several weeks to months before being painted.

Figure 4–15   Paint peels from wet plaster

Air Conditioning Units

Figure 4–16 shows a water leak in an industrial building with drop ceilings. The building had a built-up roof and was 20 years old. The owners and roof contractors hunted for months for this leak. The building is in Southern California where it seldom rains. When it does rain, most roofers are very busy and cannot get to every problem area.

The problem turned out to be a leaky air conditioner drip pan. Many air conditioning units are made up of three basic components: a controller that controls the on/off cycles, a compressor that converts heat to cold, and an evaporator that transfers the cold to the building's interior.

The point is that the evaporator is in the building for most systems. It is a series of copper coils in which very cold fluid flows. Any warm moist air, the inside air of most occupied buildings, that contacts these coils will form drops of water, condensation. This water collects in a drip pan located under the evaporator. There is a small, usually 1/2 in. diameter, drain pipe in the pan that drains the condensate to the building's drain. The drain pans require periodic maintenance, that of cleaning and the addition of bleach or commer-

Figure 4–16 Water stain on ceiling

cial mold killers. If this is not done every few months, there is a good possibility that mold will grow.

This mold blocks the drain pipe and the pan fills with condensate and overflows. The result is what appears to be a roof leak. Evaporator drip pans also are used under refrigerators and heat pump evaporators.

Improper Roof Drainage

Figure 4–17 illustrates a concrete block parapet wall above the roof line of this built-up roof. One will notice peeling paint and a white substance. The white substance is efflorescence, chemical salts from within the concrete block. Paint will not stick to salts, thus the peeling paint. I suspect that one of two things has happened here. The first possibility is that though the roof is properly draining, it is not draining fast enough. Second, the roof side of the parapet wall is probably not properly sealed and water is entering the block. I have not been on this roof so I am speculating from experience.

Another possible cause is that the parapet wall cap flashing is porous. Unpainted concrete cap block lets water seep in and minerals seep out. There does not seem to be any waterproofing or metal over or under these blocks. Either way, there is water entering these parapet wall blocks. Efflorescence, be it on block, brick, or concrete, is the result of salts leaching from within the surface. These salts require the presence of water to dissolve them and carry them to the outer surfaces.

In Fig. 4–18, it is difficult to determine whether we have a roof leak or not. Do we call in a mason? The white streaks suggest efflorescence and the large mortar cracks indicate settling. Both indicate a potential water leak into the building.

Figure 4–17   Efflorescence

Figure 4–18   Possible leak source

Chipping out the old mortar and tuck pointing in new mortar is not that difficult a job. Tuck pointing is where one uses a pointed trowel to tuck mortar into a crack. These cracks must be repaired before installing a new roof if you don't want leaks. The question is who makes the repair and who receives the money? You or the mason? In this instance, it should be you. The damage is not that great and the potential for complete disaster is small since most of the chimney is structurally sound. If it was not structurally sound, i.e., many loose or falling bricks, I would say go with the mason. The mason is experienced and insured to handle the problem correctly.

Stepped Roof Tie-in

In Fig. 4–19, stepped roof to roof tie-in has created a catch basin for leaves, insects, and mold. A possible solution to this problem is to screen this area so that the wind will blow the items off the roof. The second solution is periodic roof cleaning. Or you can leave the mess there and eventually have water leak problems.

Figure 4–19   Collection of leaves will cause roof problems

Solar Water Heaters

See Fig. 4–20. These are not skylights, they are solar water heaters. They can be potential roof leak sources—not so much the heaters, but how they are mounted to the roof. The piping into and out of the building and the method used to fasten the piping are other potential leak sources.

Figure 4–20   Items installed on roof may indicate leak sources

Poor Roof Maintenance

Figure 4–19 showed a poorly kept roof. Leaves had piled up and now have become a home to mildew, insects, and trapped moisture (see Fig. 4–21). Moisture trapped on a roof is paramount to roof destruction. Dry rot, a bacterial decomposition of material, and mildew and moss, both fungi, will eat at wood, asphalt, rock, paper, and anything else it can find. Fungi will literally eat the roof away. The thing about bacteria and fungi is that they cannot live for more than 24 hr without moisture. They die when the sun completely dries the roof. But if given a place to hide, under leaves and the like, they will survive and multiply.

Fungi protection.   Areas of high humidity and warmth produce fungus continually. Asphalt and other shingles can be discolored and eventually consumed if mold, mildew, or fungus grows on them. Certain areas of the United States are so prone to fungus growth that the shingle manufacturers add fungus-killing chemicals to their shingles. Shingles containing fungus-killing chemicals are supplied in the deep south and gulf coast states.

You may require these shingles in your area or for a specific building in your area. The shingles can be special ordered by your roofing supplier. Here are some telltale signs of fungus growth and areas of possible fungus growth.

- Existing building has green or gray-green mold growing on roof.
- Building is near the woods or a swamp.
- Building is located near a lake, river, or ocean.
- Building is located in a heavy fog-producing area.

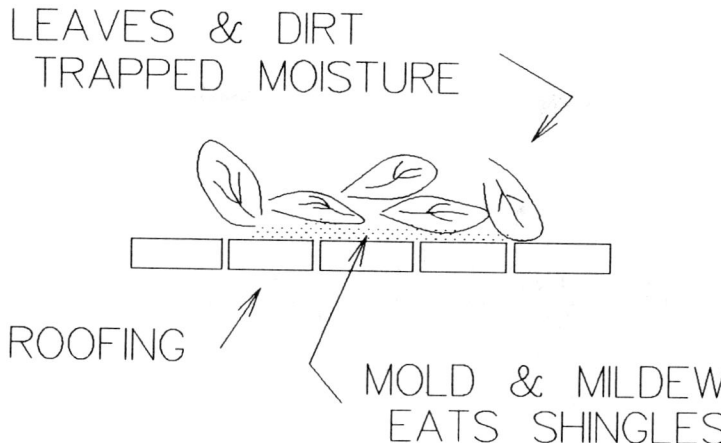

Figure 4–21   Leaves and dirt lead to mold and mildew and shingle deterioration

- Building is located in the Gulf Coast area.
- Building is located in a lower Mississippi River state.
- Building is located in Hawaii or on an island.
- Trees block the roof of sunlight.
- Roof or roof section does not receive sunlight.

Fungus has the appearance of gray to silver-gray discolorations on the shingles. Shade trees should be removed or cut back so that more sunlight shines on the roof. Fungus and mold require moisture for their survival, dry up the moisture and they will die.

Moss, Fungi, and Plants

In Fig. 4–22, moss is slowly consuming this asphalt shingle roof. There is an additional problem shown that is not yet a roofing problem, but will be in a few months. Look at the ivy growing up the side of the building. Ivy eats paint. It also eats roofing. Worse yet, ivy anchors itself under the siding and shingles. As it grows, it lifts the material and exposes the once weather-tight joints.

Still worse is that the plant traps moisture between itself and the building materials. This moisture becomes a home to bacteria and mold. You now have three different entities eating the siding and shingles. Cut the ivy back or eliminate it altogether. Spray the moss with a solution of 1 cup household bleach mixed with 1 gal of water. Do a test sample first to be sure that the roofing is colorfast.

Poor Building Locations

Sometimes, a roof never dries. The north side of buildings in the northern states of the United States are prone to this problem. It is a problem in the humid Southern Gulf states and in the coastal areas of California. These north facing roofs just do not see the sun, and

Figure 4–22   Mold is consuming this roof

moss and mildew grow unchecked. Since you cannot direct the sun to shine on these roof areas, what is the solution?

Solution A is during the material selection stage. Choose roofing materials that are not prone to attack by bacteria and fungi. This generally means asphalt shingles impregnated with algaecides. These shingles are readily available in the Southern Gulf states. They generally have to be special ordered in the northern states. Materials with a small content of copper will work. Copper is a natural killer of fungi.

Solution B comes after the roof is installed, the periodic application of algaecide (mildewcide) or bleach. Most paint stores sell liquid mildewcide in 1 and 4 oz bottles, which cost about $4.50 per oz at this writing. Mildewcide is a paint additive that is mixed with the paint at a ratio of 1 oz mildewcide to 1 gal of paint. Mildewcide does mix well with water, 1 oz to 1 gal of water, and when sprayed on a roof will kill the fungi. The problem is that the next heavy rain will wash off the additive. If mixed with paint or a water seal such as Thompsons Water Seal™, it remains on the roof, protecting it.

Treatment should be once or twice per year. Be sure to *test* the mixture before applying it to the entire roof. Apply to a small area out of sight. Wait a few days to a week to be sure that there is no color change or other detrimental results. On wood shingle or wood shake roofs, one can use CWF Clear Wood Finish™. This product will turn a darkened roof lighter as well as kill fungi. But it also will kill any grass and plants it contacts.

Bleach is an effective killer of fungi. Mix 1 cup of household bleach to 1 gal of water and apply to the affected roofing. Again, do a test on a small area first. Bleach does remove color from noncolorfast materials. It may fade colorfast materials. It must be applied regularly, once or twice per year, since rain will wash it from the roof surface.

Application of the above mixtures is by pump sprayer. Plastic pump sprayers can be purchased in most hardware stores for under $50. Be careful, wear rubber gloves, long-sleeve shirts, full-length pants, and eye protection when mixing, filling, and spraying these mixtures.

5

Roof Construction

INTRODUCTION

This chapter is a mixture of text and construction examples designed to acquaint you with the various construction techniques currently being used in the industry. The examples show construction from the top wall sills to the ridge line and are included in this text because construction will affect your job. You must be able to spot a potential problem area on paper if you are to do your job correctly and at a profit. Here are some differences between doing new construction and reroofing:

New construction work differs from doing reroofing jobs in several ways.

- First, you usually do not have the opportunity to see the actual roof surface that you will be roofing over; it is still in someone's mind or on paper.
- Second, usually your measurements are from a blueprint rather than an actual roof surface.
- Third, most of the time you will be required to use materials selected or purchased by the general contractor or architect, and these may not be materials with which you are generally familiar.
- Fourth, you may be making a bid weeks or even months before the actual job is to be done.

- Fifth, you must rely on other tradesmen to do their job correctly, the framers, the plumbers, electricians, etc. On a reroof, most problems caused by these workers have already been corrected.
- Sixth, you must know and understand the nomenclature and construction used.

It doesn't matter too much on a reroof job that you can see whether you are bidding on a flat roof or an A frame. When you are bidding from paper, it can make a tremendous difference in time, materials, setup, and techniques. Thus, we shall cover the basics of roofs and roof framing construction in this chapter.

MANUFACTURER'S SPECIFICATION SHEETS

Manufacturer's specification sheets are an important part of your business. Keep a three-ring binder or file cabinet drawer reserved for them, because you will collect a lot. Try to obtain the manufacturer's specification sheets well before starting work. You may think you understand all you need to about the materials being used, and indeed you may, but manufacturers are constantly testing their products and finding new and better methods of installation. These sheets can help you avoid costly mistakes.

DRAWING SCALE AND SYMBOLS

Architectural drawings and blueprints use symbols to represent components and materials used in construction. The scale of the drawing will often dictate the symbol to use. The larger the scale, the closer it is to actual size, the more detailed the symbol. The smaller the scale, the less detailed the symbol. Figure 5–1 depicts a view of a plywood sheet at a 1 in. to 1 ft and a 3 in. to 1 ft scale. There are dozens of other examples that I might show, but it would fill another book. I highly recommend that you enroll in a few courses in architectural design and drafting.

PLYWOOD SHOWN IN SCALE

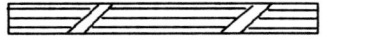

$1'' = 1'$

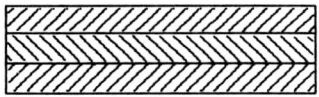

$3'' = 1'$

Figure 5–1 Symbols may change with scale used

NEIGHBORHOOD ROOFS

As you walk around your neighborhood, you will see roof lines of many various shapes and combinations of shapes. Note the various materials that have been applied. The flat roofs usually are of roll roofing or tar buildup, the steeper roofs are covered in shingles or some form of tile. The very steep peaked roofs may be clad in metal or a combination of shingles, tile, or metal.

Note the design character of the buildings. Do they look plain? Have a Spanish look to them? Or is another specific style being used?

The difference is generally most apparent in the design and covering of the roof. The same building can be made to appear luxurious, inexpensive, barnlike, super-modern, cabinlike, mansionlike, colonial, Spanish, etc., by changing the roof shape and covering. Figure 5–2 gives the basic roof shapes you will encounter during your career. You will often find combinations of these roof shapes on the same building.

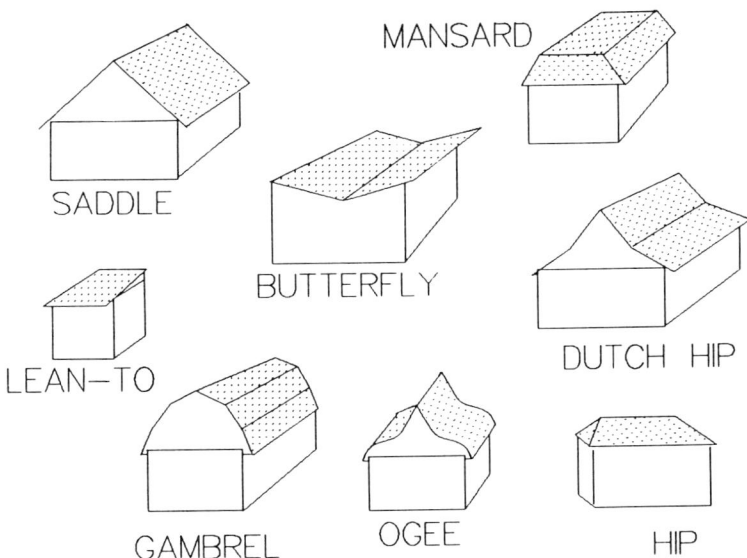

Figure 5–2   Common roof configurations

You can help yourself to easier work and a better roofing job if you can get into the act early. In other words, work with the contractor on the finer points of roof design. For example, on a mansard roof, try to insist that there is an overhang of roof sheathing where the roof changes slope. Figure 5–3 shows a roof with and without the overhang. Without the overhang area, the shingles or tiles will not seal properly and may become a source for water leaks. Tiles and shingles are designed for flat surfaces, not to be bent around corners. The leaks generated will be blamed on you, the roofer, rather than on the designer.

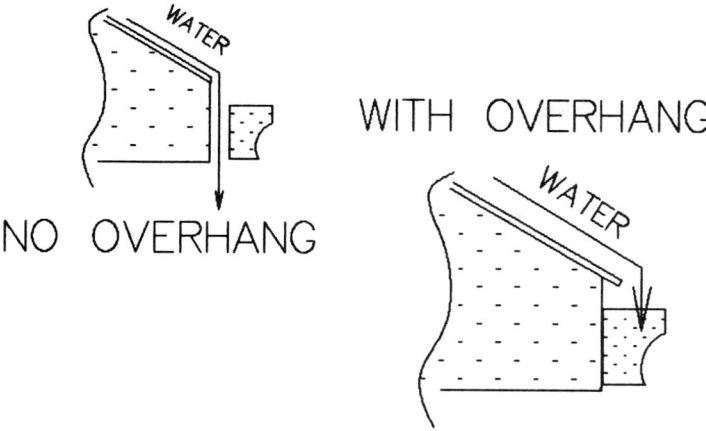

Figure 5–3 Shingle overhang is important

One place I worked on had a 'dumb' construction problem. One section of the house was built in a lean-to fashion. Again, this is fine, but the exposed rafters lack any fascia board. What happens is that rain hits and sticks to the face cut of the exposed rafters. The water then travels down the rafters to the outside wall, which becomes a source of water leaks into the building, as shown in Fig. 5–4. The builder didn't count on this and did a poor job of sealing the rafters to the wall sills. Also, the exposed ends of the beams, if not recessed or otherwise protected, will absorb water.

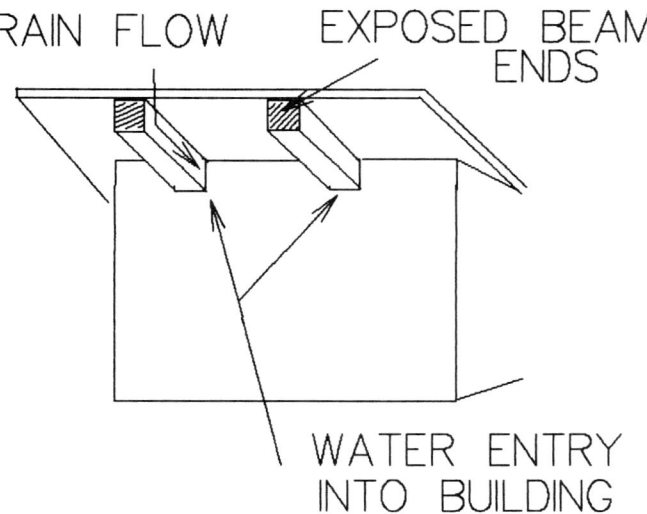

Figure 5–4 Exposed beam ends should be protected from weather

ROOF SHEATHING

Roof sheathing is essential to any roof. The sheathing holds the rafters in place and provides a secure surface for the application of the finished roofing materials. Many try to save money by using inferior quality lumber or other sheathing. This will lead to premature failure of the roof and possible destruction of the building. I advise using only quality materials that have U.B.C. approval.

Sheathing Materials

You have a selection of sheathing materials to choose from. For residential work, the T&G planking and the plywoods are best. For commercial work, you may want to use plywood, metal, or concrete for the roof sheathing. Very large commercial buildings should be sheathed in metal.

See Fig. 5–5. Most manufacturers of roofing materials require that the roof decking be free from voids and knots. Large knots should be removed and the remaining hole, or void, should be filled with an exterior wood putty. Knots or voids that cannot be removed or filled should be covered with galvanized sheet metal. This will prevent the knots from lifting out over the years. All asphalt-based roofing will eventually relax and try to conform to the shape of the roof sheathing. Voids and knots will show up if not eliminated before the roofing is applied. Knots that pop out of the wood can puncture the roofing, thus creating leaks.

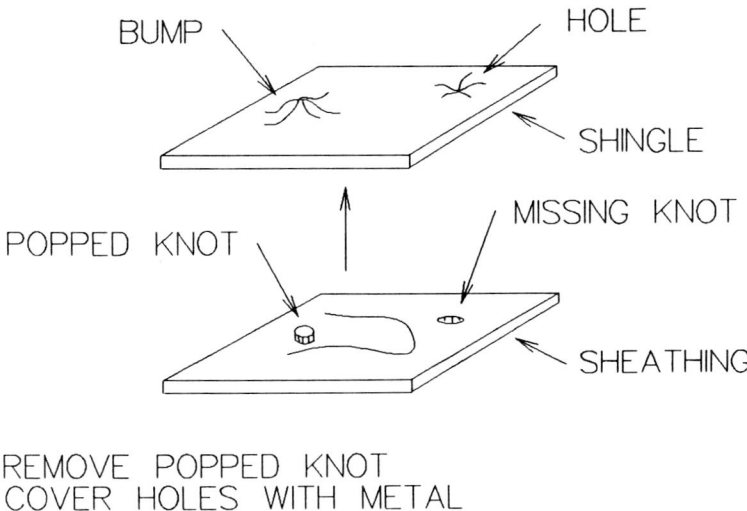

Figure 5–5   Knots and voids must be filled

T&G Sheathing

On low-pitched roofs, roofs to be covered with rock or tile, flat roofs used for patios or sun decks, or roofs requiring strong decking, the recommended decking is T&G. T&G is tongue and groove. T&G roofs are subject to people walking on them and to both rain and snow loads for extended periods of time. In the chapter on flat roofs, there are certain required construction techniques that must be used. Check with the general contractor to decide who is responsible for this construction. Sometimes it is the roofers, other times it is the framers.

Beveled Edge T&G

Use beveled edge T&G if you can find it. It makes for a nicer finished look on the underside of an exposed ceiling or deck. The disadvantage to beveled T&G is that it makes staining and painting somewhat more difficult. One has to brush coat the lumber since a roller will not get into the beveled grooves. The use of either straight cut or bevel cut lumber is fine if you can use an airless paint sprayer. Lumber should be S-dry or better yet, kiln dried. S-Grn, even T&G S-Grn will shrink with age and open the seams.

Kiln-Dried Lumber

KD, kiln-dried lumber has less than 15% water by volume. Although it is generally recognized as having less than 12% water, the U.B.C. allows some species of lumber to contain 15%. See Appendix B, table 25. S-Dry has from 12% to 19% water. S-Grn has over 19% water content. KD is the most expensive and S-Grn the least expensive. You need lumber to sheath from the end of the rafter tails to the building's outer wall line and from the building's outer wall line to the end of the rake overhangs. Many builders use lumber sheathing at both sides of the ridge lines.

T&G Thickness

Thickness of the lumber depends on the expected load and the span between rafters. Recommended thickness is 5/8 in. minimum placed perpendicular to the rafters and 3/4 in. minimum if placed diagonally on the rafters. This is for seasoned dry or S-Dry lumber (no more than 19% water content). As in all roof sheathing,the maximum deflection should not exceed 1/360th of span. This recommendation will meet or exceed code requirements. See Appendix B, table 8 for acceptable deflections. Please note that, although code may say that you can use 1/240 or 1/180th of span, some roofing material manufacturers require 1/360 of span. They may not warranty your roof if their specifications are not met.

Plywood Roof Sheathing

Plywood has thin layers of wood glued together to form a sheet. Standard sheets are 4 ft × 8 ft and can contain from 3 to 13 layers. Other sizes are available. The thicker sheets may be supplied with T&G edges and are used for flooring and some built-up roof decking. I recommend the use of 1/2 in. or 5/8 in. thick plywood for most residential roofs. To learn more about plywood, its construction, and its use, I recommend you contact the American Plywood Association (APA).

Three-ply roof sheathing is available in some areas. I do *not* recommend its use unless it is used over prior sheathing or foam insulation. It is spongy. It deflects, it bends, it sags. It was sometimes approved, but I don't understand why. This sheathing is recognizable since it is only three layers thick and it is sewn together. You can see the sewing threads across the plywood in four or five places. The sheathing is fine for roofs that have 8:12 slopes or greater. It is not advisable on slopes lower than that since it warps. Lower sloped roofs should be stronger roofs since they are subject to people walking on them and to snow loads in areas of snow.

The minimum specification for plywood roof sheathing, per the U.B.C., is four-ply, structural grade 1.

Five-ply roof sheathing is my recommendation for most roofing jobs. It is strong, provides good support, and provides a good nailing base. You should use CDX exterior plywood, structural grade 1. CD interior may be acceptable in some localities, but I still suggest CDX. I recommend using M.D.O. plywood for exposed eave and rake overhangs; it has a smooth resin-coated, paintable surface.

Particle Board or Chip Board

The use of chipboard for roof sheathing is approved in most areas. It is designed for sheathing, and it is inexpensive. It also has one problem I personally don't like. The chips can and sometimes do pop loose. The board is made from thousands of wood chips mixed with a glue binder and pressed into sheets. If any sawdust or moisture remains on the chips, if the wrong binder is used or doesn't fully coat a chip, if there are air pockets between chips, if the press wasn't set up correctly, if the chips are a mix of soft and hard wood, the chips may pop out.

You are nailing, pounding the daylights out of a roof when installing shingles. The constant pounding vibrates the sheathing and may pop loose any chips not fully secured. This is not a problem, unless it happens under a shingle ten rows back. Then you have a tear-off to do, and your big savings in material have just been lost.

Sheathing An Exposed Overhang

See Fig. 5–6. When the building design calls for exposed overhangs, the sheathing will be seen from the bottom, and this presents a problem. You must nail shingles to a sheet of 1/2 in. plywood sheathing without the nails coming through or leaving bumps on the underside of the sheathing. You could use shorter nails, but this will result in the shingles being blown off in the first good wind storm. The accepted practice is to use 1 in. nominal lumber at the overhangs. One inch nominal lumber translates to 3/4 in. thick finished lumber. 3/4 in. plywood will work, but will not look as good unless you use the recommended M.D.O. plywood. The remainder of the roof's field is sheathed in 1/2 in. plywood.

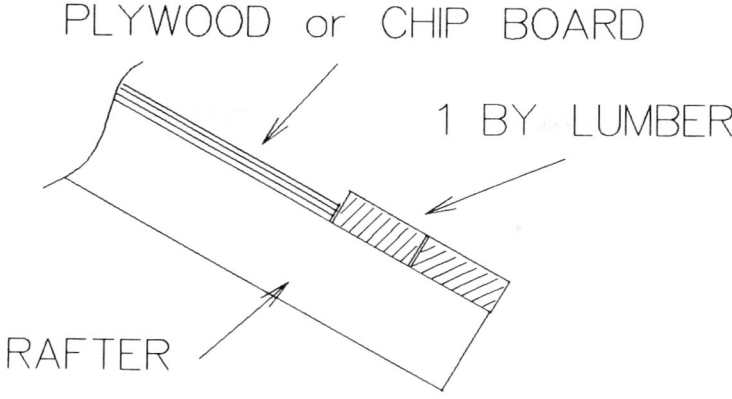

PLYWOOD or CHIP BOARD

1 BY LUMBER

RAFTER

EXPOSED OVERHANG

Figure 5–6 Overhang is built up with thick lumber

Sheathing direction.   Face grain of all plywood roof sheathing should run in the same direction (see Fig. 5–7). General practice is to run it parallel with the eave lines.

Spaced sheathing.   Figure 5–8, roof sheathing pattern for wood or wood shake shingles, is a *spaced roof.* The field sheathing is comprised of board, space, board,

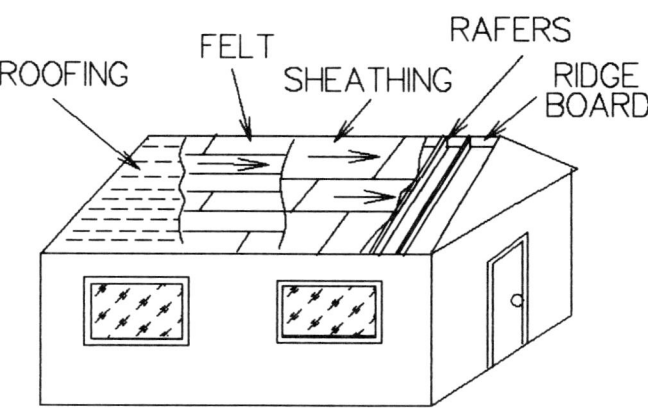

FELT AND SHEATHING RUN ACROSS
THE RAFTERS—PARALLEL TO EAVE

ROOFING   FELT   SHEATHING   RAFERS   RIDGE BOARD

Figure 5–7 Makeup of typical roof

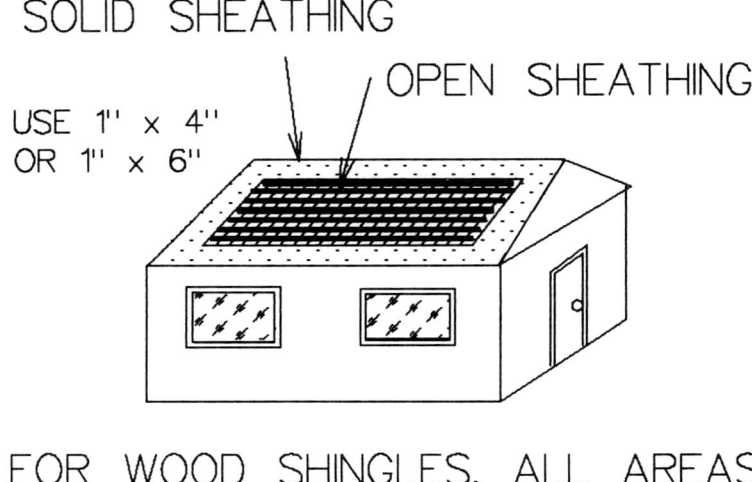

SOLID SHEATHING

OPEN SHEATHING

USE 1" × 4"
OR 1" × 6"

FOR WOOD SHINGLES, ALL AREAS
WOOD SHAKE IN COLD AREAS

Figure 5–8   Sheathing for wood shingles

space, etc. The eaves and rakes are of solid sheathing since their undersides are usually exposed to view from the ground. Spaced shingling is optional under wood shake, mandatory under wood shingles. Spaced sheathing is sometimes used under tile, but I don't recommend it.

OVERHANGS AND SOFFIT

Find out if the overhangs will be covered by soffit or not. If not, has the contractor used thicker sheathing over the overhang area? Remember you have to nail to this area. Use nails that are too short and your roofing will be blown off in the first good wind storm. Use nails that are too long and they will punch through the sheathing and be seen from below, an unsightly view (Fig. 5–9). It also makes it rough for the painters, as the nail ends will rip roller covers to shreds.

What kind of sheathing is on the building you are planning to shingle? You don't want to attempt to install asphalt shingles on an open sheathing roof or, conversely, sawed wood shingles on a solid sheathed roof. In the first instance, asphalt shingles not only will be difficult to install (no nailing surface), but if installed will warp and cup between the sheathing slats (see Fig. 5–10). In the second instance, sawed wood shingles must breathe, they need air flow under them. If installed on solid sheathing, there will be no air flow and they will mildew and rot on their undersides.

Generally it is the framing contractors job to install the sheathing, but not always.

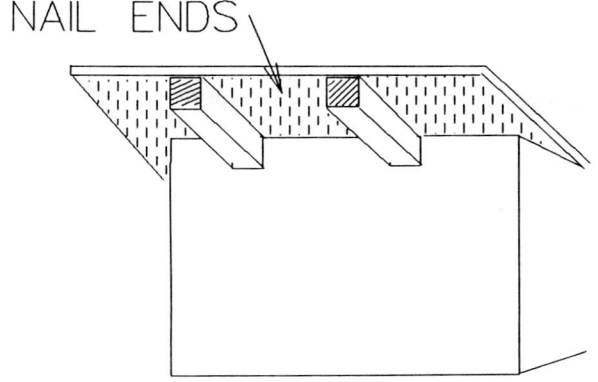

NAIL ENDS

NAILS TOO LONG AT OVERHANG

Figure 5–9   Too long nails may cause problems

There are times when it may become part of your bid package depending on the person or contractor for whom you are doing the bid. Large contractors probably have a full crew trained to install sheathing. An owner builder or small contractor may want you to do the sheathing work. There are times when the contractor did apply the sheathing but it was unsuitable for the roofing being installed. He applied a solid sheathed roof and it now needs furring strips installed so that tile can be properly installed. The strips serve a duplicate function of being footholds for the roofers.

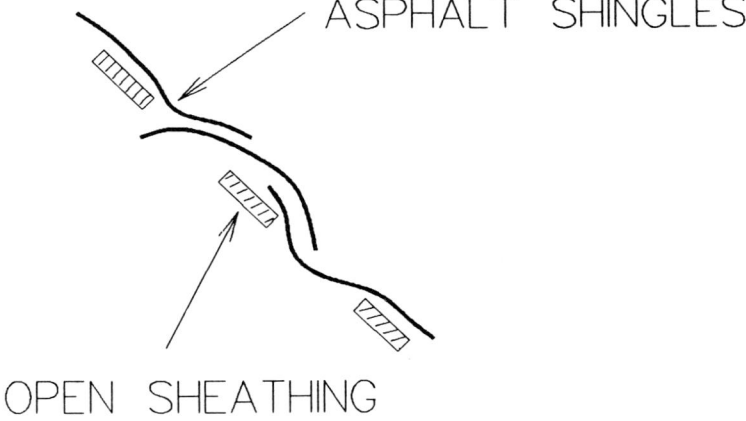

ASPHALT SHINGLES

OPEN SHEATHING

Figure 5–10   Results of using asphalt shingles over spaced or open sheathing

FASCIA

Fascia is the board used to cover the rafter's cut and exposed ends. These rafter ends are called rafter tails. By definition, fascia is a broad painted band mounted horizontally around a building.

See Fig. 5–11. Fascia is used to cover the unfinished roof sheathing edge. Fascia lumber should be reasonably free of splits, knots, and other defects since it is visible from the ground. Fascia should be painted before the drip edge is installed.

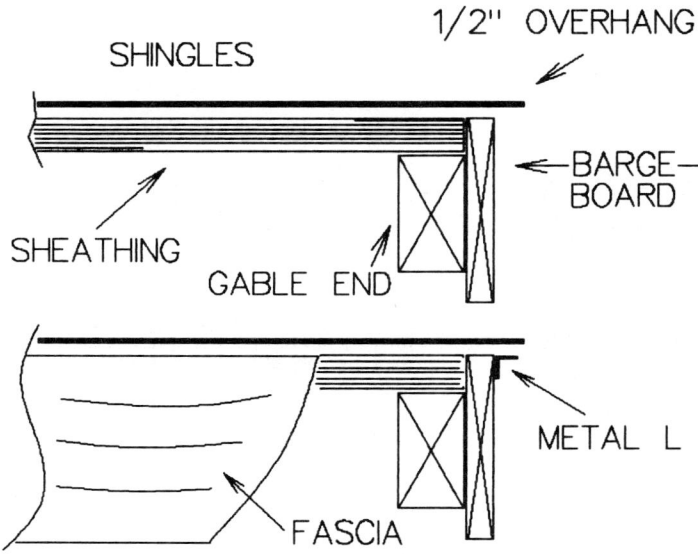

Figure 5–11   Fascia and barge-board use

The thickness of the fascia board can make a difference. To the left of Fig. 5–12 there is shown a 1 × 6 fascia installed. It bends and warps between the rafter tails.

Too thin a fascia board, a fascia board with knots and splits, a fascia board of inexpensive quality, will cause future problems. I suggest using 2 in. thick, number 2 or better, treated lumber for all eave fascia. Instead of 2 in. lumber, you can install finished 1 in. lumber over a 2 in. rough lumber fascia backer, as normally done along the rakes. The extra thickness of the 2 by lumber keeps everything dimensionally stable. To save cost, number 3 or better lumber may be used if it will be covered with vinyl or aluminum trim or with gutters.

Figure 5–13 illustrates typical fascia and soffit installation. The soffit is dressed to the siding with a *frieze block.* The frieze block supports the soffit at the siding and hides the intersection joint. These moldings can be manufactured from lumber, plastic, plaster, or metal.

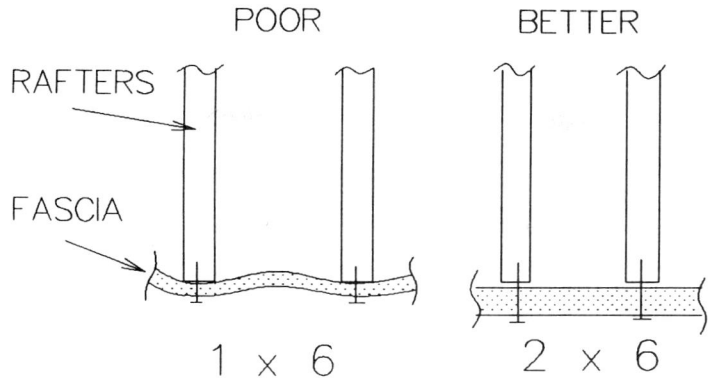

POOR   BETTER

RAFTERS

FASCIA

1 × 6   2 × 6

THICKER FASCIA LUMBER IS MORE STABLE
APPEARANCE IS BETTER & FASCIA
IS EASIER TO INSTALL

Figure 5–12   Use thick fascia or a fascia backer

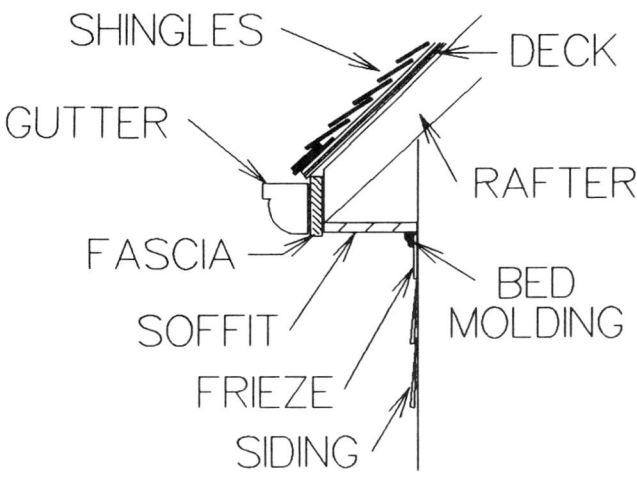

SHINGLES   DECK
GUTTER
                                  RAFTER
FASCIA
                                  BED
SOFFIT   MOLDING
FRIEZE
SIDING

CORNICE WITH RAIN GUTTER

Figure 5–13   Nomenclature of various cornice pieces

Figure 5–14   Use of molded trim and fascia

Fascia need not be dull and uninteresting (see Fig. 5–14). I believe these are molded plaster or concrete. They could be molded plastic. Low-cost molded plastic trim is allowing us to put individual style back into buildings.

Plain wood fascia can be made decorative by adding molding to its exposed face (see Fig. 5–15). Shown are two 1/2 round moldings glued and nailed to the fascia. Selection of

EXPOSED FASCIA

ADD 1/2" ROUND OR OTHER
TRIM TO DECORATE FASCIA
NAIL OR GLUE AND NAIL

Figure 5–15   Make your own fancy fascia

molding shapes and sizes is vast, thus a building can be customized easily. I do have a few suggestions on how to keep the fascia board and the molding strips from eventually rotting due to water entrapment between the two pieces. If you are attaching the molding with nails only or with nails and spots of glue, then paint both the molding and the fascia before assembly. If you are attaching the molding to the fascia with nails and full surface contact gluing, then paint the fascia after assembly. These methods will help prevent dry rot.

Bed Molding

Bed molding by definition is any molding installed below a deep projection. It is installed above the frieze board. One picture is worth a thousand words (see Fig. 5–16).

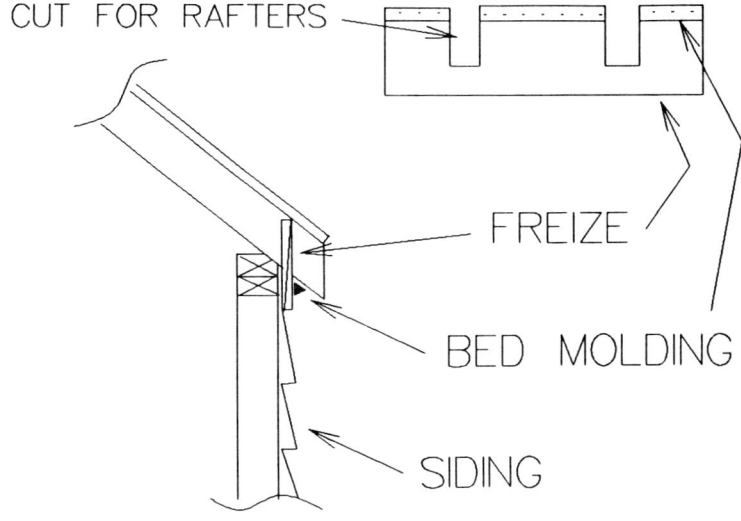

Figure 5–16   Bed molding

Shingle Molding

Shingle molding is shown in Fig. 5–17. It is installed on the fascia board just below the shingles.

Adding Gingerbread

See Fig. 5–18. This home is nothing more than a box with a roof, but then again, look at the extra care taken to add a little personal style. The fascia on the upper roof gable has molding rather than just being a plain flat board. The porch is decorated with a little gingerbread, the decorative section between the upper post. Although this home needs a little paint and some tender loving care, it still has more style than many new homes I've seen.

We have lost much of the gingerbread to the new glass box architecture. Cities like

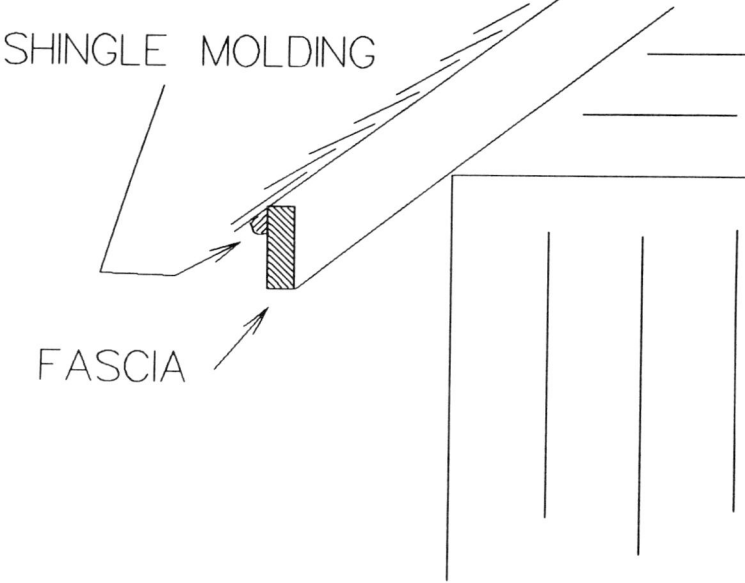

SHINGLE MOLDING

FASCIA

Figure 5–17   Shingle molding

Figure 5–18   Adding gingerbread

San Francisco had thousands of elegant-looking homes. Earthquakes, age, cost, and a new breed of architects are replacing the elegance with cold steel, glass, and concrete boxes.

See Fig. 5–19. If you ever get to Laramie, visit this place. The museum pieces are our Western history in action. The building is full of architectural fun, from the gingerbread to the spiral metal crown.

Figure 5–19 Variations in architectural style

Figure 5–20 shows that roofers of the past know their trade. Notice that the rain gutters double as walkways and work platforms. The corbels add strength and architectural style. A little lace work at the fascia, when added to the window treatment, gives the building a friendly feeling.

Figure 5–20   Double-purpose rain gutters

Overhang Rafters With Lookouts

Figure 5–21 shows a typical roof overhang design that includes a fascia board, a fascia nailer, and a lookout. The *lookout* doubles as a support for the overhang and as a nailer for a soffit. If it is to be covered by soffit, it can be face nailed to the rafter tail. If it is to be exposed, not covered with soffit, then it should be cut and nailed to the rafter tail, as shown.

Overhang Headroom

See Fig. 5–22. Physical damage has taken its toll on this corner; damaged shingles and flashing are present. This usually happens when one builds a roof overhang too close and too low to a driveway. Cars get under with no problem; it's the moving van or delivery truck that gets into trouble. Most cargo carrying trucks require 11 ft or more headroom. Some bigger ones require 13 ft or more. Design your roof overhangs accordingly.

Rafter Tail Cuts

See Fig. 5–23. I love this one and I have seen it frequently. The rafter end cuts are 90 degrees to the roof plane. Then, a gutter is installed, a gutter that should catch water runoff and carry it

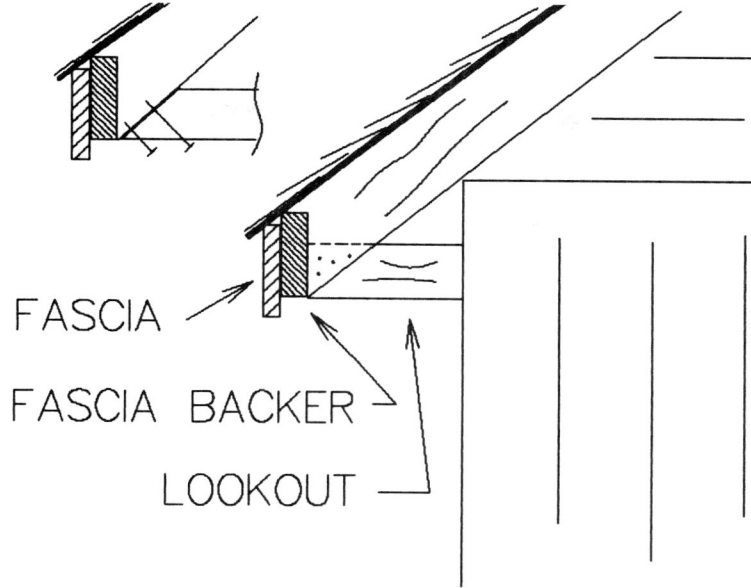

FASCIA

FASCIA BACKER

LOOKOUT

Figure 5–21   Look out for the lookout

Figure 5–22   Ensure proper ground to roof height

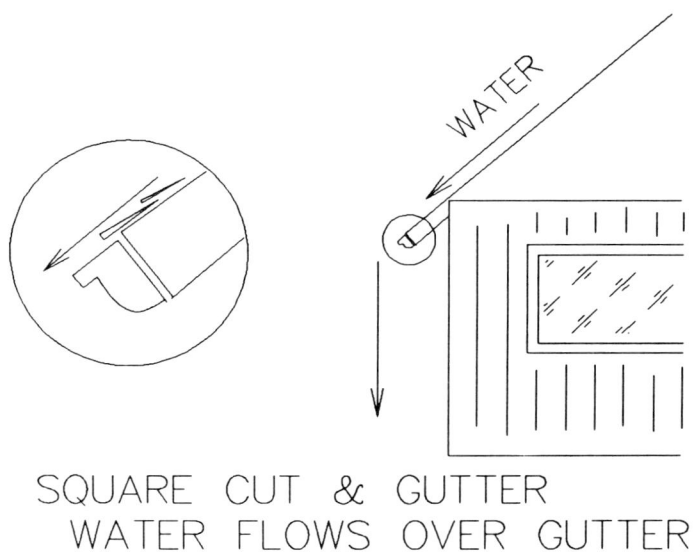

SQUARE CUT & GUTTER
WATER FLOWS OVER GUTTER

Figure 5–23   Improper rafter cut for installation of gutters

away. On 8:12 and greater slope roofs, the water will flow over the gutter and on to the ground. This ground water seeps into the earth and eventually ruins the building's foundation.

See Fig. 5–24. The rafter end should be cut to accept mounting of the gutter so that the gutter is as shown. The water flow from the roof is to enter the gutter and be carried to a point out and away from the building.

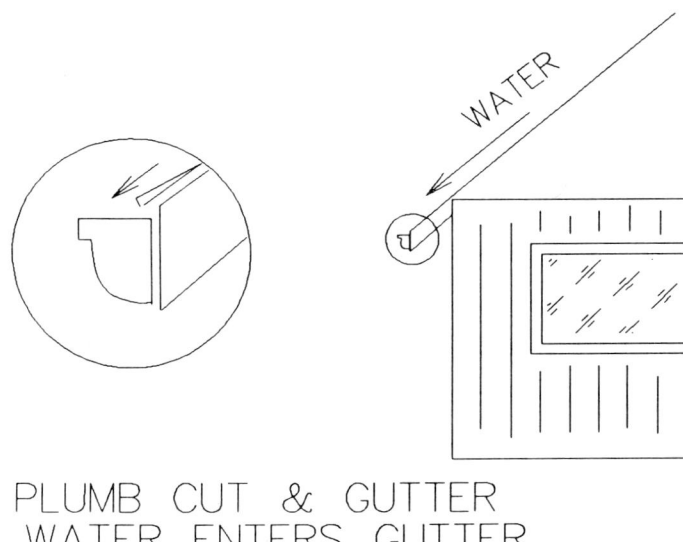

PLUMB CUT & GUTTER
WATER ENTERS GUTTER

Figure 5–24   Proper rafter cut for installation of gutters

Figures 5–25 to 5–32 give possible rafter tail end cuts for various visual and architectural effects. The templates should be enlarged to full scale of the lumber being used. Once drawn, a cardboard, plastic, wood, or metal template can be made. Some designs will accept a rain gutter, others will not. Some will accept fascia and soffit. All can be shop formed before installation. The straight cut ones can be cut after the rafters are installed. The rounded rafters are best cut with a band saw. They should be cut before being installed.

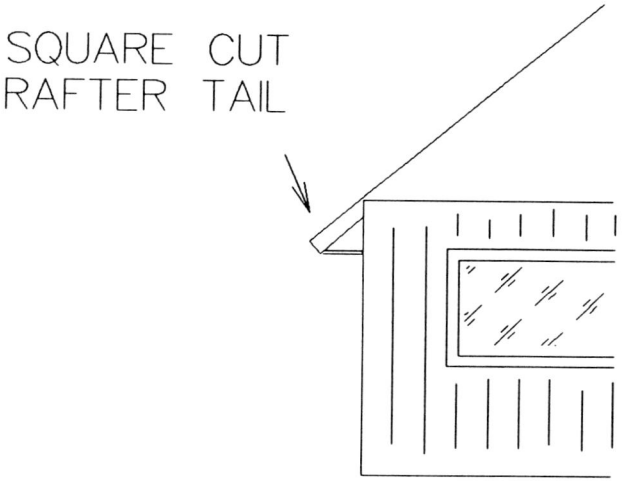

Figure 5–25 (to 5–32)   Various rafter tail cuts

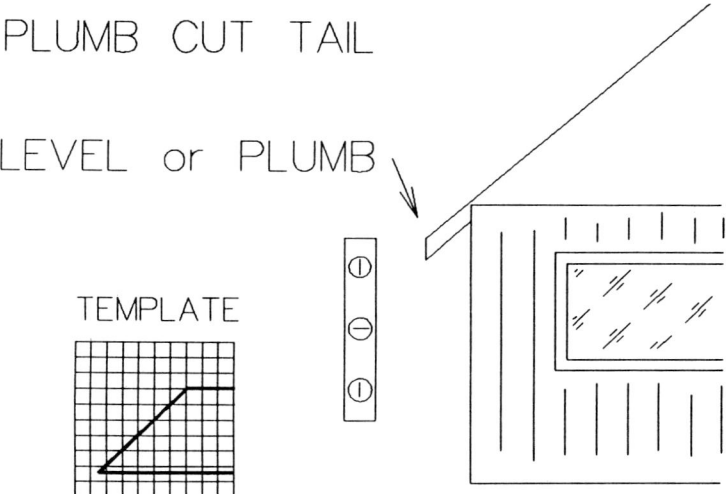

Figure 5–26

GUTTER REPLACES FASCIA
NO SOFFIT

TEMPLATE

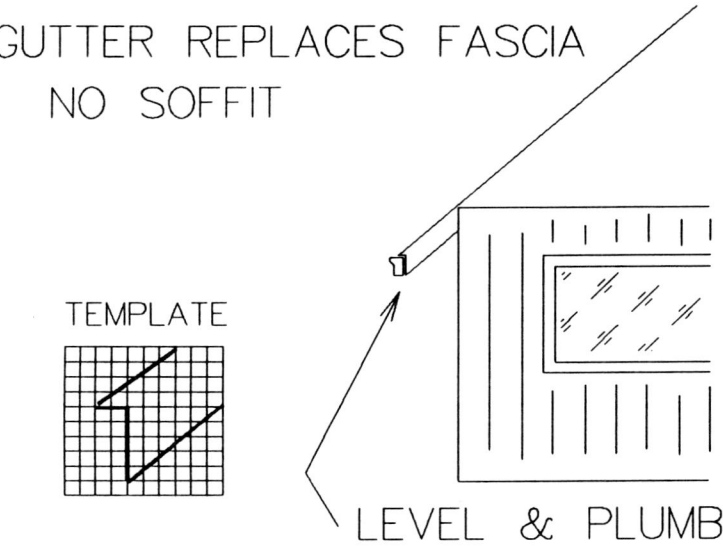

LEVEL & PLUMB

Figure 5–27

NO GUTTER & NO SOFFIT

TEMPLATE

Figure 5–28

FOR GUTTER & SOFFIT
INSTALLATION

TEMPLATE

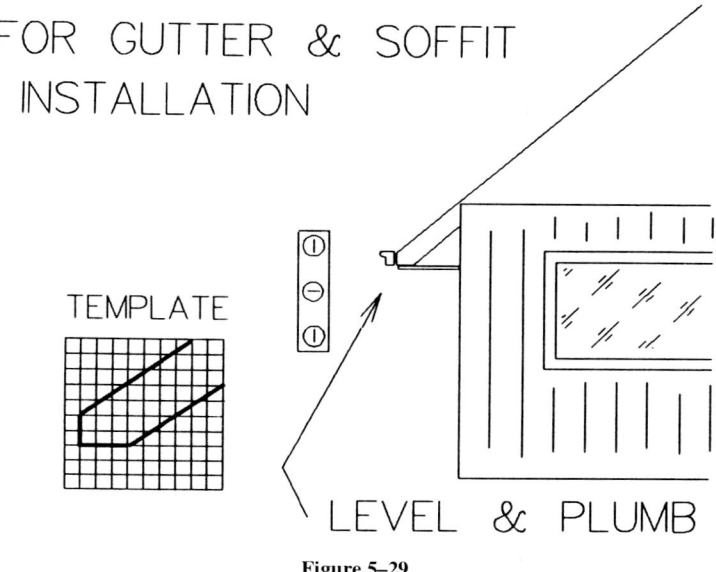

LEVEL & PLUMB

Figure 5–29

QUARTER CIRCLE

TEMPLATE

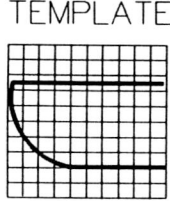

Figure 5–30

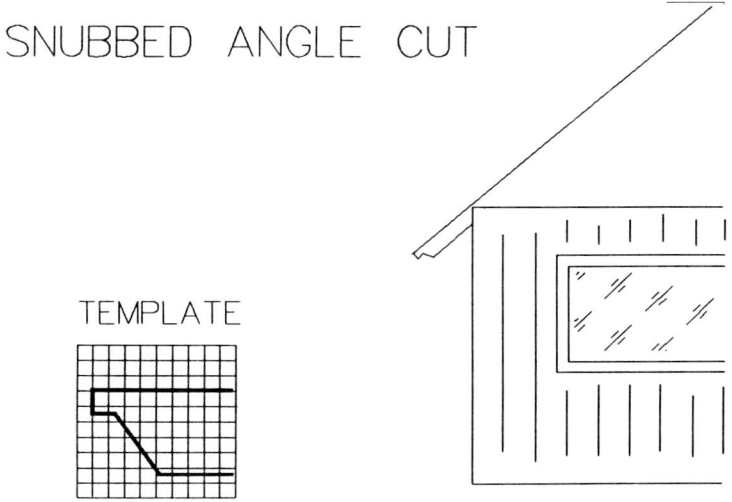

SNUBBED ANGLE CUT

TEMPLATE

Figure 5–31

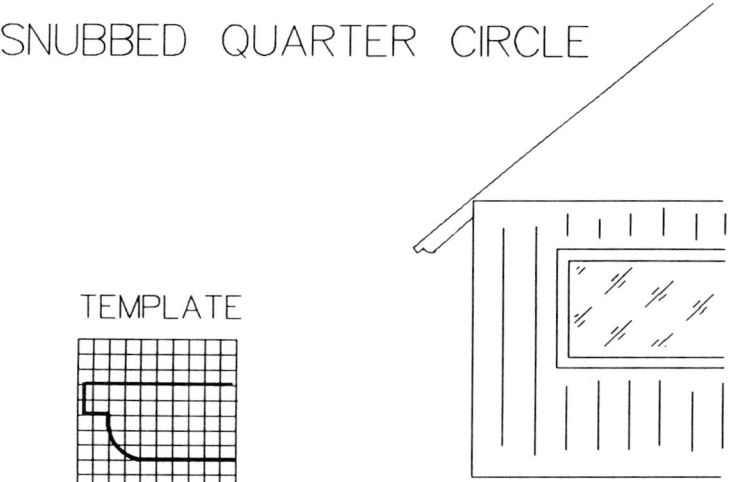

SNUBBED QUARTER CIRCLE

TEMPLATE

Figure 5–32

ROOF SHAPES AND SIZES

Roofs come in all shapes and sizes (see Fig. 5–33). This roof is 18 ft square. It provides protective covering to those persons trying to find their door keys while juggling five bags of groceries in a rain storm. The roof and its decorative column provide a pleasant change to what otherwise would be a dull entryway.

Figure 5–33 Roofing jobs come in all sizes

Figure 5–34 is a built-up flat roof in an apartment complex. The overhang, a little narrow for my taste, contains emergency exit and walkway lights. The poured concrete walkway is the roof for the lower floor. It, too, contains lighting for the safety of tenants.

See Fig. 5–35. So who says a roof has to be rectangular? This creative home is the highlight of the neighborhood. A more conventional Hip roof is shown in Figure 5–36.

This figure also shows interesting architectural design created by various roof lines. The stain under the window is caused by the upper roof drain pipe emptying onto the lower roof.

Figure 5–37 illustrates locking shingles on a Mansard roof. The separation between upper and lower floors is the rain gutter system.

Figure 5–38 shows an interesting roof line. Figure 5–39 is a variation of a Dutch Hip roof. See Fig. 5–40. A little fun and style are shown here. Different color shingles and different roof angles add interest to this public building.

In Fig. 5–41, the owner took a plain, dull house and turned it into his personalized home. The shingles are asphalt. As shown in Fig. 5–42, with proper design and planning, a roof need not be square and boxy.

See Fig. 5–43. This barn has a loft for storage of hay and feed. The ridge board was left extended for a purpose. It provides a solid anchor for a block and tackle hoist.

Figure 5–34   Flat roof overhang protects stair and walk areas

Figure 5–35   Custom-designed roof

Figure 5–36   Use of different roof lines adds character to building

Figure 5–37   Variation of a mansard roof

Figure 5–38   Unusual roof line, may have been a second-story add-on

Figure 5–39   Variation of a Dutch hip roof

Figure 5–40   Varied shingle color used for effect

Figure 5–41   Obtaining drainage on a near flat roof

Figure 5–42   Roof curvature

Figure 5–43   Extended ridge board has a purpose

Roof Nomenclature

Figure 5–44 illustrates the main portions of a roof: the left rake, ridge, right rake, field, and eave.

The Shed Dormer

Figure 5–45 illustrates a shed dormer.

The Gable Dormer

Figure 5–46 illustrates a gable dormer.

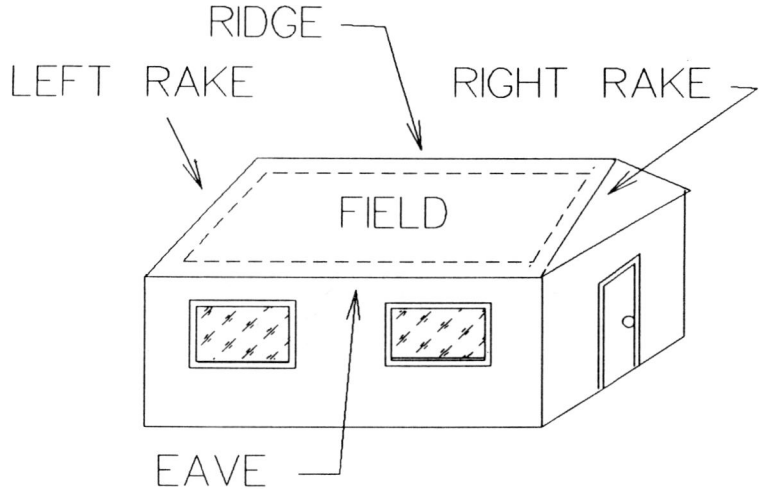

Figure 5–44   Parts of a roof

SHED TYPE DORMER

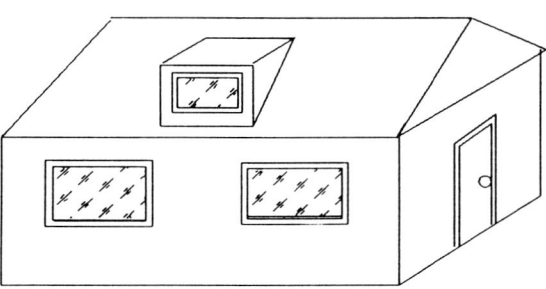

Figure 5–45   Shed dormer

GABLE TYPE DORMER

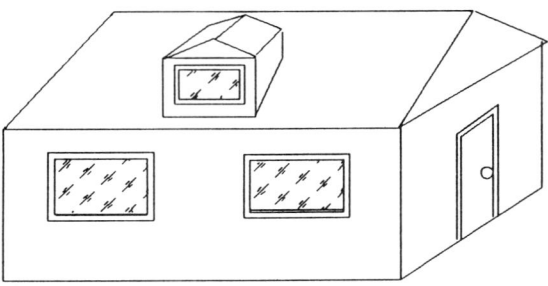

Figure 5–46 Gable dormer

ROOF TRUSS DESIGN

What kind of truss is being built? Figure 5–47 shows several different truss designs that are currently in use. Why should this matter to you? Well if you are asked to install clay tile, you will soon find out. The ridge area of the trusses must be built up with 2 by 4 lumber in order for you to have a surface on which to nail (Fig. 5–48).

Without this extra board, you can't finish the job. If the trusses are 2 × 3 in. lumber, or the *king post* design, they will not hold the weight of the tile. I'll go into more detail in

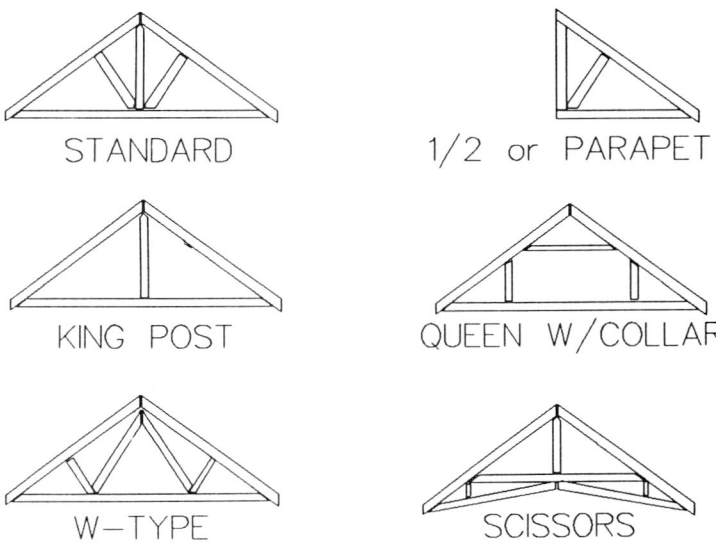

Figure 5–47 Roof trusses in general use

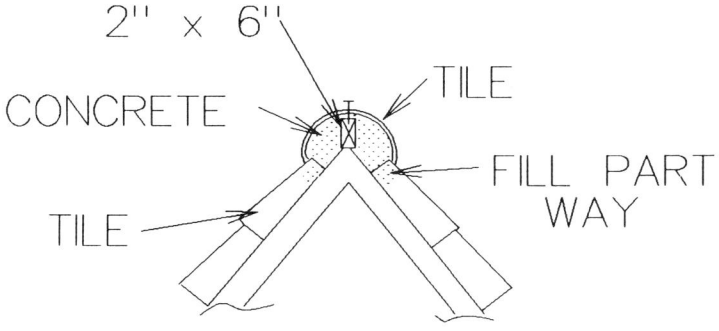

TYPICAL MISSION TILE
RIDGE TREATMENT

2" x 6"

CONCRETE

TILE

TILE

FILL PART
WAY

Figure 5–48   Truss design affects the roofing material used

the chapter on tile. Roof variations will affect the methods and materials you use for each roof.

The designs also will affect the method and placement of the insulation installed and the ability of the building's owners to expand their floor space. The *queen post* with collar beam is the best design for expansion, but requires the insulation be installed above the collar beam and to the outsides of the queen post. The *scissors* design is for when one wants to install a vaulted ceiling. The *parapet* is normally used to provide a covered walkway in front of a building, and may require flashing at the junction of the roof to the wall.

Roof trusses have changed with the arrival of modern manufacturing, glues, and presses. Figure 5–49 is a truss made from plywood, used to support flat roof decking. It is dimensionally stable since it is not a solid piece of S-dry or S-grn lumber, or a 2 × 12. Lumber can split, crack, warp, twist, shrink, and sometimes be a nuisance to use. These plywood trusses do none of the above. They are lighter in weight than solid lumber and are easier to handle on the job site. They are prescored every foot or so for knockouts, that pop out with a strong blow of a hammer. Knockouts provide spaces for running pipes (sprinkler systems, water) and conduits along the roof or under floor. These trusses are designed for and are used for floor joists as well as roof rafters. For more information, one manufacturer is: Trus Joist Corporation, 9777 West Chinden Blvd., P.O. Box 60, Boise, Idaho, 83707, 1-800-338-0515.

Additional information on roof truss designs can be obtained from: The Small Homes Council, 1 East Mary Road, Champaign, Illinois, 61820, and from: American Plywood Association, 1119 A Street, Tacoma, Washington, 98401. You also may obtain valuable information from: Construction Specifications Institute, Washington, D.C., 20036, or CSI, Alexandria, VA, 22314. The CSI has developed a system of preprinted (fill-in-the-blanks roof specification) forms, called Manu-Spec™. One such form is used for specification when doing BUR work.

PLYWOOD
TRUSS or
JOIST

Figure 5–49   Plywood truss or joist

Stanley Bostitch, East Greenwich, Rhode Island, 02818, has developed truss assembly methods that you can use at the job site or indoors (Fig. 5–50).

FABRICATED BEAMS

Fabricated or laminated beams are coming of age. They are used for residential, commercial, and industrial buildings where large spans are required or elegant open beam structures are called for. The beams are made of various core lumbers with pine, oak, cedar, or other lumber as the exposed facings. Face lumber can be sanded and prestained at the assembly plant. Straight or curved beams can be fabricated to fit any building architecture.

The lumber used is finished sized lumber, 3/4 in. for curved beams, 1-1/2 in. for straight beams. When ordering, order by number of laminations and finished lumber size. Example: If you want a straight 40 ft beam, 15 in. deep, order a 40 foot × 10 lamination.

RAFTER TO RIDGE BEAM CONNECTION

Figure 5–51 illustrates a typical site-built rafter to ridge beam installation. The ridge beam is built and metal joist hangers are installed (can be installed on the ground before being placed, if desired). The rafters are then inserted and nailed. Depending on the manufacturer

Compact, Walk Through Truss Jig With TFCN Kit

You get all truss jig hardware including saw guides, hardened-steel tabletop clincher anvils and detailed construction plans for six track-mounted movable nailing tables (2 each heel and web, 1 peak and 1 splice). Six tables accommodate practically any size or style truss up to 40-foot (12.2 meters) span and 7-in-12 pitch. Just add extra tables and track to handle any size pitch and span trusses! "Walk-through" spaces between tables simplifies cutting, nailing and unloading trusses. Truss tables and track may be installed indoors or out.

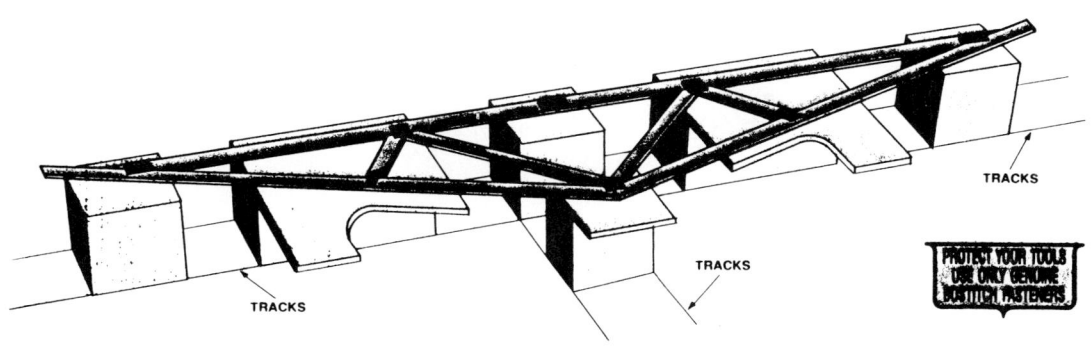

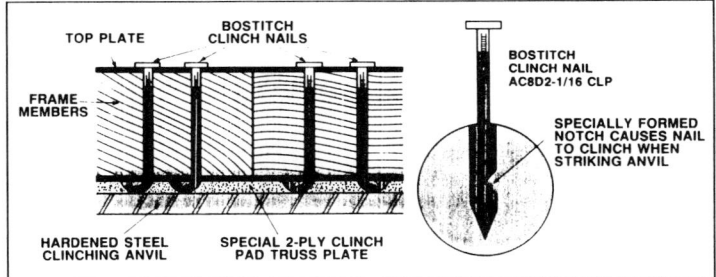

The Heart of Bostitch Clinch-Nail Truss System is the special two-ply truss plate and companion clinch-point nail. Each special nail is power-driven completely through the joint and is automatically clinched on the opposite side. Two-ply construction of Bostitch lower clinch pad truss plate insures precisely controlled clinching action.

Figure 5–50   Roof truss jig (Reprinted with permission of Stanley Bostitch)

of the hanger, 12 to 16 nails per hanger are generally required, 6 to 8 in the beam, 6 to 8 in the rafter. The nails are specially designed for maximum performance; they are short and thick.

See Fig. 5–52. To prevent the building from separating at the roof peak, it is advisable to strap the rafters together. This metal strap is not required for most residential construction. It is required in some localities for commercial and flat roof construction. It may be required in your area, so check with your building inspector. I feel it is good security and worth the few extra dollars spent, especially in earthquake areas. The strap is required on slopes of 21:12 or more. Collar beams may be installed in place of the straps, if needed.

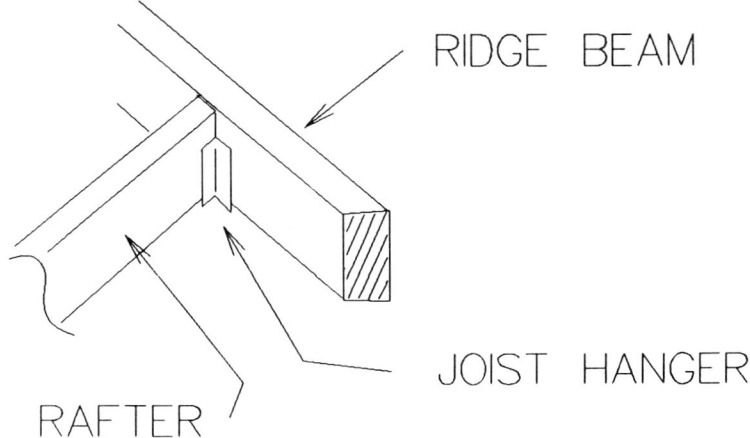

Figure 5–51   Beam to rafter intersection

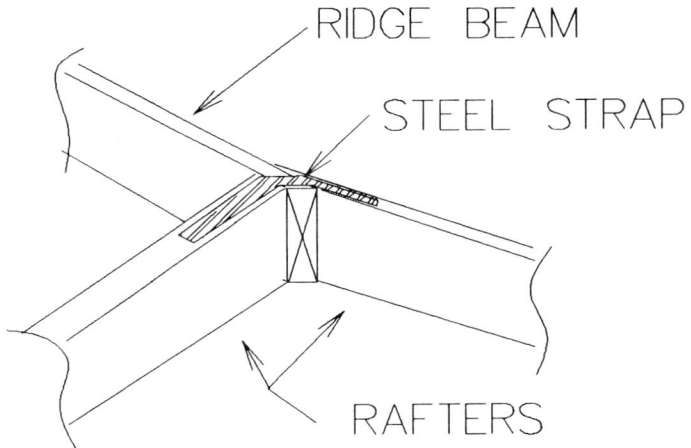

Figure 5–52   Metal strap is required on some roof installations

ROOF TO WALL SECUREMENT

Figure 5–53 is a cross-sectional view of how the typical wall to roof joist/rafter intersection is made. Shown is the insulation with its vapor barrier properly installed toward the building's interior. The roof joists sit on the wall sill plates and are fastened to the wall by any of several methods. The blocking pieces go between each joist. Blocking keeps small animals, birds, etc. out of the attic area. It keeps the insulation in the attic area and it helps

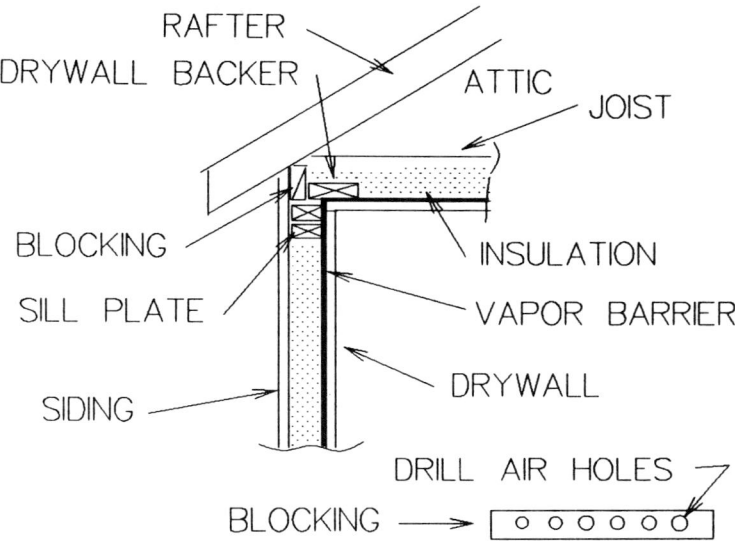

Figure 5–53 Typical wall to roof intersection

prevent the joist from twisting. Any joist that twists would destroy the ceiling. Blocking pieces are sometimes drilled to allow ventilation into and out of the attic. The holes drilled must not exceed one third the width of the lumber. A piece of screening is stapled over the holes on the interior side to prevent insect and animal entry.

METAL HANGERS

In light framing, residential work, the roof trusses can be toe nailed to the wall sill plates (see Fig. 5–54). Many areas of the country will accept this method, others will not. California, Florida, and other states where the building will be subjected to high winds or earthquakes require seismic hangers. These are metal strips designed to hold the truss work to the sill plate. These metal strips are *plate anchors* but are frequently called *seismic anchors*.

Eight nails are required for each anchor. Four anchors are required for each truss, one on each side of the truss where it attaches to the sill plates.

You are not required to paint these anchors since they are covered by the siding and the interior walls. If used in a situation where they are exposed, such as a carport or shed, then I suggest you do paint them before installation. Dip them in mineral spirits for a minute to remove oils. Then dip in metal primer and allow to dry. When dry dip or spray them with a good grade of nonchalking exterior latex enamel.

Figure 5–55 illustrates galvanized sheet metal hangers laid out on a plastic tarp for painting. Not shown is the cost of the hangers. We made a few phone calls and found a local

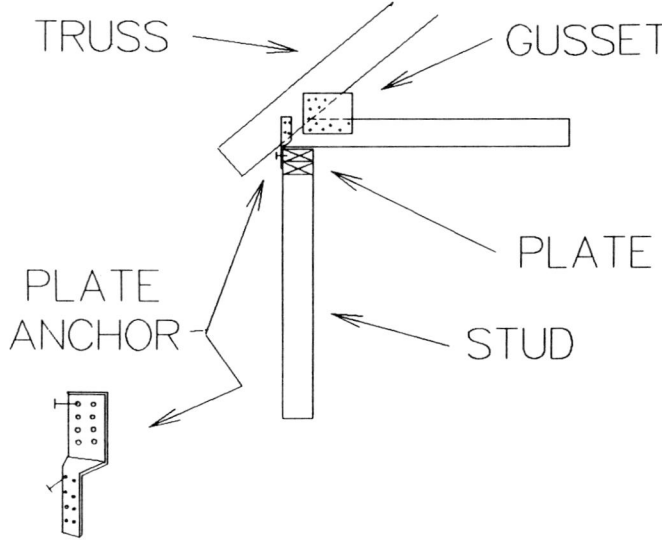

Figure 5–54 Plate anchors may be required

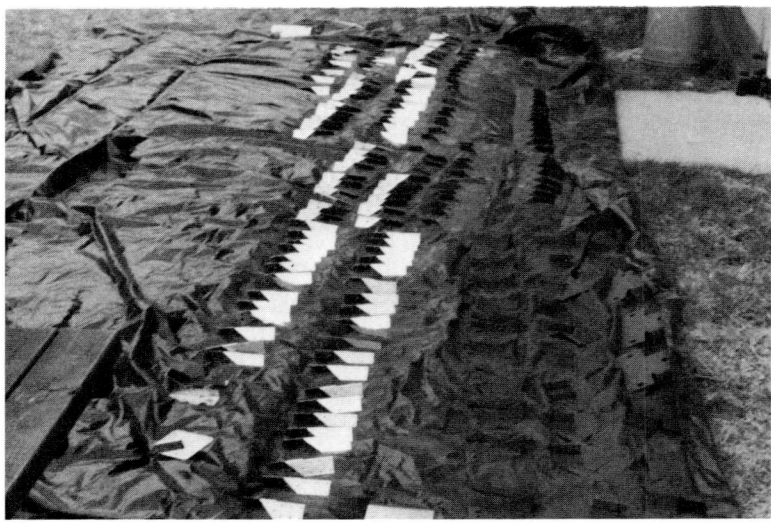

Figure 5–55 Paint metal before installing

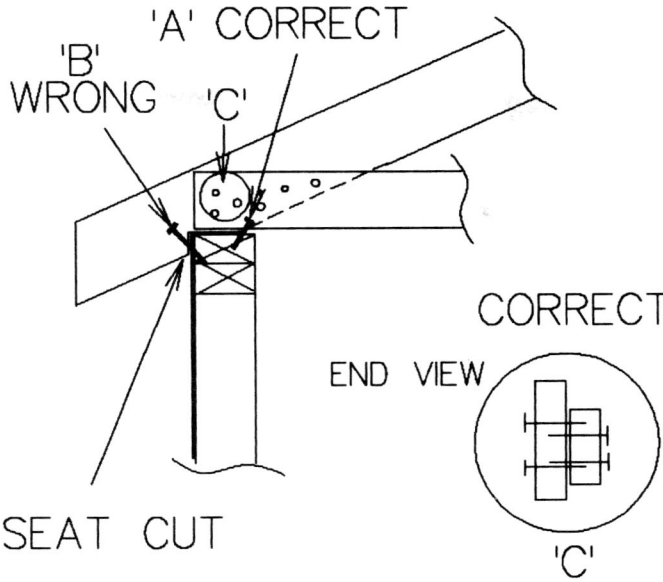

Figure 5–56   Joist to rafter tie in

distributor willing to sell to us in bulk. The savings were nearly 45% from what we had been quoted by the lumber yard.

Figure 5–56 illustrates a simple tie-in of joist and rafter to a wall sill plate. Nailing is done at point A. Nails placed at point B will not hold since they will be with the grain of the rafter lumber and will split the lumber. The face nails at C should come in from both sides, two from the rafter to the joist, two from the joist to the rafter. This method gives maximum holding.

Figure 5–57 shows a typical 1-1/2 story tie-in of joist and rafter to the wall. This only applies to a stick- or site-built building. Stick-built infers that all components are built from individual sticks, lumber, rather than being factory built and shipped to the erection site. The construction shown is used where the joists are used as a floor for an attic room. The extra sill plate then becomes a stop for the flooring (dotted line).

Figure 5–58 shows one method of securing the roof rafters to the sill plate. This is popular on *post and beam* or *barn* construction where the building's main supports are posts set on 48 in. or greater centers. The rafter, the sill plate, and the post are drilled to accept a wood or metal dowel rod. Dowels should be a minimum of 12 in. in length and a minimum of 1/2 in. in diameter. Exterior glue is used to keep the dowel in place. For added strength against uplift, a second and third dowel are used. These dowels are placed at points a and b. They pass through the vertical dowel and lock it in place. This construction was used well before the invention of nails and is still in use today. Most of the homes and barns built in or before the 1800s used pegs and dowels to keep them together. Most are still standing a hundred or so years later.

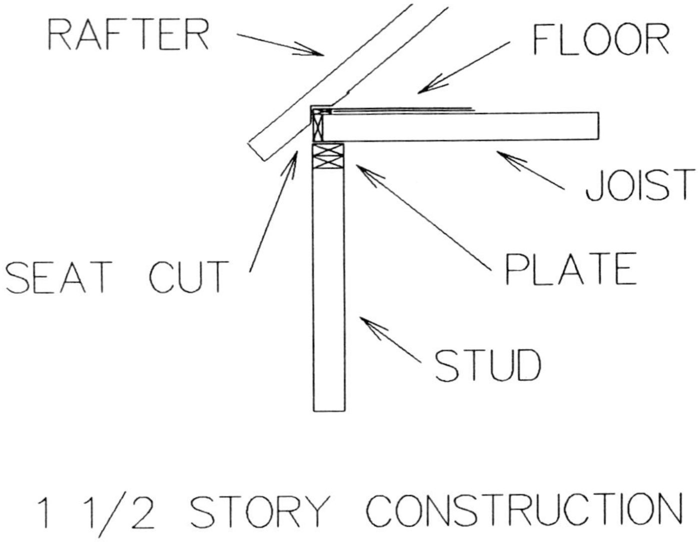

RAFTER → FLOOR

JOIST

SEAT CUT PLATE

STUD

1 1/2 STORY CONSTRUCTION

Figure 5–57 Joist to rafter tie-in

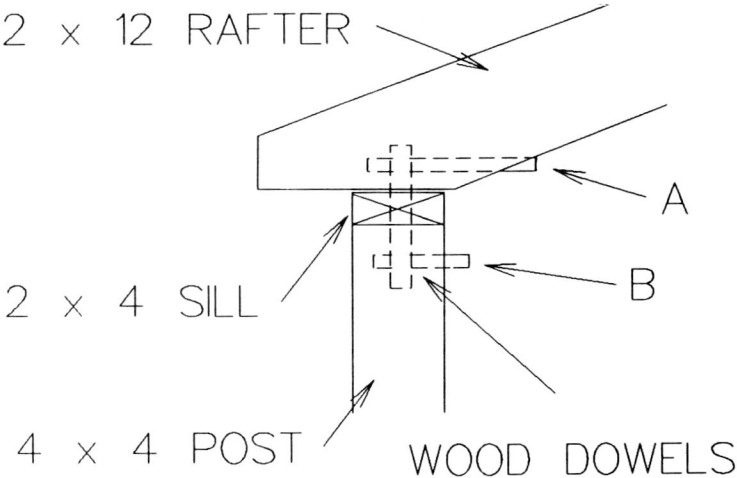

2 × 12 RAFTER

A

B

2 × 4 SILL

4 × 4 POST WOOD DOWELS

Figure 5–58 Post and beam construction

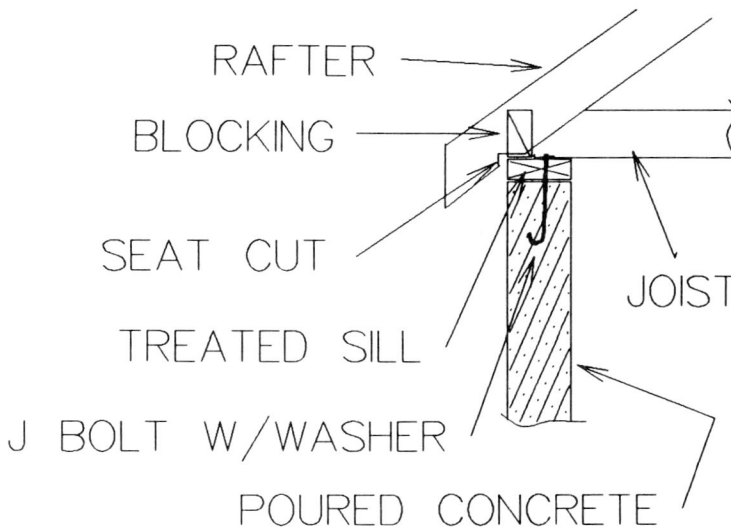

RAFTER

BLOCKING

SEAT CUT

TREATED SILL

J BOLT W/WASHER

POURED CONCRETE

JOIST

Figure 5–59 Joist to poured concrete wall tie-in

Figure 5–59 illustrates an acceptable method of attaching rafters and joists to a poured concrete or hollow masonry block wall. The treated wood sill plate is bolted to the wall via anchor bolts embedded in the wall. The joists and rafters are then fastened to the sill with nails, seismic anchors, or metal L brackets. Shown is a *seat* or *bird's mouth* cut in the rafter. This cut helps in the alignment of the rafters to the sill and helps distribute some roof weight to the sill. The treated lumber sill plate is required by code. Lumber, in contact with earth or masonry, must be treated lumber.

The anchor bolts are J-shaped bolts designed to be inserted in wet concrete or mortar. The J hook locks the anchor in place. These anchors are usually 1/4 to 1/2 in. in diameter, 12 to 18 in. long, and supplied with a flat washer and nut. Cost is from one to three dollars each. Spacing between anchor bolts is per building code, one for each 6 ft of sill length and one within 18 in. of all ends, corners, and openings. Fiberglass insulation is sometimes used between the sill and the masonry to help prevent air leakage into the building. This fiberglass can be purchased in rolls approximately 1 in. thick and 4 in. wide.

Inexpensive strap anchors can be used to secure the joists to the sill plate (see Fig. 5–60). These anchors are embedded into the concrete or mortar during construction of the wall. The joists are slipped into them and nailed. Some designs wrap up and over the top of the joist for maximum protection against uplift.

See Fig. 5–61. Bond blocks, channel blocks, and lentil blocks can be used as top sills for some buildings. Check local codes.

The anchor bolts or straps can be cemented into these blocks the same as in Fig. 5–60. The major difference is that we can now make our tie-in all the way to the building's foundation. The channel formed by these blocks allows us to run horizontal rebar around the

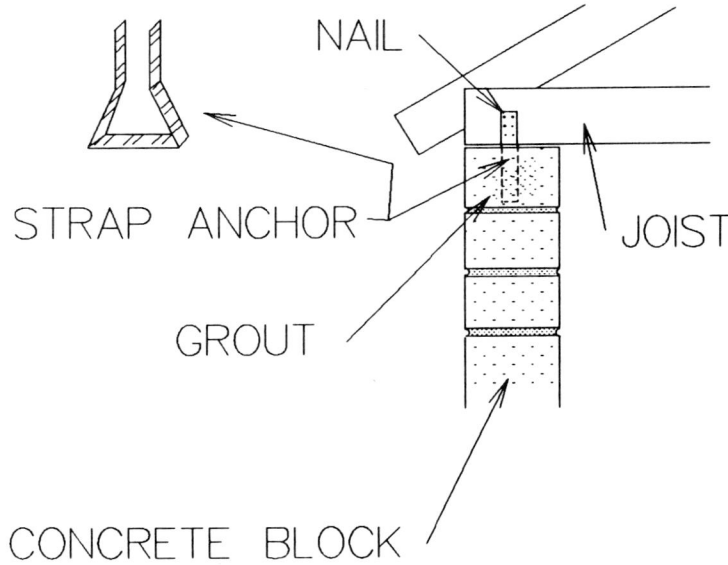

Figure 5–60   Joist to hollow block wall tie-in

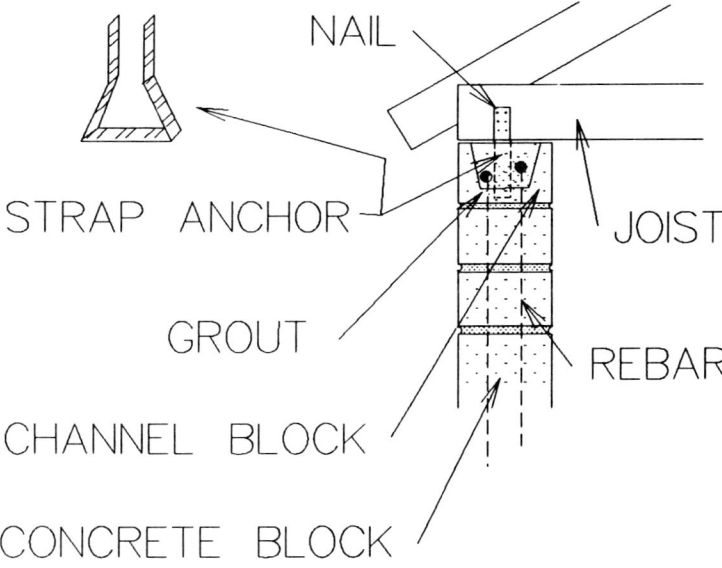

Figure 5–61   Use of channel block for joist to wall tie-in

upper perimeter of the building walls. This horizontal rebar can be tied to the vertical rebar coming up from the foundation. This method of tie-in prevents the walls from spreading outward, tilting. I have seen builders bring the vertical rebar from the foundation up and over the joist. This provides extra protection from uplift but is not a substitute for standard anchors.

DUTCH HIP ROOF BEING ASSEMBLED

Figure 5–62 illustrates Dutch Hip trusses being placed on my father's home. The telescoping lift makes the job easy and reasonably safe. The trusses are kept bundled until used and laid on the main supporting wall upper plates (see Fig. 5–63).

The gable end trusses are moved into position and secured to the outer wall sills with metal straps (see Fig. 5–64). A string line is strung from gable peak to gable peak. The string line will establish placement of the common truss peaks.

One by twos are used as temporary braces from truss to truss before sheathing is installed (see Fig. 5–65). The vertical plumb of the trusses and truss to truss spacing must be held in careful alignment.

Hip, valley, and jack rafters were fabricated on site (see Fig. 5–66). The building is U-shaped and made it difficult to calculate the proper cuts involved. The truss manufacturer complicated things by refusing to build trusses for the purpose. They had never built Dutch Hip roof trusses, much less tried to design hip and valley trusses for one.

Figs. 5–67 and 5–68 show the effect of different widths of the building. These different widths required slightly different roof slopes and a break-point area on the roof.

Figure 5–62   Modern lifting equipment makes the job easy

Figure 5–63   Bundle is laid across bearing walls

Figure 5–64   Gable ends are raised into place

Figure 5–65 Hips are added. One-by lumber holds everything in place

Figure 5–66 Valley rafters are site fabricated

WIDTH CHANGE = SLOPE CHANGE

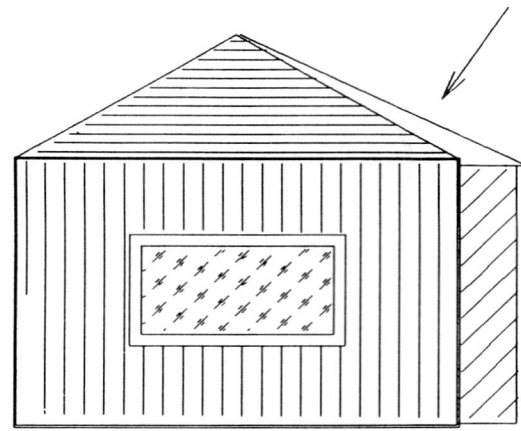

Figure 5–67 Building's width change will change roof slope

EXTENDED ROOF

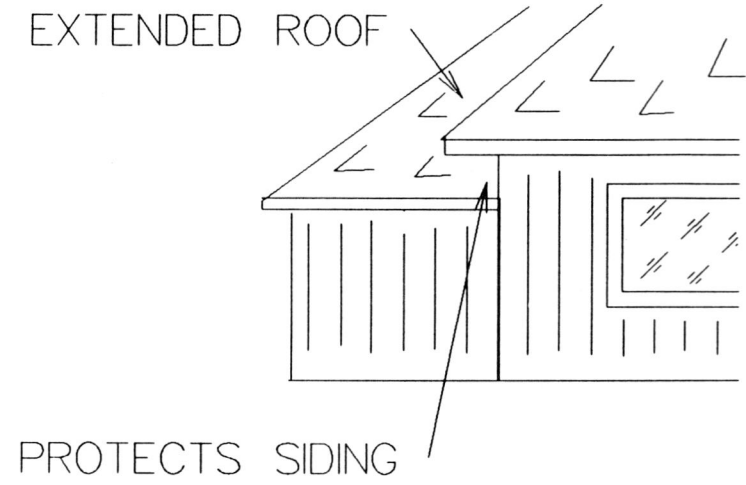

PROTECTS SIDING

Figure 5–68 Overhangs used to protect siding above roof line

Figure 5–69   Green lumber split upon drying

See Fig. 5–69. The bang would have awakened the dead. This upper cord literally exploded from internal stress. I found that California, where this building was constructed, is a difficult state in which to purchase KD or S-Dry lumber. Most lumber is S-Grn and when used as framing or trusses, it must be sheathed very quickly. The hot sun speeds drying of the wet lumber and the lumber warps quickly. Sheathing keeps the moisture evaporation at an acceptable level to allow the lumber to adjust. The sheathing also keeps the lumber from twisting and warping. In this instance, the lumber being unsheathed and being fastened to the wall sills lacked a place to relieve this warping.

See Fig. 5–70, a new roof being installed. I would normally be working from the bottom up, sitting on the already installed shingles. The problem incurred was heat, 105 plus degree heat from 10 AM to 8 PM. The shingles literally melted under you.

See Fig. 5–71, the completed building. Note the gable end vents. The opposite ends of the building are of hip design and, therefore, no gables exist. Automatic power roof ventilators were used to overcome the problem. See the chapter on venting.

There are gutters installed around the perimeter of the entire building. We used brown vinyl gutters and matched the trim stain to the gutter color. The gutter then became an intergral part of the trim.

CATHEDRAL CEILINGS

Figures 5–72 to 5–75 show various acceptable construction techniques used for cathedral ceilings. The rafters usually become the exposed beams or a base for the exposed beams.

Figure 5–70   The author wanted to get his picture in this book

Figure 5–71   The completed roofing job

Nailer rafters should be used if ceiling drywall and insulation are used. Insulation can be fiberglass but is usually foam. Foam, for its thickness, provides higher R-values then fiberglass. Kraft or foil faced fiberglass can be installed from within the building on all designs. Foam sheet insulation is easiest installed from the outside of the building. Install the drywall and then, from the roof, drop in the foam sheets. This outside-in method depends on good weather conditions. Finish the job with the roof decking and roofing.

   In all instances, the vapor barrier is to face the inside of the building. 1/4 round trim is used to finish the drywall to beam intersections.

   Figure 5–73 uses low-cost foam plastic beams. These beams can be added to a room that was not designed to have beams. The nailer lumber should fit the width of the plastic beams purchased. Beams can be nailed or glued to the nailer lumber.

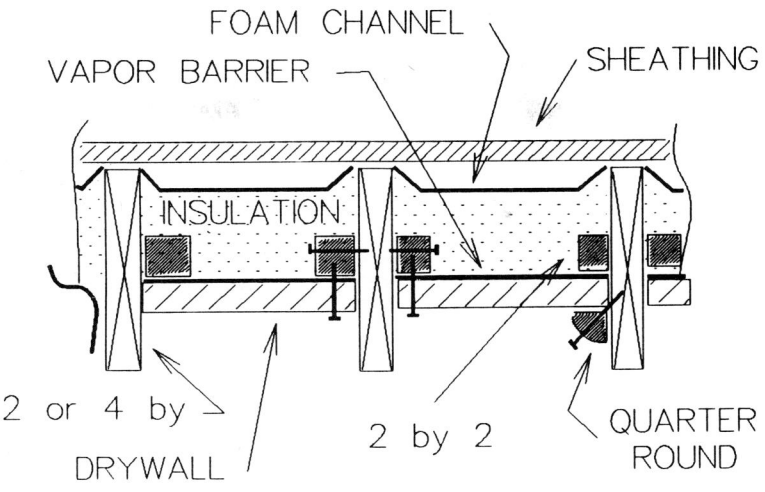

Figure 5–72   Possible cross section of a cathedral ceiling

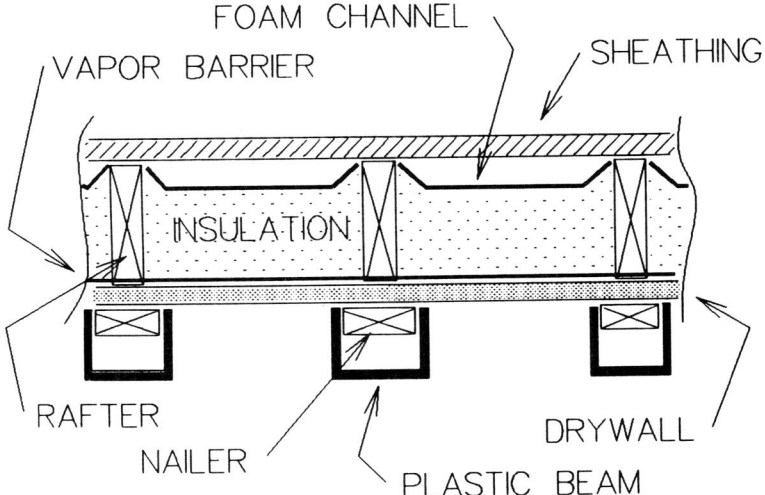

Figure 5–73   Possible method of securing plastic beams

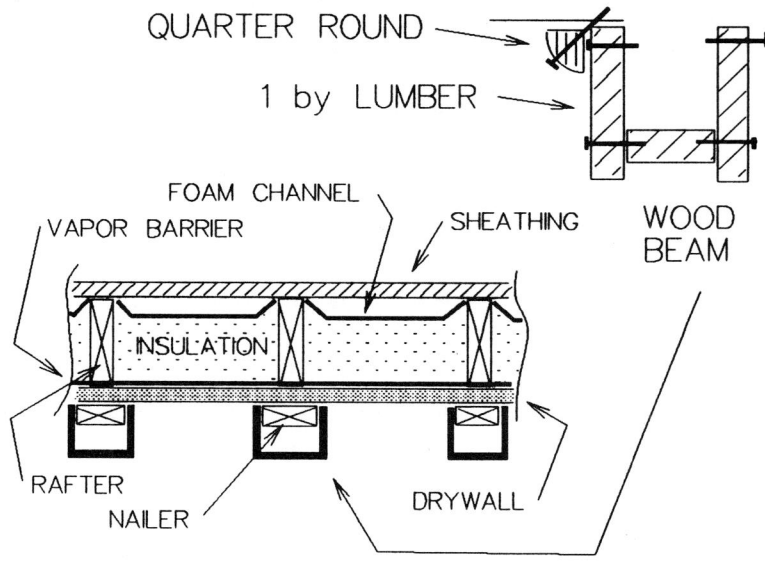

QUARTER ROUND

1 by LUMBER

WOOD BEAM

FOAM CHANNEL

VAPOR BARRIER

SHEATHING

INSULATION

RAFTER

NAILER

DRYWALL

Figure 5–74   Fabricated wood beam construction

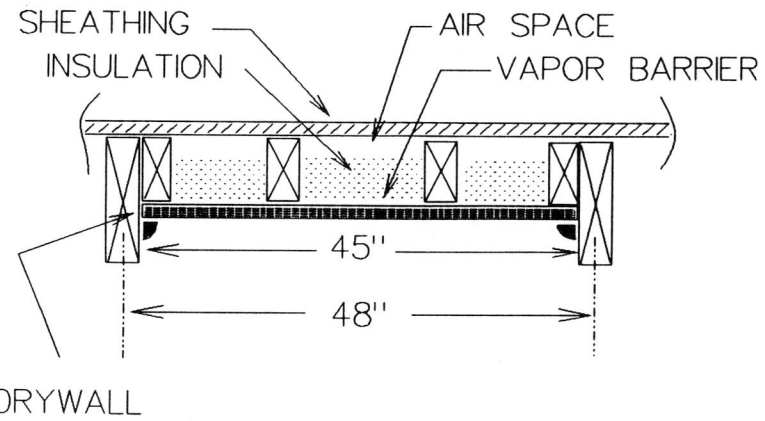

SHEATHING
INSULATION

AIR SPACE
VAPOR BARRIER

45"

48"

DRYWALL

BEAM CEILING DESIGN

Figure 5–75   Possible cross section of a cathedral ceiling

Here are several inexpensive methods of fabricating decorative beams. Figure 5–74 uses 1 × lumber over the rafters to obtain a finished beam. Figure 5–75 shows rafter construction for 48 in. on center beams. Rafter spacing is 16 in. on center.

ROOF VENTING

This is a quick overview of roof venting. Detailed information is contained in the chapter on flashing and venting.

Figure 5–76 illustrates types of vents you find on a roof. Little details can make or break you. One such detail is the number, size, and location of vent pipes. You, as a roofer, are generally required to supply the flashing for these roof vents. Do you have to prime and paint them as well as supply and install them? This can add up to many extra hours of work.

Vent Pipes for DWV

Why are those vent pipes there? Vent pipes are a necessary component of the plumbing package. To make a drain function correctly, one must allow air into the drain line after the trap. A trap is that S-shaped pipe you find under your sink. It is used to keep rodents and sewer fumes out of your house by trapping water in it. Water acts as a barrier to both. Without the vent pipe, air would be trapped between the S trap water and the sewer and that air would prevent the sink from ever draining. Every drain in a building *must* have a vent to

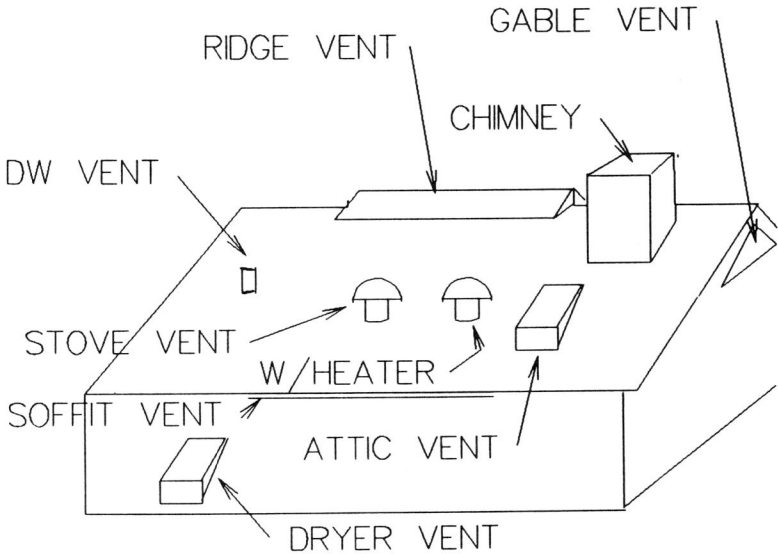

Figure 5–76 Various vents you may find

atmosphere. Since this vent is venting sewer gases, it must be ended above the building where the wind is in motion.

Sewer vents are generally 1-1/2 in. or 4 in. in diameter and are not capped off. Caps are not needed as any water entering the pipe drains to the sewer system.

Reventing

Some builders, such as myself, dislike having a conglomeration of pipes sticking through the roof. We use what is termed *reventing* (Fig. 5–77).

Reventing is where these sewer vents are connected to one large vent in the attic area and then only the large vent exits the roof. It is legal, it works, and it makes for a nicer looking building while cutting down on roof leaks. It does take extra pipe and elbows and cost more. The cost is under $100 for the average home, just about the cost of all those leak-prone roof vent flashings. Reventing will make your job much easier, as there will be far less vent flashing to install.

Other Venting

You may encounter several types of vents on a roof. If the building is gas heated or has a gas water heater, you will have a 6 in. or 8 in. flue vent to flash. You may have 6 in. range vents even if it is an all-electric building. Most ranges have a vent fan over them.

Attic Air Vents

Attic areas must be vented and you should be prepared to install roof vents. Turbine vents,which turn by wind power or rising internal attic heat, are inexpensive and used in Southern climates. Turbine vents are called rotary wind vents in some localities.

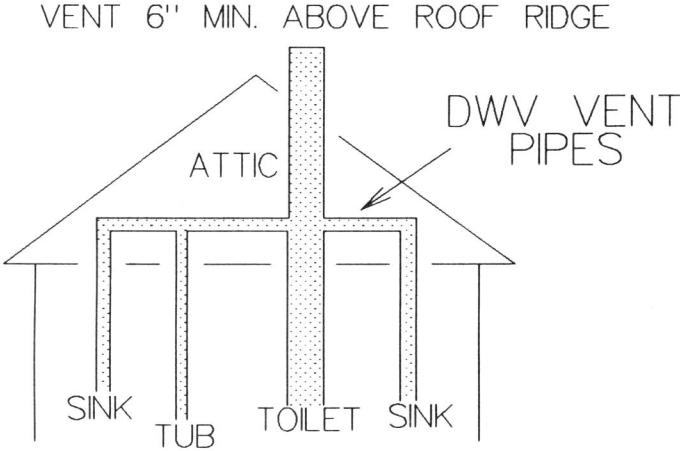

Figure 5–77 Reventing eliminates many pipes from roof

Power vents are becoming common in many areas. These are fans that turn on or off by a preset thermostat built into them. When the attic space reaches, say, 140 degrees, the fan starts. If you are asked to install these, you must know something about electrical hookups. Be sure the contractor has provided an electrical junction box, with an on/off switch, near the location of the proposed vent. Another name used for power vents is mushroom vent.

You may be asked to install ridge venting. This is a continuous vent that doubles as the ridge cap. Many designs are not very suitable for use on clay tile roofs, some are. Most are very suitable for asphalt roofs.

Flush-to-the-roof, gable, and soffit vents are also available. Whichever vent is used, you should obtain copies of the manufacturer's recommended specifications for installation and proper sealing.

Fireplace and Chimney Flashing

There is one other venting problem you may face. If the building has a fireplace, then you have to schedule some of your work with that of the brick mason. It is very difficult to install roof flashing around a chimney after the mason has already installed the counter flashing. The cricket, the roof flashing, and the counter flashing should be installed during the brick laying.

Fire Protection Around Chimneys

For proper fire protection around a chimney, there should be a 3/4 in. gap between the chimney and the roof sheathing (see Fig. 5–78). A 2 in. gap is left between the rafters, blocking lumber, and the chimney. Check with your local building inspector as this may be different in your locality.

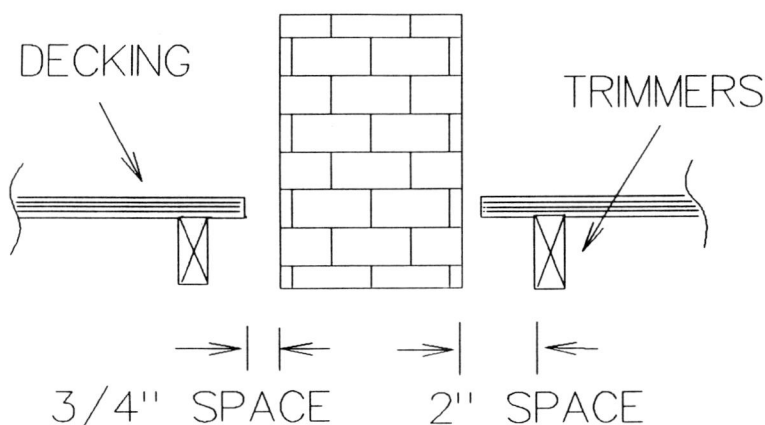

Figure 5–78   Code required spacing around chimney

Upper Area Flashing

Does the building have dormers or an offset second story? Again, you must work closely with other contractors, namely the framing and siding contractors. The roof connection, from the addition to the lower roof surface, must have flashing installed. This is a very difficult job to do if the upper area siding has been installed.

ROOM ADDITIONS

Figures 5–79 and 5–80 show second-story additions to older homes. There are potential roofing problem areas when a second-story addition is added to a single-story home. The

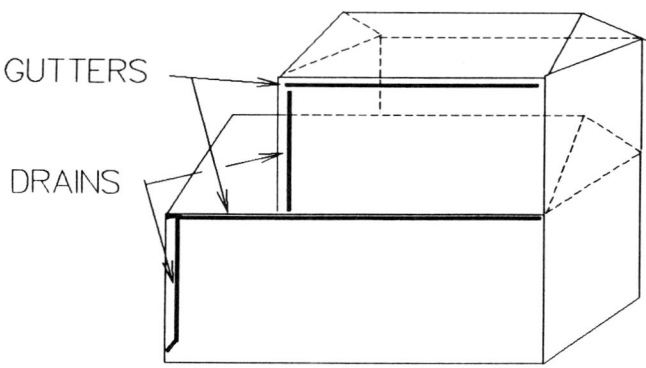

Figure 5–79 Typical second-story gutter system

Figure 5–80 Second-story room addition

roofer must work closely with the framer, siding installer, drain installer, and other contractors. Each contractor should know the problems of the other and work together as a team.

If all work is not properly coordinated, there will be water leaks. There will be unhappy people and much finger pointing. Example: The siding is put on before the wall to lower roof flashing. The roofers must now remove and reinstall siding to do their job properly. The siding contractor then does not want to warranty the siding because the roofer removed it. The customer sues the siding contractor, who sues the roofer, and so on.

If there is a general contractor involved, then have them set up a meeting with all the contractors before any work commences. Let the subcontractors and the general contractor know what you require. Listen to what they require of you. Schedule efforts so each can do his job and each will be satisfied.

DEALING WITH THE PAINTER

You must also schedule work with the painters. A building contractor wants to close-in his building as fast as possible against the effects of weather. Thus, the roof goes on when the external wall framing is in place. It may be days, weeks, or even months later before the painters get to work. You have installed this beautiful tile roof and along comes the painter, who pulls out his airless and starts dumping on the paint. You now have three rows of red Spanish tile that are painted white. It is the painter's fault for not being careful, and the contractor will make the painter pay for the damage. But, you did a good job, and don't want to see it ruined.

Painting the Eaves

Try to get the eaves and overhang areas painted before you install the roofing material. It can prevent you from having to redo your roof.

Painting has a few other positive uses. First is that the drip edge you installed will get a coat of protective paint, thus extending its life. Second, most reroofing jobs require new sheathing in several areas of the overhangs due to moisture rotting out the wood. This area is prone to damage from rain, snow, industrial fumes, etc. I apply a coat of paint on the sheathing and up the roof for a few feet before installing the roofing. It adds years of life to the total job. Since the painter is spraying up there anyway, have him spray this area as well.

Some will argue that the painter now has to make two trips to the job site, one for this, the second to come back and do the siding. But, consider this. You will save time since you can have the painter spray all your flashing, installed or not. The painter will save time because he doesn't have to be super careful about getting paint on your roofing material, thus reverting to roller or brush work. The builder saves since neither you nor the painter had to spend extra time doing, or redoing, the job. The customer or building owner saves by having years of life added to the property and by needing less preventive maintenance.

COMMERCIAL ROOF CONSTRUCTION

Before we get too far along, let's discuss some roof construction. Most residential buildings use plywood over a framework of rafters. We then apply underlayment and shingles.

In flat roof work, we may indeed find the same type of structure, especially in residential work. We have the rafters sheathed with plywood and we apply tar and roll roofing. But plywood has a small problem, it often relaxes. As it does, it sags between the rafters, forming dips. On a peaked roof, this is of little consequence, because it will still get proper drainage. On a flat roof these dips can form puddles and areas of rot. The use of heavier plywood or the use of T&G roof sheathing is recommended to avoid this problem.

Many commercial and industrial buildings use corrugated metal for the decking. The rafters are often not rafters, but are trusses due to the spans encountered. The average small commercial building is 20 ft across; therefore, the span is 20 ft. An industrial building can be 50 to 150 ft across its width. Laminated beams can be used in place of rafters and sheath these beams with steel corrugated decking.

Here are a few decking and sheathing methods used for commercial and industrial buildings. The items shown are for BUR or Built-Up Roofing systems. A licensed civil or mechanical engineer should be involved in the construction and designs. The roofing supplier's installation manuals must be obtained and complied with since most BUR work is very specific and improper decking can kill an otherwise fine job.

Concrete Trusses

Prestressed, precast concrete beams are becoming popular for commercial construction (see Fig. 5–81). Shown are single and dual T concrete deck trusses.

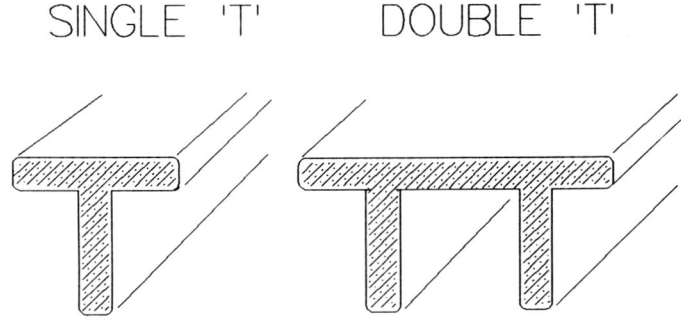

Figure 5–81   Two types of precast concrete roof trusses

Concrete cracks as it cures or gets old. These cracks are small and if repaired in time will not affect the general strength of the concrete. The key here is repaired in time. Prestressed concrete repairs the concrete before it is broken. Before the concrete is poured into the mold, several cables are inserted lengthwise in the mold. These cables are then tensioned, stretched tight end to end. The concrete is poured and allowed to cure. The cables are then disconnected from the tensioner machinery and the mold removed. As the concrete cracks, the stretched, tensioned cables try to shrink back to original shape. They pull the concrete cracks together. Friction then keeps the cracks from splitting even more.

Concrete beam trusses can be fabricated to lengths of 80 ft or more, the length restrictions being the required length for the building being covered and the local "Over the Road" laws. Most states and communities have restrictions on length and weight of objects transported on their roads. Practical length is about 65 ft.

These beams weigh several tons each and several must be used to cover a building. It is imperative that the building be constructed to carry this weight load. Steel-framed buildings must often be used.

Prestressed concrete roof deck trusses are not fabricated to be interlocked into a watertight deck (see Fig. 5–82). They require another coating to be installed. Usually that other coating is "poured in place" lightweight insulating concrete. To this coating, an additional protection of a built-up, felt/asphalt roof may be required. Instead of the built-up roof, a single membrane or a liquid membrane roof may be installed, described in the chapter on built-up roofs.

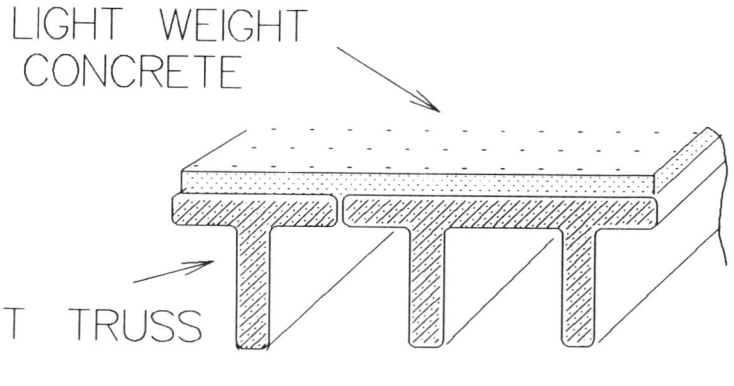

LIGHT WEIGHT
CONCRETE

T TRUSS

50 POUND LIGHT WEIGHT
INSULATING CONCRETE
BUR DECKING

Figure 5–82 Lightweight concrete used as a roof deck

Steel Roofing Decks

Steel roof decks are usually installed on steel trusses (see Fig. 5–83). This allows the deck to be welded in place, making the deck and the trusses into a unit. The open web truss is the most commonly used truss for steel deck roofs. It is easy to manufacture, very strong, and reasonably inexpensive.

One welds the steel decking to the open web truss (see Fig. 5–84). This decking is then covered with foam insulation or lightweight insulating concrete, shown. The steel

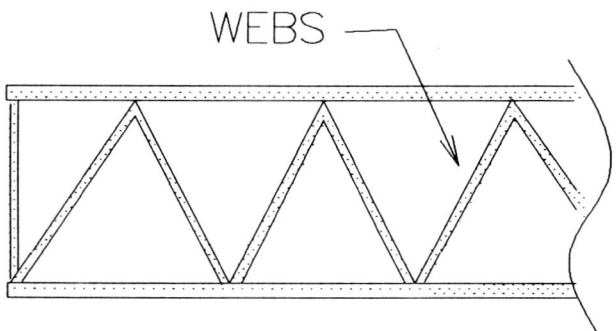

WEBS

OPEN WEB TRUSS

Figure 5–83 Open web truss

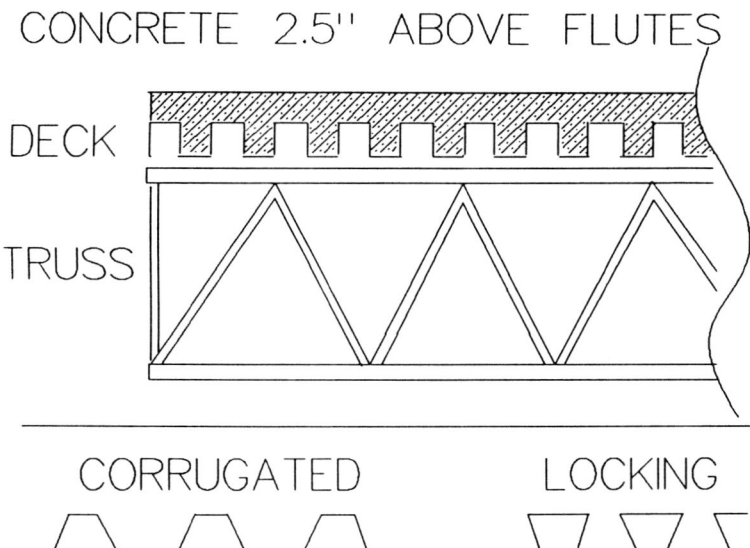

CONCRETE 2.5" ABOVE FLUTES

DECK

TRUSS

CORRUGATED LOCKING

Figure 5–84 Concrete deck on open web truss with metal decking

decking can be of conventional corrugated design or of a locking design. The locking design is preferred for poured concrete as it locks the concrete in place.

MISCELLANEOUS TOPICS

Heavy-Duty Sawing

For sawing decking at the rakes or for sawing large-diameter lumber, a chain saw works well. Use an electric saw if possible, it makes less noise and shouldn't bother the neighbors.

Cutting Rafters to Length

When cutting rafters to their proper length, don't forget to deduct for the ridge beam. A 2 by beam is 1-1/2 in. thick and you must deduct 3/4 in. from the length of the rafter for a properly fit rafter.

Illumination of Work Area

I recommend that all attic and crawl spaces be wired with light ON indicating switches and lights. Beats trying to work with a flashlight. Light indicating switches light up at the switch when the switch is in the on position. This way you don't leave an attic light on for days or weeks at a time. Install the switches in a location that will be seen by normal household traffic.

Skylights

Skylights are windows to the sky. They can be openable or fixed, flat or domed, flush mounted or raised. Skylights can be single, double, or triple paneled. The dome types use heat-formed clear plastic. The plastic is acrylic, Plexiglass™, or polycarbonate, Lexan™ or comparable brands. The flat-glazed skylights are of tempered glass or special Low-E glass. Low-E glass lets in light, but keeps U/V out. Skylights made with Low-E glass are double-glazed and the space between the two panes is filled with argon gas. The argon gas provides a certain amount of insulation value for heat loss prevention. See the chapter on flashing for flashing recommendations.

6

Nailing Instructions

PROPER NAIL SELECTION

The selection of the proper type of nail is important for a long-lasting roof. Use large head, hot dipped galvanized, or blue ring shank nails for asphalt roofing and plywood sheathing. The roughness of the nail shank gives the holding power needed. Blued nails have been heated to a temperature at which the metal has turned blue in color. This is an oxide that prevents the nail from rusting. HDG (Hot Dipped Galvanized) nails are nails dipped into a hot zinc or tin/lead/zinc mixture.

I prefer HDG or hot dipped galvanized nails since the coating is very thick and very rough. I believe the HDG nails will last longer and provide better holding power. The disadvantage of the HDG nails is the lack of handling ease during installation of the shingles. They do have sharp surface protrusions that can cut your fingers. The other possible disadvantage is the quality of the coating by some manufacturers. A poor-quality coating will not protect the steel and the nail will rust.

EG, electro galvanized nails are steel nails with a very thin electroplated coating of zinc or tin/lead/zinc. These are the most commonly used galvanized nails. I feel that the HDG nails are superior. In my estimation, the coating on the EG nails is too thin to afford proper long-term protection. Plus, the coating is smooth and does not provide a good firm grip. My experience with EG nails is that they will scratch and rust.

Another commonly used nail is the cement coated nail. Cement-coated nails are those with an amber or greenish color. They are supposed to cement themselves in place when nailed into wet lumber since the moisture content of the lumber activates the glue coating. Roof decking should be dry when installed, thus, cement-coated nails will not work properly. I do not like cement-coated nails. My experience has been that the cement holds well to the wood, but not to the nail itself. You should make your judgment, as I may have gotten a few batches of nails that were defective.

Nails supplied for roofing guns are normally EG nails, but Stanley-Bostitch claims to have a superior grade. Their coating contains chrome and should last longer than standard EG nails.

Aluminum nails are recommended for use whenever possible and for cedar and redwood use. Aluminum nails should be used only on aluminum flashing, not with steel flashing. Use aluminum nails whenever the nail heads will be exposed,such as during the installation of the last ridge shingle. Use gasketed aluminum nails for securing fiberglass corrugated roofing, gasketed galvanized nails for corrugated steel roofing.

NAILS ARE DANGEROUS

One or two other comments about nails before we continue. You will be applying roofing felt to the roof before you shingle. The felt can be held in place with T-50 staples, HDG nails, or with the newer square, stress plate nails. These are sheet metal plates about 1-1/2 in. square that have a nail punched through them. I recently visited two work sites. The first was a new construction and the roofers were using the stress plate nails. Dozens of these were allowed to fall to the ground. Most landed with the steel plate on the ground and the nail pointing straight up. This is a very dangerous situation for all the workers. The second site was an apartment building covered with Cedar Shake and I had to do some work on the roof. The roofers, several years before, were sloppy and left hundreds of loose nails all over the roof. Not only were these nails rusting, thus leaving unsightly rust stains, but walking was hazardous. It is hard enough to keep from falling off a roof without someone putting what is the equivalent of roller bearings all over the place.

NAIL AND STAPLE SIZES

Nails are measured in D numbers. The starting point is 1 in. (2D) and each D number adds 1/4 of an inch to that figure, up to 10D. See Appendix B, table 10. For example, a 4D nail is 1-1/2 in. long and a 6D is 2 in. long. Proper nail size selection is controlled by the U.B.C., section 2510, tables 25G, 25Q, and section 3203, table 32B. Please note that staples are not permitted for asbestos-cement shingles, flat or curved tiles, and that they must be approved by the building inspector for all other types of shingles.

NAIL STRIPPER BOX

You will read about nail stripper boxes in many manuals of roofing. The nail stripper box is fast becoming a part of a lost art. Newer roofing technology uses nail guns and staplers.

These power-driven machines contain a magazine in which coils or strips of nails are placed. It is much faster and easier than the old hammer away methods of the past.

NAILING PROBLEMS

A nail that is too short is the common problem in high wind prone areas. This is especially true at the eaves. The nail doesn't penetrate deep enough into the sheathing to hold the shingle. Winds of 30 to 100+ miles per hour are common throughout most areas of the United States. Wind blows sideways and gets under the edges of shingles and tiles at the eaves. A 30 or 40 mile per hour wind pushes strongly. Picture this force under a shingle that only weighs a pound or two. I've seen entire roofs (shingles) peel back and leave a roof in a wind storm. One roof sailed into a field over a quarter mile away. When one shingle starts to tear off, the rest follow like a kite. Remember, when selecting nail length, that you are not always nailing through one thickness of shingle. Shingles overlap at several locations and the nail length must go through two or more shingles before hitting the sheathing. This is common in areas such as the starter strip, tie-ins, roof hips, roof caps or ridges, and reroofs. Wood shingles require a nail length that will go through three thicknesses of shingle. Also remember that you are nailing through a layer of roofing felt and sometimes metal flashing. The penetration into the sheathing should be a minimum of 1/2.

Most of today's builders use 1/2 in. thick sheathing with 5/8 to 3/4 in. sheathing at the eaves. This is so you, the roofer, can use longer nails at the eaves without penetrating through to the underside of the sheathing, an unsightly condition. I prefer to use long nails that do penetrate the sheathing and then I cover the lower side of the eaves with fascia and soffit. The soffit not only hides the nails but gives a better general appearance to the building. Do not use long nails for the field shingles, as it may create future problems.

Cover Exposed Nails

Be sure that all nails are covered with another shingle. Exposed nails are a sure sign of shoddy workmanship and will lead to premature roof failure. Often you have to place a temporary nail to have a place to hang things, i.e., part of the safety equipment, or for anchoring a chalk line. After these nails are pulled, you *must* go back and fill the nail holes with sealer. A night-time inspection of the roof is the easiest way to locate these holes. Light up the attic or crawl space with a few bright lamps pointed upward. Go on the roof and look for any spots of light shining through. Place a 4 or 5 in. wooden dowel in the hole or use another way of marking. Then you can locate and repair the holes the following day.

Use of Extra Long Nails

It is common practice to secure shingles with extra long nails (see Fig. 6–1). The thought is that if the nails extend through the decking, the shingles will not be blown off in heavy winds, and this is in fact true.

But, nails extending through the field sheathing may collect moisture from within the

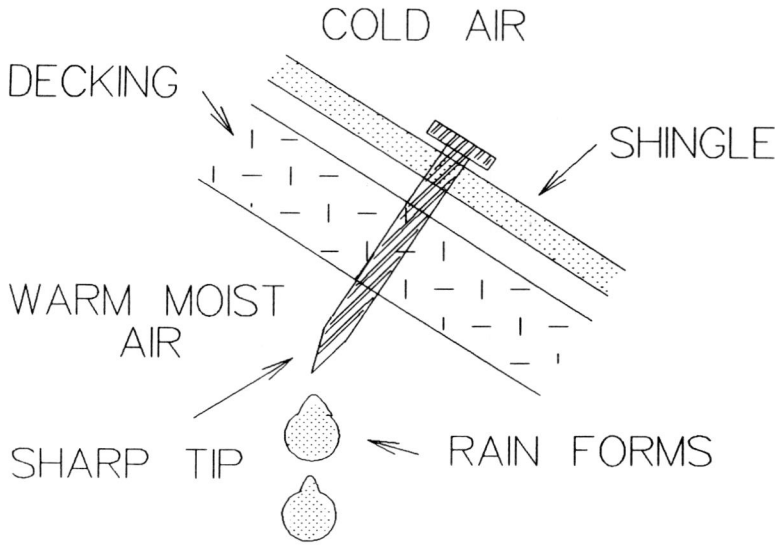

COLD AIR

DECKING

SHINGLE

WARM MOIST
AIR

SHARP TIP ← RAIN FORMS

Figure 6–1   Use of long nails may cause water problems

building. This moisture is warm and when it contacts the cooler nails it condenses into water. The water then drips off the tip of the nail and into the building insulation or on to the top of the ceilings. If the moisture dips into the insulation, it will destroy the R-value of the insulation. If the moisture drips on the ceiling drywall, it will destroy the drywall. The problem can be minimized with proper attic venting.

There are two other reasons for not nailing through the field sheathing. First is that the underside of the sheathing becomes a pin cushion, with thousands of protruding nails that make it dangerous for anyone working in the attic space. The second reason is that it makes it very difficult to remove the shingles when it comes time to replace the roof. Proper sized nails can be removed with the shingles using a shovel. Over-length nails must be pulled one by one with a claw hammer or nail puller.

NAILING T&G SHEATHING LUMBER

Figure 6–2 illustrates proper nailing of T&G plank sheathing. Decking that is over 3 in. in thickness should be predrilled for the nails. Nail length should be twice as long as the thickness of the plank.

Figure 6–3 shows T&G roof decking properly installed with its tongue facing toward the ridge. The reason is that you will be nailing through the tongue as you proceed up the roof. Nailing the tongue of each board, before face nailing the board, squeezes the board into the prior one. The first T&G board is rip cut its full length to provide a clean edge at the eaves. See Appendix B, table 3 for nailing requirements.

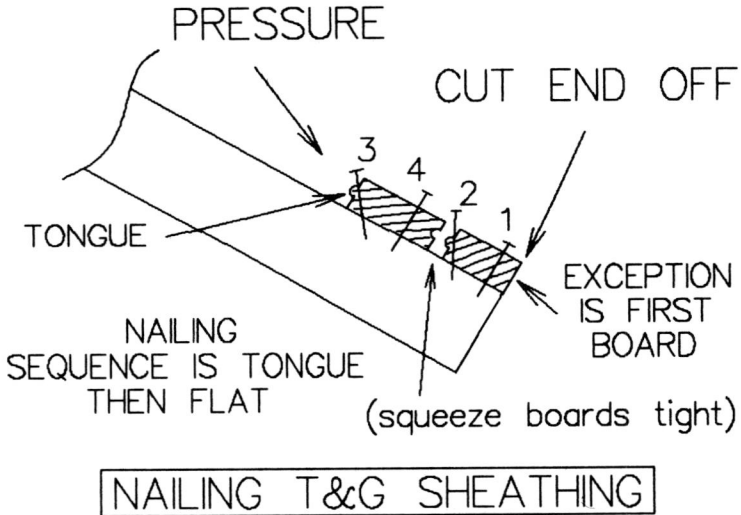

Figure 6–2   Proper nailing sequence for T&G decking

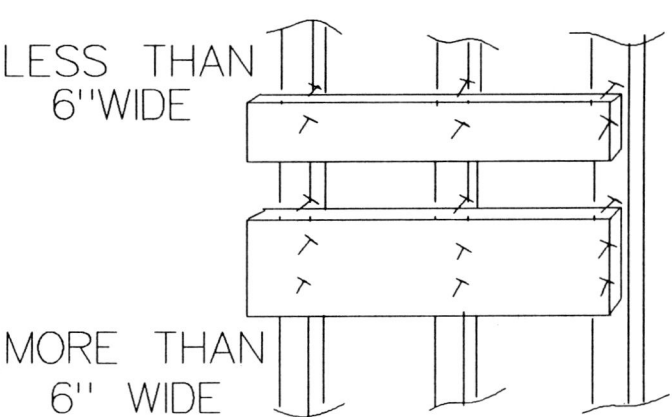

Figure 6–3   Nailing techniques change with T&G board thickness with width changes

Four inch lumber should be drilled through the edge and nailed. The edge nailing requires drilling a hole through the lumber from one edge to the other. The lumber is then nailed to the proceeding piece with long thin straw nails.

NAILING PLYWOOD SHEATHING

Figure 6–4 shows the correct end to end, side to side spacing for plywood sheathing. Also shown is the code approved nailing pattern and the minimum number of nails per 4 × 8 ft sheet for various rafter spans. You can use more nails but not less. The clips or other blocking must be used if the span exceeds 24 in. Do the inspectors check all this, the number of nails, the spacing, the distance from nail to nail? You bet they do. See Fig. 6–5; the tape measure belonged to the inspector; he didn't want to be photographed though. The man spent more time on this roof than I did.

PLYWOOD EDGE CLIPS

Plywood edge clips are to be used wherever two sheets of plywood butt together and are unsupported for 24 in. or more (see Fig. 6–6). These clips are available at most lumber yards.

The common name for plywood edge clips is H-clip due to its H shape. Another name is plywood framing clip. Some clips are manufactured of stamped sheet metal, others are cut from aluminum extrusions.

H-clips are code required in most areas. The theory is that two sheets of plywood held

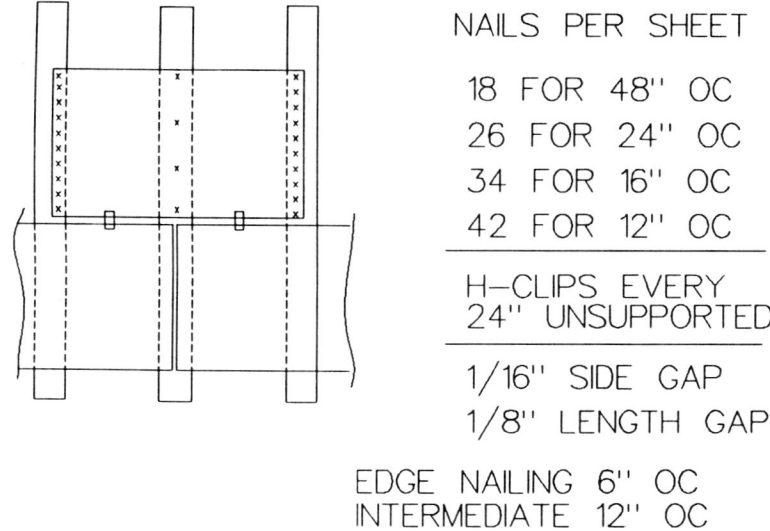

NAILS PER SHEET

18 FOR 48" OC
26 FOR 24" OC
34 FOR 16" OC
42 FOR 12" OC

H–CLIPS EVERY
24" UNSUPPORTED

1/16" SIDE GAP
1/8" LENGTH GAP

EDGE NAILING 6" OC
INTERMEDIATE 12" OC

Figure 6–4   Code approved nailing pattern for plywood roof sheathing

Figure 6–5   Building inspectors may check the nail locations

EDGE or 'H' CLIP

REQUIRED EVERY 24" ON
UNSUPPORTED SHEATHING

Figure 6–6   Plywood can be supported with edge clips

with an H-clip will not sag, or if they do sag, they will sag in unison. Each sheet of plywood gains strength from the other.

Use 6D ring shank nails for plywood less than 1/2 in. thick. Use 8D for plywood 5/8 to 1 in. thick. Nails should be galvanized or aluminum.

NAILING PLYWOOD GUSSETS

When nailing plywood gussets to form a T or a V connection, use the following length nails:

> 3/8 to 1/2 in. thick plywood: use 4D nails
> 1/2 to 7/8 in. thick plywood: use 6D nails

For maximum strength, gussets should be glued with a construction adhesive and then nailed.

FASTENERS REQUIRED FOR SECUREMENT OF INSULATION SHEETS

Insulation fasteners are determined by the insulation used, the insulation's thickness, the type of roofing system being applied, and the building's decking material. The manufacturer's product insulation manuals are very specific about the type of fastener and its placements. In general, each sheet of 4×4 insulation sheet must have a minimum of 4 fasteners, 1 each at each corner to pass FM-60. A 2×4 sheet of insulation foam must have a minimum of 2 fasteners, each near the 2-ft edge and along the center line. A 3×4 sheet must have a minimum of 3 fasteners, two in opposite corners along the 4-ft edge and 1 on the other 4-ft edge, centered between the 2 corners.

To meet FM-90, the 2×4 sheet must have 4 fasteners, 1 in each corner. The 4×4 sheet must have 8 fasteners, 4 staggered rows of 2 from end to end or 3:2:3. The 3×4 is the same as the FM-60 requirement. I suggest that you treat 8- and 9-ft long sheets as two 4-ft sheets laid end to end. Fasteners on 8- and 9-ft boards are to be from 18 to 24 in. apart and 12 in. from the edges. A minimum of 8 to a maximum of 16 fasteners are required on 4×8 boards. The distance from the edges to the fasteners on all other size boards is 6 in. minimum to 12 in. maximum. (See Fig. 6–7.)

See Fig. 6–8. Foam insulation is fastened to metal decking with metal plate fasteners sometimes called *cap fasteners* (Fig. 6–9). The fasteners are of two types, the driver nail and the self-piercing screw. Figure 6–10 shows each. The driver nails are special barbed nails designed to be struck with a rubber mallet. The tips pierce the corrugated metal and the barbs lock in. Self-drilling screws have a cutting tip and a screw cutting shank; the head can be phillips, common, spine, or hex cut. The spine and hex cut are preferred by most roofers since they stay in the power screwdriver bit better than the phillips or common head screw will. The screwdriver bit should be magnetic if steel screws are used.

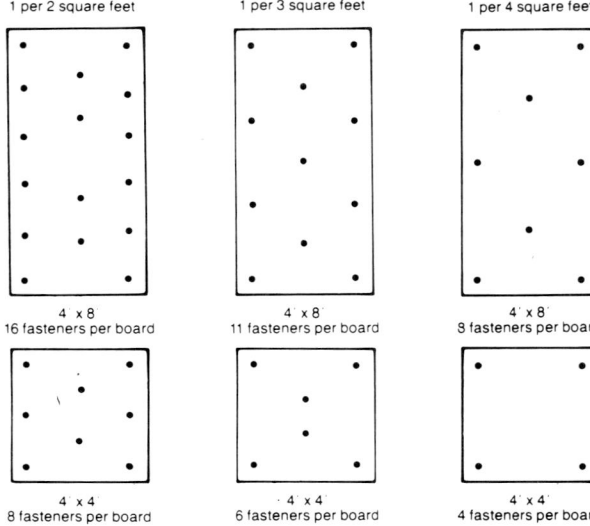

Figure 6–7 Recommended fastener patterns for foam sheet insulation (Reprinted with permission of U.S. Intec, Inc.)

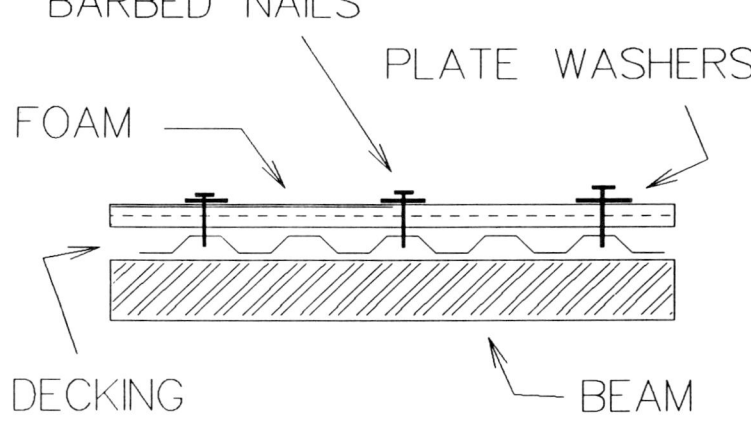

BARBED NAILS

PLATE WASHERS

FOAM

DECKING

BEAM

ATTACHING FOAM INSULATION
TO STEEL DECKING

Figure 6–8 Plate washer

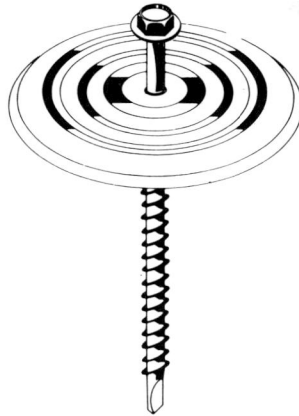

Figure 6–9 One brand of self-drilling roofing screw (Reprinted with permission of U.S. Intec, Inc.)

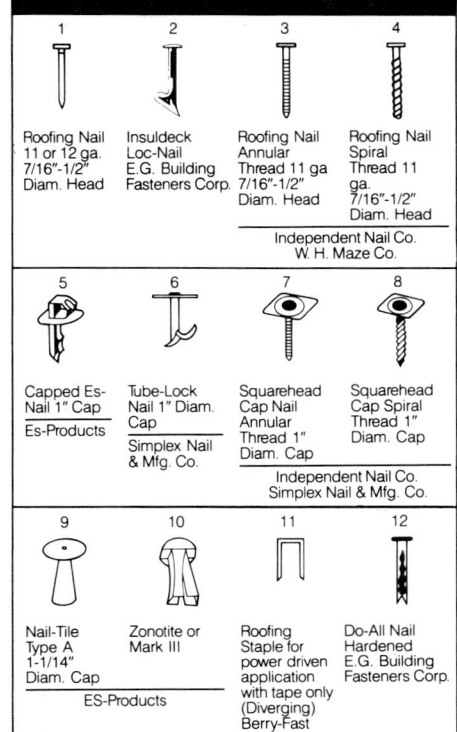

Note: Nails with less than 7/16″ diameter heads shall be nailed through tin-discs.

Figure 6–10 Roof fastener guide (Reprinted with permission of U.S. Intec, Inc.)

Foam insulation is installed to insulating concrete with tube nails (see Fig. 6–11). These nails are specially made with controlled bending properties. The tube part keeps the nail from bending too soon and acts as a nail head stop.

NAILS AND FASTENERS FOR BUILT-UP OR FLAT ROOFING

I've shown you various nails and fasteners used for installing flat roofing. Here are a few others (Figs. 6–12 and 6–13).

See Fig. 6–14. The reason that any nail or screw type fastener holds two items together is that the nail or screw head's bottom surface is wide. It is wide enough so that the material being held cannot pull through it. If the material being held is brittle, soft, porous, or of poor structural strength, a nail by itself will not hold. Asphalt roof felts and foam insulation board used for roof deck insulation are materials that will not hold. This problem is overcome by distribution of force. By using metal plates under our nail or screw heads, the downward holding force of the nail is distributed over a greater area. It is now very difficult for the felt piles or the insulation to lift and tear through at the nail head.

Metal distribution plate nails are available in many different shapes and sizes. The foam insulation or built-up roofing manufacturer specifies the approved fasteners for their product. Use of other than specified fasteners can void the manufacturer's warranty.

FASTENER PENETRATION

Fasteners are to penetrate a minimum of 1/2 in. into the sheathing or decking. Stress or distribution plates are to be used, 2 1/2 to 4 in. diameter plates are recommended. A vapor barrier may be required by the insulation manufacturer. On solid wood decking, the fasteners should be sized to penetrate three quarters of the way into the decking.

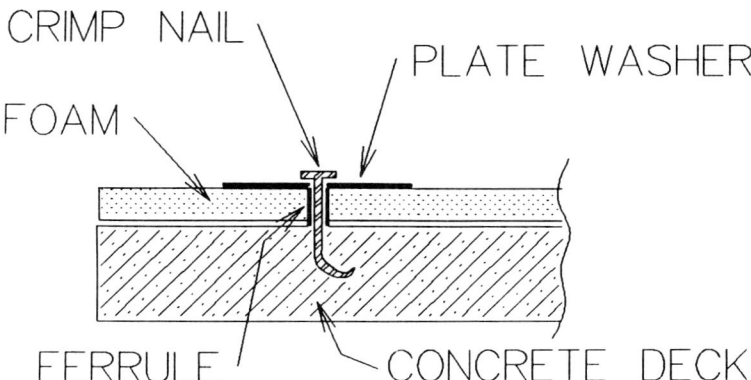

Figure 6–11   Securing foam insulation to a concrete deck

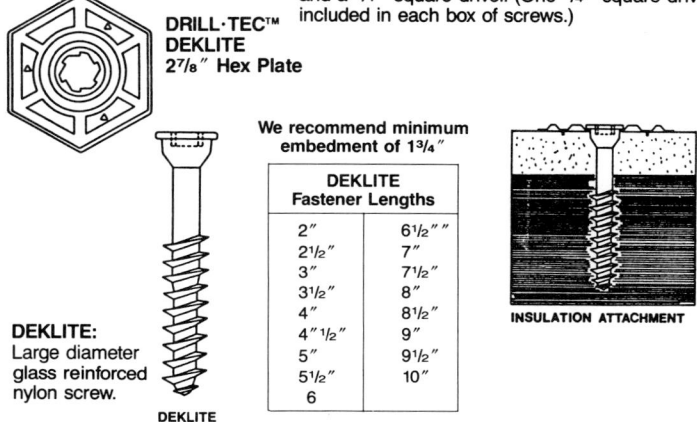

DRILL·TEC™ DEKLITE FASTENERS (DL)
Low Cost, Speedy Installation

Self-embedding DEKLITE fasteners are easily driven into Tectum, Gypsum or light-weight concrete deck material.

When using DEKLITE Fasteners with Gypsum Deck just pre-drill a pilot hole for the DEKLITE fastener and see the benefits of fast, easy installation and holding power in fastening insulation to gypsum roof deck. Installs with a standard electric screw gun and a 1/4″ square driver. (One 1/4″ square driver is included in each box of screws.)

DRILL·TEC™ DEKLITE 2⁷/₈″ Hex Plate

DEKLITE: Large diameter glass reinforced nylon screw.

DEKLITE

We recommend minimum embedment of 1³/₄″

DEKLITE Fastener Lengths	
2″	6¹/₂″″
2¹/₂″	7″
3″	7¹/₂″
3¹/₂″	8″
4″	8¹/₂″
4″ ¹/₂″	9″
5″	9¹/₂″
5¹/₂″	10″
6	

INSULATION ATTACHMENT

Figure 6–12 Drill-Tec™ Deklite fasteners (DL) (Reprinted with permission of U.S. Intec, Inc.)

CORRECT NAIL SELECTION FOR ASPHALT SHINGLES

The correct nail for asphalt roofing is a 7/16 in. roofing nail. This will penetrate two 1/8 in. thick shingles and the sheathing. Most sheathing is 1/2 in. and minimum penetration of the nails or staples is to be halfway into the sheathing. For each single layer of shingle added to the standard roof, add 1/8 in. to the nail length.

Please take note of this typical problem: You are on the roof and nailing away. You get to the ridge cap shingles and find you only brought one length of nail with you, 7/16 in. You look around, no one sees, so you use what you have. Such a pretty job, it looks great. That night the winds hit 45 miles per hour. The next morning you have an irate customer, as all the ridge shingles have blown off. This time you grab nails of the correct length, 7/8 in. or longer.

Even if you do the job correctly the second time, your reputation as a roofer is suspect. The entire job is now suspect by the customer (or by your boss if working for someone). They will now find all sorts of other problems, real or not. This is time and money consuming. Plus, people have a habit of always badmouthing someone who made a mistake. You don't generally hear too much about the people who do their jobs correctly the first time. It is expected of them and will be expected of you.

Figure 6–15 illustrates correct nailing points for three-tab asphalt shingles. All nails

DRILL·TEC™ CONCRETE DECK FASTENERS (CF)

Benefits

• Corrosion protection meets the rigid Kesternich test DIN 50018 SFW 2.0
• Prevents costly installation breakage
• Pilot end provides alignment for fast and easy perpendicular entry
• Thread design provides high pull-out values and easy self tapping
• Compatible with DRILL·TEC™ metal stress plates
• Meets Factory Mutual 4470
• ³/₁₆″ standard masonry bit is required

Features

• Low profile #3 recess Phillips head
• Powder coated
• Alloy selected to provide high torsional strength
• Pilot end for fast alignment in pre-drilled hole
• Deep 10 threads per inch
• Available in 11 lengths, 1¼″ to 8″
• #3 phillips bit is included in each box of screws

Figure 6–13 Drill-Tec™ Concrete deck fasteners (DF) (Reprinted with permission of U.S. Intec, Inc.)

or staples should be covered. Exposed nails are a source of leaks and possibly rust (Fig. 6–16).

See Fig. 6–17. When nailing shingles, *look* at what you are doing. Look at what your workers are doing. Nails set at an angle, not perpendicular to the roof plane, or nails not driven in far enough lead to callbacks. The shingle directly above the improperly seated nail will eventually punch through. This small crack or hole formed by the nail head will lead to water entering under your shingles. Also, the lack of sufficient penetration of the nails into the sheathing can cause the shingles to blow off during a wind storm.

Nails driven in too far will lead to punch through of the lower shingle (see Fig. 6–18). This can result in blown off shingles, even during a moderate wind.

A properly seated nail is perpendicular, and forms an angle of 90 degrees to the roof plane (see Fig. 6–19). It is driven in to the point that it just starts to dent the shingle being nailed. It is of sufficient length to penetrate a minimum of halfway into the roof sheathing. Three quarters in would be better.

If you are using an electric or pneumatic stapler or nailer, check frequently for proper drive pressure. You don't want to redo your shingling work just because something changed and you didn't catch it in time.

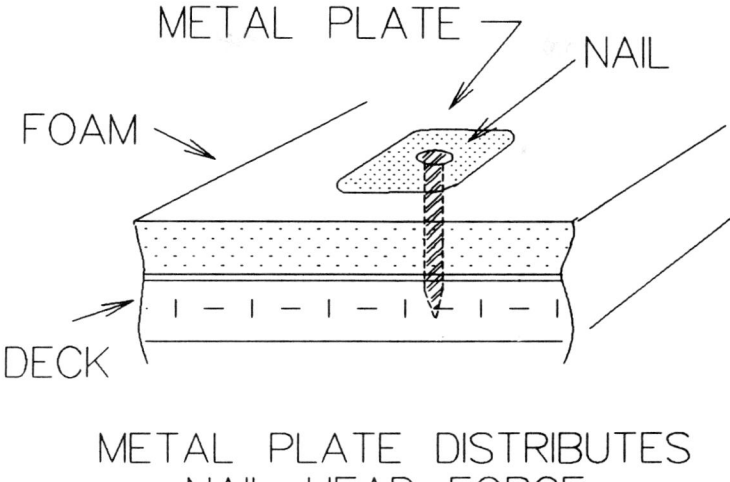

METAL PLATE DISTRIBUTES
NAIL HEAD FORCE

Figure 6–14   Plate washers used to distribute nail head force

NAILING PROCEDURES FOR WOOD SHAKE AND SHINGLE

Both wood shingles and shake are attached by nailing or stapling to the roof sheathing. If using staples, wide crown staples are recommended. The staple must penetrate through the shingle and at least halfway into the sheathing. Use aluminum, copper, stainless steel, or brass, but do *not* use steel or coated steel nails or staples. Figure 6–20 shows the nailing points for wood and wood shake shingles.

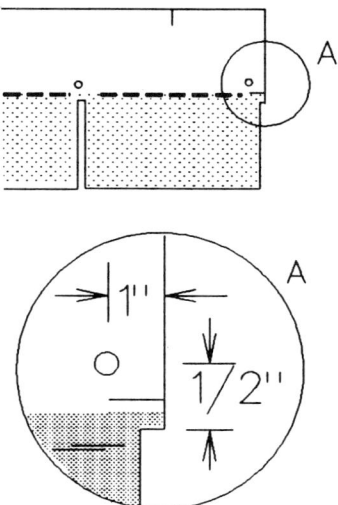

Figure 6–15   Nailing locations for asphalt shingles

Note: No more and no less than two nails or staples per shingle are to be used. Attachment is to be 3/4 in. in from the wood shingle's edge and 2 in. above the weather exposure line. Attachment of shake is 1 in. from the edge and 2 in. above the weather exposure line.

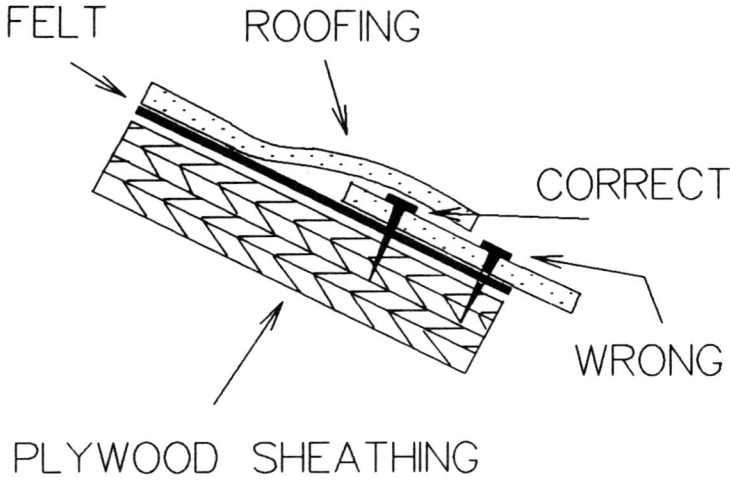

Figure 6–16 Nail heads should be covered

Nailing at the Rakes

Wood shingles and shake should extend past the rakes by 3/4 to 1 in. Nailing is to be as close to the rake as possible to prevent edge curling.

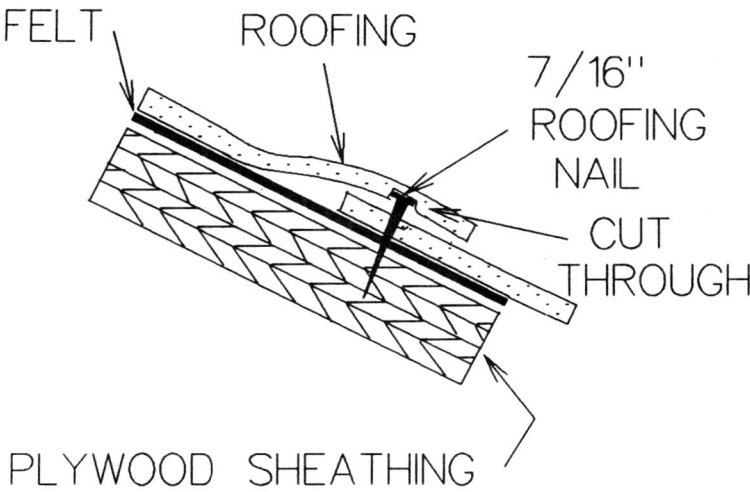

Figure 6–17 This will result in damage to shingles

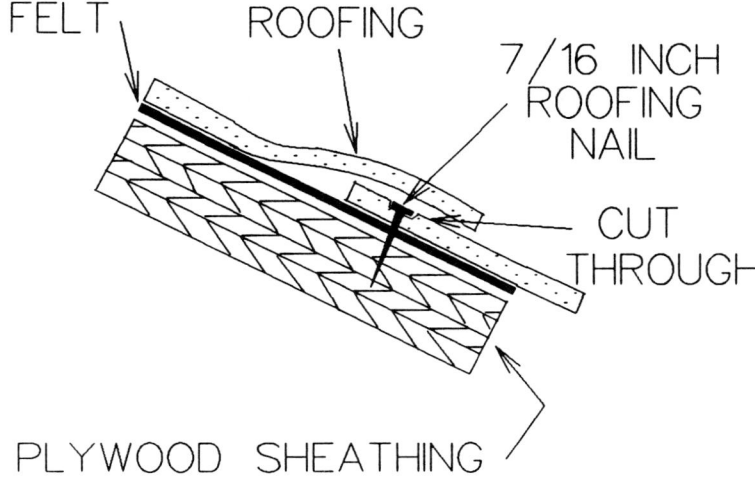

Figure 6–18   This may result in wind blowing off the shingles

Wrong Nail Location Problems

Wood will warp if not supported or nailed properly. If you nail too far from the shingle's edges, the shingle will curl on both edges. If you nail too far above the weather exposure line, the shingles will curl with the butt end up. Using one nail will allow the shingles to twist or turn from side to side. Using more than two nails is a waste of nails and may cause lengthwise shingle splitting.

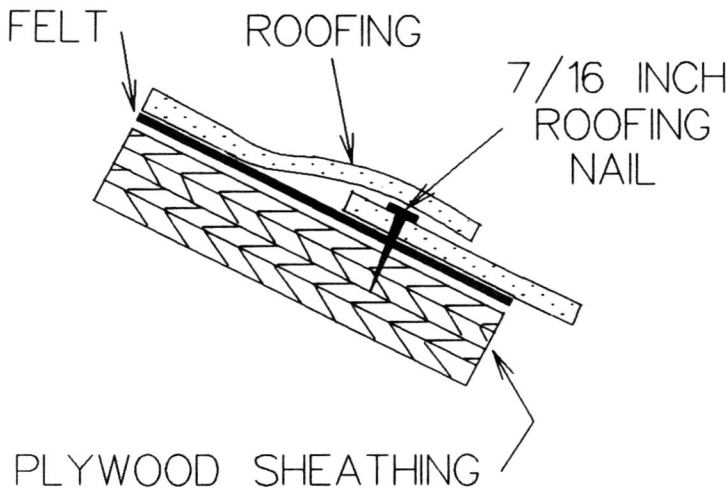

Figure 6–19   Proper nailing of asphalt shingle

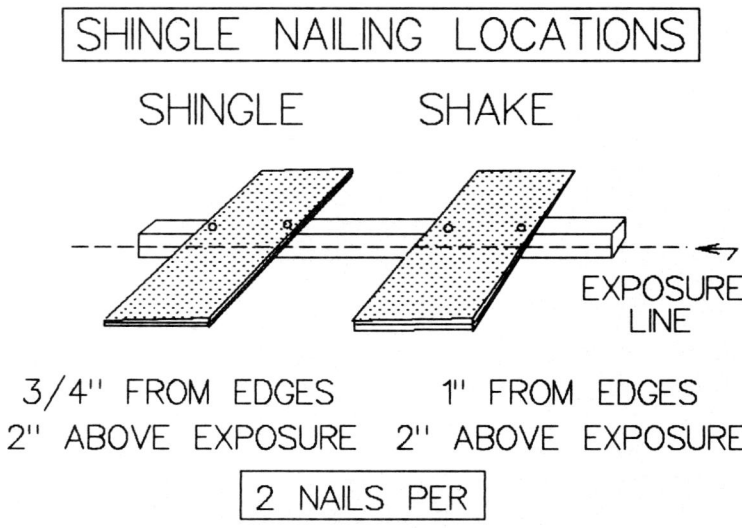

Figure 6–20   Nailing locations for wood shingles and shake

Here are the recommended nail lengths to use when shingling over a new roof or a roof where the old shingles have been removed.

- For wood shingles of 16 or 18 in. length, use 3D nails.
- For wood shingles of 24 in. length, use 4D nails.
- For wood shake of all lengths, use 6D or 7D nails.
- For hip and ridge shingles or shake, use 8D nails.

Reroofing nail sizes.

- For reroofing over existing wood shingles or shake, use nails that are two sizes longer than for a single shingle.

NAILS AND FASTENERS FOR CORRUGATED ROOFING

Nails or screws used for corrugated plastic, fiberglass, steel, or aluminum roofing should be gasketed (see Fig. 6–21). The gasket is usually a neoprene rubber washer that when compressed makes a tight water seal. Shown are three possible designs. The least expensive consists of the nail and the washer. This design is adequate for most outdoor shed-type buildings. The second design incorporates a flat metal washer between the gasket and the nail or screw head. This washer usually protects the gasket from destruction during the installation process. The metal washer distributes the nail head's pressure evenly without

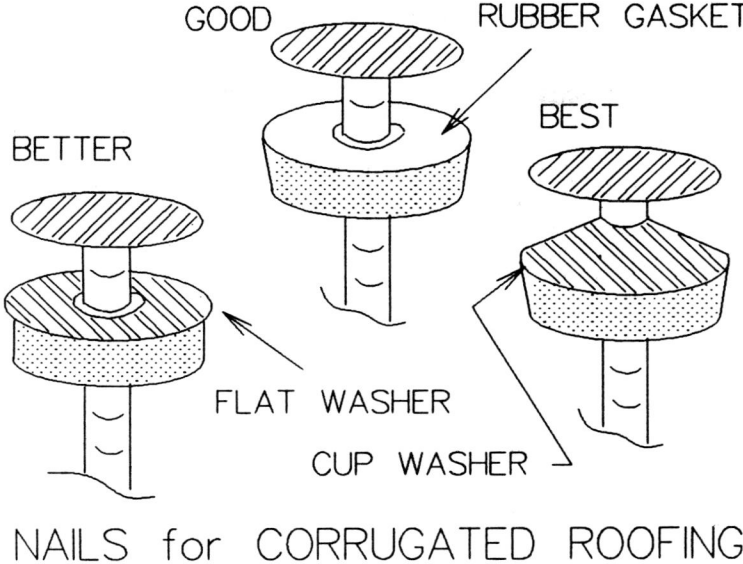

NAILS for CORRUGATED ROOFING

Figure 6–21   Gasketed nails

cutting into the gasket. When used with screws it prevents the turning screw head from cutting into the gasket washer. Vertical alignment of these two fasteners is somewhat critical. If driven in at an angle, they will not seal properly. To overcome this problem the cupped washer gasket is used. The metal washer is cone shaped. This type of fastener will form a leak-tight seal even if not driven in perfectly straight.

NAILING TO CORRUGATED DECKING

Corrugated decking that will be covered with foam insulation or other deck sheathing is attached as shown (Fig. 6–22). Rubber gasket nails or fasteners are not used. All attachment is through the valleys, not the ribs or flutes. Do not nail so tight as to distort the corrugations.

Corrugated sheet roofing that will not be covered with another material should be nailed through the ribs of the corrugations (see Fig. 6–23). Gasketed nails should be used to prevent water leakage. Wood nailers may be used if necessary to strengthen the corrugations. Use care in nailing, do not nail too tightly. Too much downward pressure will distort the corrugations and sealing gaskets. This will make it very difficult to keep the roof panels in alignment across the roof field. It also may cause leaks. Do not nail in the corrugation valleys. Nails in the valleys are subject to vast amounts of runoff water. The protruding nail heads can collect dirt and leaves and form dams that hold this runoff water. Leaks are sure to develop.

CORRUGATED DECKING

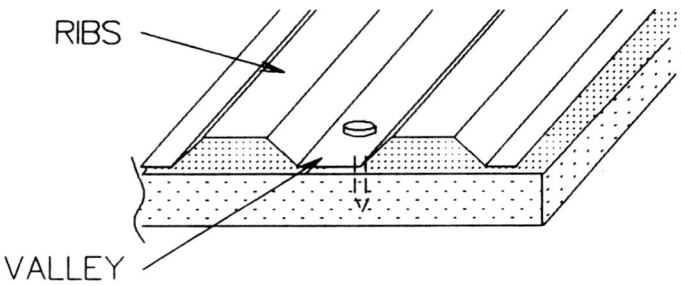

DECKING WILL BE COVERED
BY INSULATION SHEETS
NAILING IS IN VALLEYS

Figure 6–22   Nailing of exposed corrugated decking

NAILING INTO TILE

Tile is not especially forgiving when it comes to nailing it to roof sheathing. A too strong hammer blow on cement, slate, or clay tile will create several pieces of irregularly shaped tile. Metal tile will easily bend or dent. Plastic tile will crack and split.

It is for this reason that most manufacturers of tile have drilled the tile for you. Plus,

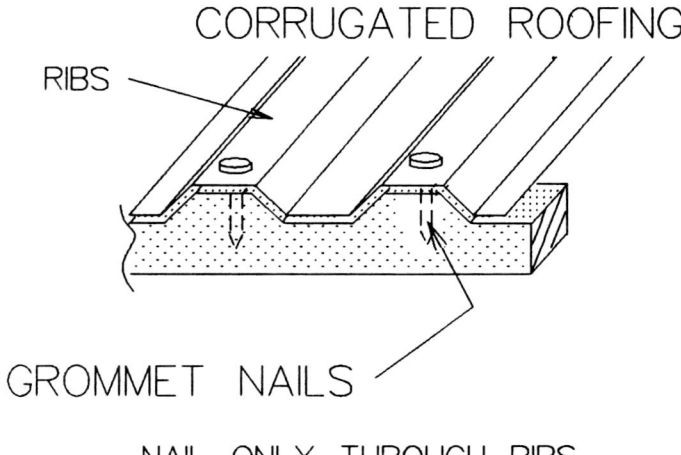

NAIL ONLY THROUGH RIBS

Figure 6–23   Nailing of concealed corrugated decking

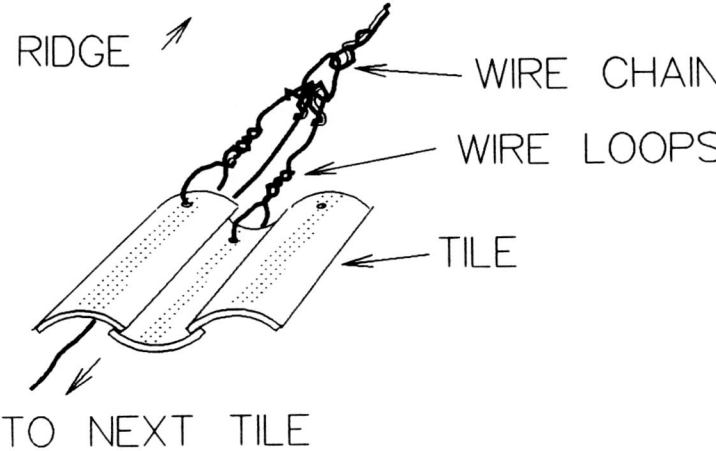

Figure 6–24 Wire used to secure clay tiles

many manufacturers supply special clips, tabs, and other securement items for installation of their tile products. It is highly recommended that you use the manufacturer's recommended fasteners.

Slate, clay, and cement tiles are normally wired in place on the roof sheathing (Fig. 6–24). Wire gauge is to be a minimum of #14 gauge, per code. Many tiles, when used on low slope roofs, do not require any securement; their weight and a built-in lip keep them in place.

The building code states that at least one nail or hanger be used if the tile is narrower

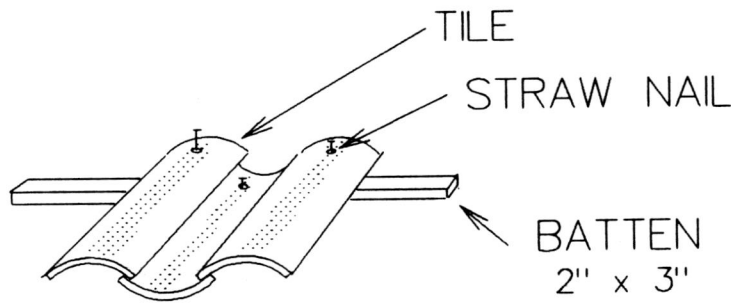

NAILING SPANISH TILE

Figure 6–25 Nails used to secure tiles

PURCHASE PREDRILLED or DRILL WITH A
CARBIDE TIPPED DRILL BIT

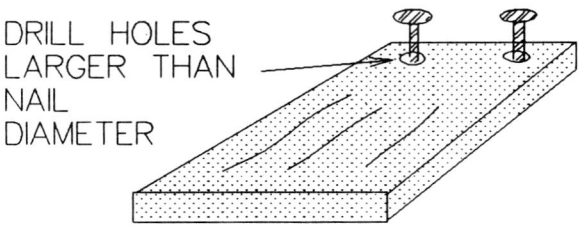

DRILL HOLES
LARGER THAN
NAIL
DIAMETER

NAILING OF SLATE or
CONCRETE FLAT TILE

Figure 6–26   Nailing of flat tile

than 16 in. and the installed weight is greater than 7-1/2 lb, and the slope is less than 7:12. All other tile is to be secured with two wire ties, nails, or clips.

In my installation instructions where I show tile being nailed, do so with care. Use nails that are smaller in diameter than the hole in the tile. Hammer the nails to within 1/16 to 1/8 in. of the tile surface. *Do not hammer flush* with the tile surface. Tile will expand and contract, a too-tight nail will not allow for this expansion and contraction and the tile will crack.

Note: If you are using an automatic nail gun, be sure to check the depth calibration often. It will prevent costly rework due to loose or bent tiles.

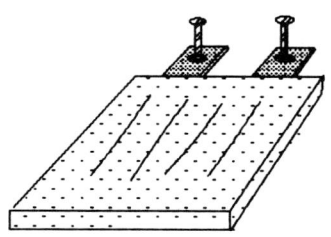

SOME ALUMINUM SHINGLES
HAVE NAILING TABS

Figure 6–27   Nailing tabs are sometimes provided

Here are the typical nailing locations for a variety of tiles. See Appendix B, table 17 for the recommended nail types.

Figure 6–25 illustrates typical nailing points for barrel or Spanish clay tile roofing. Holes must be drilled if not supplied already predrilled.

Figure 6–26 illustrates the nailing pattern for flat concrete or slate tiles.

Some brands of steel, vinyl, or aluminum shingles have nailing tabs on their rear or covered edge (see Fig. 6–27). Plastic and metal tiles will have the required nailing pattern printed on the carton or wrapper. Some use two nails at the tail edge, others must be nailed at the tail and across the butt edge facing. Follow manufacturer's direction for proper nailing and nail sizing.

7

Insulating a Roof

INTRODUCTION

This chapter is about insulation used on roofs and its interaction with roofing. Most roofs are insulated in one form or another. The insulation may be in direct contact with the roof decking or it may be between the ceiling joist. Insulation is used to keep heat or cold in a building so that the occupants of the building remain comfortable. A well-insulated building will require far less heating and cooling energy expenditure than a noninsulated or poorly insulated building. The roofing contractor should inspect for proper insulation during a roofing job. The poorly insulated building should be reported to the building's owner.

A well-insulated building will last longer than a poorly insulated building. There are less or slower variations in temperature and structural members tend to expand and contract less. There will be less moisture escape from the enclosed rooms and, therefore, less damage to the building's components.

Please read the chapter on flashings and vents to see how proper venting interacts with insulation values. Insulation is rated by a system of values and formulas that show the effectiveness of the insulation.

R-, C-, K-, AND U-VALUES

The resistance to heat flow through an object is its R-value. The C-value is the amount of conductance of heat through an object and is the reciprocal of the R-value. The U-value is the total of the R-values of a composite surface. For example: Typical wall construction is siding, insulation board, air space, insulation, air space, and drywall. The sum of the R-values of each of these items is the U-value. The resistance to heat flow on any of the items is the R-value. The passage of heat through any of the items is the C-value. The K-value is the amount of heat in BTUs that will pass through one square foot of a homogeneous material 1 in. thick, in 1 hr. The better insulations have high R-values and low C-, K-, and U-values.

R-Values in a Building

Figure 7–1 and Appendix B, table 26 depict the thickness and R-values of various attic insulations. Thicker material will have a higher R number and there will be less heat loss from the rooms. Today's modern, energy-efficient homes are built with R-values of 30 or more. To the roofer, a higher R-value means a longer lasting roof. There is less temperature difference between the attic air and the outside air, thus less moisture collection. There is better control of expansion and contraction of the roofing materials.

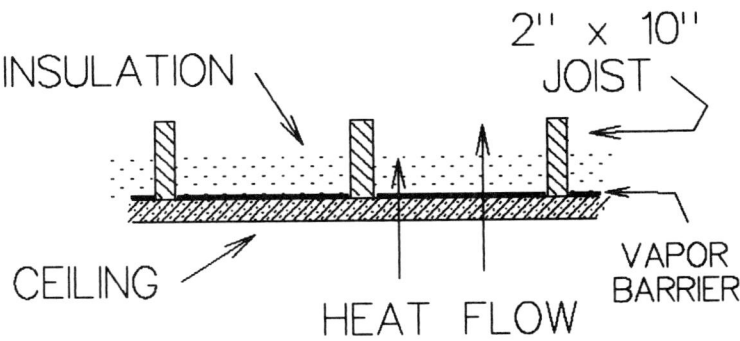

ATTIC AREA

INSULATION

2" x 10" JOIST

CEILING

HEAT FLOW

VAPOR BARRIER

SEE D.O.E. RECOMMENDATIONS

Figure 7–1 Typical room to attic construction

DOE

The Department of Energy publishes a paper titled "DOE Insulation Fact Sheet (DOE/CE-0180)." This is available from the Department of Energy, Technical Information Center, P.O. Box 62, Oak Ridge, TN, 37830.

In essence, the fact sheet suggests the proper thicknesses of ceiling insulation for various types of heating systems and insulation zones. An insulation zone is a portion of the country that should have a minimum of 'x' amount of insulation installed. The lowest ceiling insulation, R-value, recommended by the DOE, is R-30. You should use R-30 in the Gulf Coast states and southern California and Arizona. The DOE recommends R-49 in the mountain and Northern states. These include Nevada, Utah, Colorado, Wyoming, Nebraska, Ohio, Illinois, New York, and all states north. All other states should use R-38 ceiling insulation. The R-values given are for homes with electric resistance heating. For oil or gas heating, the R-value may be reduced slightly.

How to Determine Existing R-Value

To determine the R-value of existing fiberglass insulation, measure its depth. Multiply the measurement, in inches, by 3.23. The resulting answer is the R-value of the insulation. For example, 3.5 in. insulation multiplied by 3.23 equals 11.30 or R-11.

R-11 is the minimum recommended R-value for exterior walls. If you include the interior drywall, the vapor barrier, the fiberglass batt, the outer sheathing, the air spaces, and the siding, the R-value will increase by one or two.

Ceiling Insulation R-Values

In insulating a ceiling, you have the ceiling drywall and the vapor barrier only, thus, the R-value is small. To get the DOE-recommended ratings, use the following. Use 9-1/2 in. thick fiberglass insulation for a R-30 rating. Use 12 in. for a R-38 rating, and 15 in. for a R-49 rating. See Appendix B, table 26 for other insulation R-values.

Currently, with the exception of special 7-3/4 in. thick attic insulation (Certain Teed™, rated R-25), you cannot purchase fiber glass in the thicker sizes than 3-5/8 in. or 6-1/2 in. To obtain the R-values needed, you must stack the insulation. Place the first layer between the ceiling joists and a second layer over the first. The second layer should be placed perpendicular to the first. If a third layer is needed, it should be placed in the same direction as the first. Appendix B, table 26 gives the insulation batt thickness to combine for the desired R-value.

Caution

Be sure to fill all cracks and voids. Do *not* block the eave vents or any other vents. Do *not* cover light fixtures or electrical appliances or junction boxes. Building and fire codes require that these items be kept clear of insulation so that they can disperse heat. Many building inspectors will require you to build an open top box around these fixtures. The box prevents insulation from inadvertently being moved to the top of the fixtures. Blocking these fixtures prevents them from dissipating heat and may cause fixture damage or a fire; check local codes.

DETERMINING INSULATION NEEDED FOR A CEILING

To determine the amount of fiberglass insulation needed, measure the length and width of the area to find the square footage of insulation required. Then, use one of these multipliers. If the insulation is placed between joists that are on 16 in. centers, use .90. If the joists are on 24 in. centers, use .94. If the insulation is above the joists, use 1.0. Example: You must insulate a 10×10 ft area. The insulation will be installed between ceiling joists placed at 16 in. centers. A second layer above the joists will be required: $10 \times 10 = 100$ sq ft $\times$ .90 = 90 sq ft of insulation material required for the first layer: $10 \times 10 = 100 \times 1 = 100$ sq ft for the second layer. The total is equal to 90 + 100 or 190 sq ft of insulation.

INSTALLING A SERVICE WALKWAY

If the insulation thickness exceeds the ceiling joist depth, you will have difficulty walking in the attic area (see Fig. 7–2). Thus, you should install a service walkway. A service walkway in the otherwise unused attic is required since you need access to wiring, vents, TV lines, air conditioning, etc. Usually, these items are installed in the attic area in most homes.

COMPRESSING INSULATION

When it comes to insulation, more is sometimes not better. Many feel that the more insulation they cram into a space, the better insulated the space becomes. This is not true. Insula-

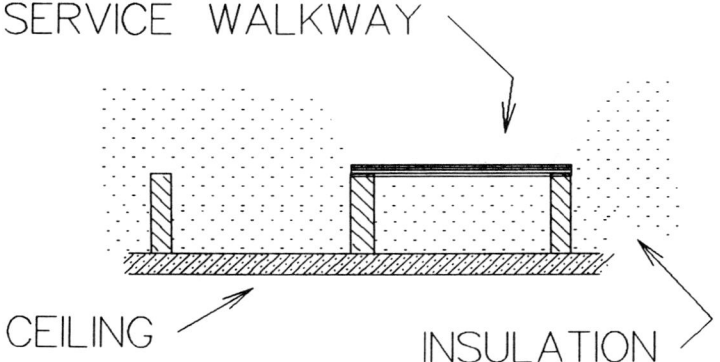

Figure 7–2   Service walkway

tion is effective because it contains dead air spaces between the fibers or cells of the insulation. When the air is squeezed out of the insulation, it defeats the design intention. Dead air is air that does not circulate freely. It provides a good portion of the insulation materials R-value.

INSULATING CORNERS OF ROOMS

Ceiling insulation that is not installed at a constant thickness can create problems (see Fig. 7–3). This is especially true at the eave and rake areas. The outer corners of most outer rooms are cold. The insulation installer either did not install the insulation all the way to the corner or he compressed the insulation into the corner.

Compressed insulation can lose up to 90% of its insulation value. It is better to use a thinner insulation, say R-19, than a highly compressed R-49 at the eaves and rakes. But, since the insulation is thinner at these corners and, therefore, the corners are colder, the warm moist room air will try to exit the building at these points. That which looks like a roof leak may only be condensation. This is true at the inside ridge of a room with a cathedral ceiling. One leak I observed appeared to be a leak, but was in reality coffee. The coffee pot generated steam that rose to the peak, cooled off, and condensed into droplets.

The heat loss at the point shown in Fig. 7–3 can be of some benefit in that it does help minimize ice and snow buildup on your roof. To help prevent moisture damage to the roof sheathing, top sill plates, ceiling joists, rafters, and metal strapping, install a vapor barrier. The vapor barrier is installed across the wall to ceiling junction and should be continuous for best blockage of moisture.

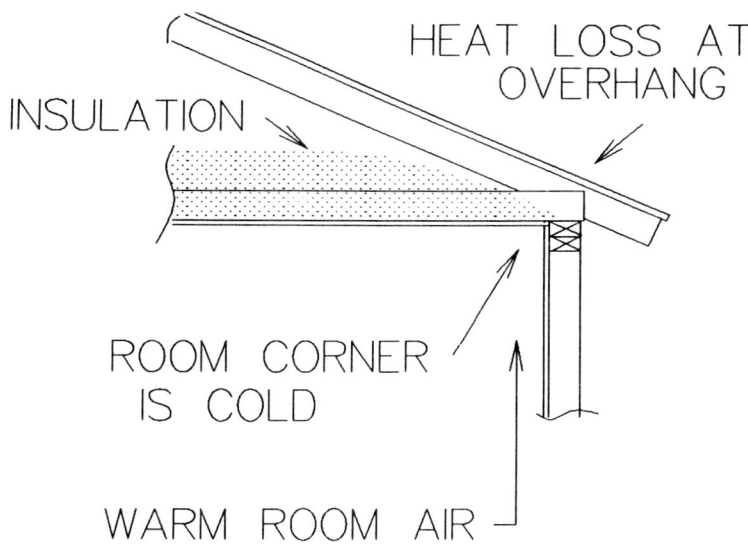

Figure 7–3   Insulation at building's corners

POOR WORKMANSHIP

Many general contractors feel that installing insulation is a nonprecision job. After all, you just tuck the insulation between the joist and staple it in place. The general contractor, therefore, hires personnel who are not really installers. This is wrong, as you shall see.

See Fig.7–4. Several years ago I was doing a job in a very expensive home in Malibu, California. I wish to thank publicly the person who left the beer can on one of the ceiling joists, buried 6 in. under the insulation. I didn't break a leg or kill myself, but I could have. I did fall through the ceiling. I did get very bruised and I did lose 2 days work fixing the ceiling and repainting the room. I almost lost a customer. After all, I did destroy their bedroom ceiling and had chunks of drywall all over their bed. But, they understood. I do not! Was this an oversight by the contractor? Just forgot to clean up after himself? Or, was this a big joke? Why was he drinking on the job in the first place? I trust he is out of business. I wonder how many of his fellow contractors he has injured with his unthinking, uncaring attitude? Any guesses? My point to you, the roofing contractor, is be careful; there are obstacles everywhere. And clean up after yourself; leave the drinking and practical jokes at home.

There are other mistakes made by untrained or unskilled insulation installers. These mistakes include:

- Lack of an air channel between the insulation and the roof sheathing.
- Overcompressing the insulation into openings.
- Not insulating all openings.
- Installing the insulation's vapor barrier in the wrong location.
- Not installing a vapor barrier.
- Installing all layers of insulation in the same direction.
- Insulating the wrong area of the attic space.

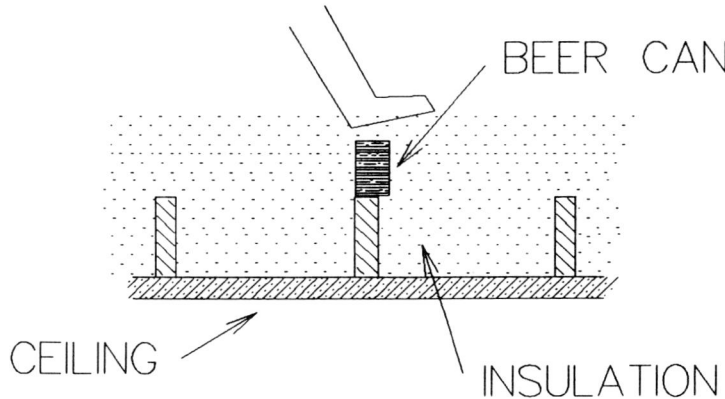

Figure 7–4 Poor work practice

- Not providing proper ventilation of the attic space.
- Miscalculating the required amount and thickness of insulation.
- Using insulation that derates the building's fire rating.
- Using insulation that derates the building's wind rating.
- Not sealing cracks before installing the insulation.
- Insulating too close to electrical fixture boxes.
- Using wet insulation.
- Using insulation that is banned in the community.
- Using crushed or damaged sheet insulation.
- Using the wrong number of fasteners for sheet insulation.
- Using the wrong fastener pattern for sheet insulation.
- Using the wrong fasteners for the application.
- Fastening sheet insulation too tightly to the decking.
- Installing sheet insulation so the joints match the decking joints.
- Not fastening sheet insulation to the raised ribs of metal decking.
- Applying heated asphalt to sheet insulation not designed for it.
- Installing sheet insulation directly on new concrete.
- Improperly venting sheet insulation.

That is a partial listing of the things that can go wrong while installing insulation. I'm sure the building inspector can find fault with a few other items. Excessive moisture or heat under your roof will deteriorate the roof's interior surface. Flashings will rust, sheathing will become soft and sag, shingles will dry rot or warp and twist. Blisters will form or worse, the roofing will blow away. See the chapter on problems.

TOOLS OF THE TRADE

Insulation installing does not require a truckload of tools. Here are the common items you will need.

1. For roll or batt type insulation:
   - A stapler. Manual, hammer type, or power driven.
   - A razor knife with easy change blades.
   - A tape measure.
   - A 2- or 3-ft straight edge.
2. For loose fill insulation.
   - A caulking gun with sealer.
   - A few buckets of various sizes.
   - A few funnels of various sizes.
   - Blower with hopper (optional).

3. For pour-in-place foams.
   - Mixing and dispensing equipment.
   - Tape measure.
   - Power drill with properly sized large hole cutters.
   - A caulking gun with sealer.
   - A power sander.
4. For sheet insulation.
   - A hand saw for cutting thick sheets.
   - A powered fastener gun or drill.
   - A razor blade knife with easy change blades.
   - A 4 ft or longer straight edge.
   - A tape measure.
   - A rubber mallet for some types of fasteners.
   - A hammer.
   - A carpenter's saw.

There will be other tools depending on the insulation manufacturer's requirements, but these should get you going.

SAFETY EQUIPMENT

Note: Proper safety equipment is required. Most of the insulation materials that you will be using are fibrous or produce somewhat hazardous dust when cut or handled. Dust mask, eye protection, and skin protection are recommended. Obtain, read, and heed the manufacturer's PSDNs before starting work.

MATERIALS USED FOR INSULATION

The materials used for the manufacture of insulation falls into four categories:

- Rolls or blankets, mats or batts.
- Loose fill. Granular or puffed.
- Liquid foam-in-place. One or two part mixes.
- Sheet foams. Faced or unfaced.

I'll start with the blankets and batts. These are the most common materials.

Fiberglass Blankets and Batts

Fiberglass is the main component. As the name implies, the material is glass fiber random laid and cut into strips 15-1/2 in. wide. Fiberglass is fire resistant, mildew resistant, and provides no food value to insects or rodents. Fiberglass insulation is available as unfaced, single-sided faced, and double-sided faced. The facings can be Kraft paper or aluminum foil. The unfaced insulation is used for walls that have a separate vapor barrier or for ceiling insulation where a second or third layer is required. The first layer of insulation is between the ceiling joists and should be faced. The vapor barrier facing faces toward the heated room. Double-faced insulation should have the foil side toward the room, the paper side toward the outer wall surface. Double-sided insulation is not recommended for ceilings. Blankets are sold as rolls of insulation containing 'X' amount of square feet. Batts are sold as precut insulation with 4 ft and 8 ft lengths available. Thicknesses vary and are determined by R-value rating.

Another batt type insulation is mineral wool. Mineral wool is described as a "vitreous, fibrous insulation material." Loosely translated it means a "glass-like fiber." Rock, slag, glass, and at one time asbestos are used to make mineral wool. The R-value of mineral wool is usually higher than that of fiber glass for the same insulation thickness.

Loose Fill Insulation

Loose fill insulation is used for insulating cavity walls, masonry block walls, and for ceiling insulation. Loose fill is sold in bags and poured or blown into place. Being loose fill, it does tend to flow downward. All cracks and holes must be sealed if you do not want insulation pouring into the building's rooms.

In accessible areas, the insulation is poured into place. In closed-in or inaccessible areas, it must be blown into place. This will require a motor-driven hopper with feed hoses.

The materials used for loose fill are:

- Fiberglass
- Rock wool
- Perlite
- Mineral wool
- Cellulose
- Vermiculite

Perlite and vermiculite are the smallest particles and flow easily. It is for this reason that they are not generally used for ceiling insulation. They are frequently used for masonry block insulation. Both do absorb water and both must be protected from water contact.

The remaining insulations are fairly large and tend to bond together by fiber interlocking. Thus, the problem of flow is not as great as with the smaller insulation. There is a possibility of the larger materials settling or matting down. This settling is caused by the materials' weight and may result in loss of insulating value.

Liquid Foam-in-Place Insulation

There are one- and two-part urethanes that are sold as a liquid. When exposed to air or when the two parts are combined, the liquid expands into a solid closed cell foam. The expansion rate is about 10 to 1. During expansion, the foam can generate great pressure and if solidly contained, or improperly measured, can cause damage to the item to which it is installed. Liquid foam-in-place insulation is used for insulating the walls of older uninsulated buildings. Liquid foam-in-place insulation is banned in many communities due to its potential health and fire hazards. It does burn and give off toxic fumes and does require a fire-resistant barrier. Installers of liquid foam insulation may have to be state licensed in your area.

To install liquid foam-in-place insulation, one drills holes in the building's siding. Every cavity between wall studs must be drilled, including the cavities on both sides of horizontal fire blocking. The liquid is then pumped into the cavities and allowed to expand. The holes are then plugged with wood, plastic wood filler, or other means. The building will require repainting after the insulation is installed. The outer interior walls may require an additional layer of drywall or plaster to meet fire codes. Foam-in-place insulation is seldom used for ceiling insulation.

Sheet Insulation

Sheet insulation is used for siding underlayment, burial in or under concrete slabs, and for BUR deck insulation. It may be used for standard attic insulation, but usually is not due to fire codes. It is frequently used for cathedral ceiling and recreational vehicle roof insulation.

The materials used for sheet insulation are:

- Fiberglass
- Phenolic
- Polystyrene
- Cellulose
- Wood fiber
- Perlite
- Polyurethane
- Polyisocyanurate
- Mineral fibers
- Combinations of all

It is not unlikely to find sheet insulation manufactured as combination sheets. The base or core will be urethane or isocyanurate. The outer skins will be wood fiber board, wood chip board, perlite, fiber glass, aluminum foil, or phenolic plastic board. The outer skin protects the closed cell structure of the foam plastics from physical damage while providing a smooth base for the roofing. An outer skin may be needed to increase the fire resistance rating, especially for the plastic base insulations.

Some insulation may contain perlite or glass fibers within the foam. This adds to the insulation value and to the foam's structural support.

Foam board installation.   In most installations, the foam board is mechanically fastened to the roof decking with nails or screws and stress or distribution plates. Some BUR installations require that the foam be asphalted or otherwise cemented in place. Perforated foam board is recommended for these applications. The perf holes trap the liquid cements and bond the roof plies to the insulation board. Some installations of BUR will require that the insulation be vented to remove excess moisture. See the chapter on flat roofs.

Foam Board Sizes.   Sheet insulation comes in thicknesses of 1/4 in. to over 4 in. with R-values of less than 1 to over R-25. Sheet insulation is available in 2 × 4, 3 × 4, 4 × 4, 4 × 8, and 4 × 9 ft sheets. Foam boards with a thickness of over 2-1/2 in. are usually a special-order product.

The weights of foam board range from approximately 1/4 lb to 3-1/2 lb per sq ft. Weights vary according to type of board and R-value.

Sheet insulation R-values.   Appendix B, table 2 shows some typical R-values obtained with foam insulation. These values are aged 6 month values. Foam insulation does not reach its full R-value for up to 5 years after manufacture.

The large variance in R-value range for a particular thickness is due to the composition of the product and the foam used. The highest R-value numbers were that of Owens-Corning's, Energy Shield™, sheathing. This sheathing is a polyisocyanurate foam covered with a foil-kraft-foil facing on both sides and is rated as Class I insulation when properly installed.

The ratings are the aged R-values as the materials leave the factory. The actual retention or in-service R-values may be much less. For example: Foam used for foundations where there is much moisture present may lose up to 60% of its initial aged R-value. If properly installed and kept dry, with age, it may increase in R-value. This is typical of a roof installation situation.

Bead board and Styrofoam™.   Bead board and Styrofoam are both polystyrene products. The difference is in how they are manufactured. Styrofoam starts with a styrene resin and foams the styrene in a mold. This produces a tight, closed cell foam board with a tough outer skin. Bead board starts with styrene pellets that are expanded with steam and heat into the mold. This produces a semiclosed cell product with approximately 24% less insulation value than Styrofoam (per Dow Chemical Co., the makers of Styrofoam).

Bead board and Styrofoam weigh about 28 oz per cu ft and are light enough to float on water. Both products are water resistant but can be made watertight with a thin coating of epoxy or Wilhold™ cement. Do not attempt to cement plastic sheet to galvanized metal using epoxy, because the epoxy will not adhere properly.

T&G insulation board.   Sheet insulation may be fabricated with straight or T&G edges. The T&G foam boards will give better resistance to air infiltration than the straight-edged products. Both products should incorporate a vapor barrier, either integral to the product or installed separately.

DOUBLE UP THE INSULATION

Air infiltration into a building accounts for 38% of the building's heat loss. Heat short circuits can account for 17%, and ceilings account for 5% of the heat losses. These figures are for attic type buildings. Flat roof and cathedral buildings have higher heat losses, due to the absence of an attic space.

Air infiltration is through holes and cracks while heat short circuits are through nails, joist, studs, rafters, and all other uninsulated items that join the building's interior with its exterior. When you install a single layer of insulation, you install it with fasteners. These fasteners become a heat short circuit. The cracks between the individual insulation sheets are a source of air infiltration. It is better to install two thin sheets of insulation that, when added, equal the thickness of the single sheet you wanted to use. The second sheet is installed perpendicular to the first and covers the nails and seams of the first sheet. This reduces the air infiltration and heat short circuits to near zero.

FOAM INSULATION FIRE PRECAUTIONS

Note: Sheet and foamed insulation will burn if exposed to fire. All sheet and foamed insulation must be protected with a vapor barrier and a fire-resistant covering. Coverings may be perlite, gypsum plaster, cement, concrete, metal, lumber, or drywall. While burning, all sheet and foamed insulation will give off fumes which contain chemicals that are hazardous.

Chimneys and Heat-Producing Fixtures

Do not insulate within 3 in. of heat-producing equipment or light fixtures. (Check local building code. Some require that a wood box frame be used around each fixture such as a stove pipe or chimney.) Insulation should not lie against a fireplace or stove chimney—a 3 in. to 6 in. clearance is needed (check local code and manufacturer of chimney for specific instructions).

CATHEDRAL CEILING INSULATING TECHNIQUES

See Fig. 7–5. Sometime during your roofing career you will be required to roof over or install a cathedral ceiling. Where is the insulation? It is sandwiched between the ceiling drywall or interior lumber planking and the roof sheathing. See the chapter on new construction.

Sometimes, the insulation is installed by the insulation contractor to the underside of the roof sheathing. Other times, the insulation is installed on the top of the roof sheathing by the roofing contractor.

I inspected one house that had a room add-on with a cathedral ceiling. The contractor who did the work has long since left the area. It is a shame that the taxpayers paid out good

CATHEDRAL CEILING

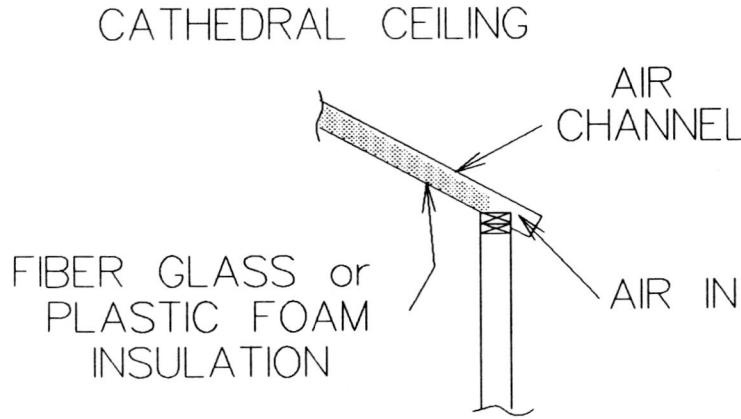

Figure 7–5   An air channel is required

money for the job. The job was financed by a state government grant program and was less than a year old. Each rafter could be seen from the room's interior. There are black stains on the ceiling drywall from the walls to the peak, due to a moisture problem.

The contractor made four serious mistakes.

• The first was to use S-Grn lumber and then quickly cover it with the drywall. The 19% plus moisture from the lumber and the drywall spackle's moisture were trapped between the ceiling drywall and the roof sheathing. This moisture effectively killed the R-value of the insulation.

• The second mistake made was the absence of a vapor barrier. The homeowner lives in this room and has it decorated with many moisture releasing plants. The paint was in poor condition and in several places the drywall seams were coming apart. The insulation and the rafter lumber were soaked with moisture. This caused several rafters to warp and crack the ceiling drywall.

• The third mistake was the venting system used. The contractor used soffit vents, but did not use a ridge vent. Soffit vents by themselves do little. See the chapter on flashing and vents.

• The fourth mistake was the contractor's selection of insulation thickness. The rafters are 2 × 10 lumber with a finished depth of 8.5 in. The insulation used was R-38, 12 in. thick. This compressed the insulation and reduced its R-value considerably. The compressed insulation prevents air circulation above the insulation, which in effect, prevents the later installation of a ridge vent. Ridge vents require soffit vents and an air channel from the soffit to the ridge. This air channel can be made from low-cost foam ventilation chutes available at roofing supply houses. The ventilation chutes are inserted between the insulation and the roof decking.

Back to our problem home. It will cost this customer, or the government, thousands of dollars to make the necessary repairs. The ceiling drywall and the existing insulation must be removed. The proper solution is to install a ridge vent, new insulation, new drywall, and new paint. The temporary solution was to clean the surfaces, repair the drywall tape seams, and level out the bumps caused by the warped rafters. Then the cracks were filled, the drywall primed with a shellac base primer, and then repainted with semigloss paint. The homeowner was advised to sue the contractor for the cost of this temporary repair and require him to make the correct fixes.

INSULATING AN ALREADY INSULATED ROOF

I do *not* recommend that you apply foam insulation to a roof that is already insulated at the room's ceiling joist. This new insulation will be a waste of money since you will be insulating an outside vented area. Plus, the insulation will act like a vapor barrier and it will create high levels of condensation at the roof sheathing to foam interface and within the attic area. This condensation will wet the ceiling insulation and decrease the overall R-value.

USING SHEET FOAM ON A BUR

There are times when you will be requested to provide insulation when roofing or reroofing a flat roof. There are several materials that can be used for this purpose. These materials can be applied to the topside of the decking and then covered, a built-up roof system.

Built-up roofing systems are tightly regulated by the building code, the manufacturers, and Factory Mutual Engineering. Factory Mutual (FM) is an engineering company that specializes in commercial properties. They do testing and specifications for built-up roofing products and installation techniques.

FACTORY MUTUAL REQUIREMENTS

Built-up roof insulation is light in weight and flammable. The proper installation techniques must be used in order for a building to pass the FM Class I, Class II, FM-60 or FM-90 Factory Mutual ratings. FM-60 means the roof must not blow off in a 60-mile per hour wind. FM-90 means it must not blow off in a 90-mile per hour wind. FM Class I is their best fire rating. FM Class II is their worse fire rating, with Not Classified as not fire rated and possibly uninsurable. See the chapter on testing and specifications for more information.

CORRUGATED DECK FLUTE SPANS

When installing foam insulation on corrugated metal decking, the maximum flute to flute distance for 1/2 in. thick and under insulation is 1-5/8 in. For 1/2 to 1 in. thick insulation, it is 2-5/8 in. For insulation over 1-1/5 in. to 2-1/3 in. thick, it is 3-5/8 in. For insulation over 2-1/3 in. thick, the flute to flute span is 4-3/8 in. NOTE: Insulation sheets of less than 1 in. thickness may not be FM approved. This is because the thin material can break through the fasteners in a high wind situation.

MAXIMUM WIND UPLIFT

The air currents at the edges and corners of a building produce extreme uplift currents. For this reason, it is recommended that the sheet foam insulation be fastened to FM-90 standard plus 50% more fasteners at all building corners and along the building's edges. This applies even if the rest of the insulation is installed to FM-60 specifications.

Parapet Walls and Wind Uplift

You will find that flat roofs with parapet walls can produce severe uplift that may damage the roof surface. As the wind travels over the parapet, it forms a vacuum over the roof. This vacuum is not unlike the vacuum formed over the wings of a 747 jetliner. If it can hold a 747 in the air, it certainly can rip off a poorly attached roof (see Fig. 7–6).

BALLASTED ROOF INSULATION

Some manufacturers of roofing and roof insulation require the use of ballast. Ballast is gravel or slag used as the final roof covering. You must remember that the sheet insulation

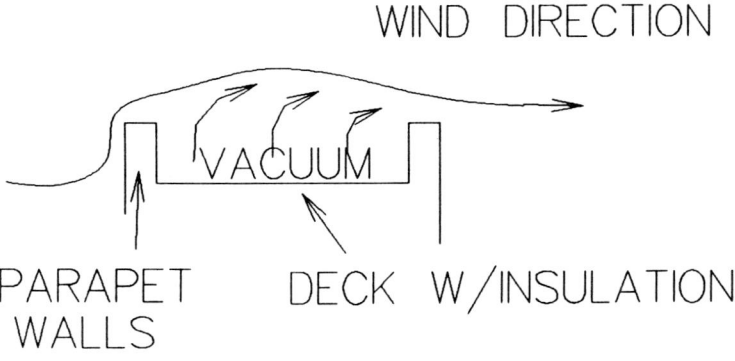

Figure 7–6   Parapet walls may lead to roof being sucked off decking

fasteners are the only things holding the roofing to the roof sheathing. The remainder of the roof plies sit on top of this insulation. If the insulation fastener system fails, the roof can be blown off. Ballast weighs from 300 to 400 lb per roof square, which is sufficient weight to hold everything in place.

ROOF INSULATION VERSUS ROOF COATING

For the most part, roof insulation does a necessary job, but there is one problem of which you should be aware. On BUR roofs, the insulation tends to keep heat in the roof plies. This causes an unbalanced temperature gradient from bottom to top surface of the roofing plies. This causes premature destruction of the roof due to blistering, delamination, and drying.

Foam Insulation versus PVC Single Ply Roofing

Several foam insulation sheets are not compatible when in direct contact with PVC material. A divorcing layer must be used to prevent chemical interaction of the materials. Consult the manufacturer's specification sheet for the insulation being installed.

Service Temperature

The service temperature of a roof is the temperature extremes that the roof might be subjected to. Most roof insulation is good for a service temperature of −100 to +250 degrees Fahrenheit.

Dimensional Stability of Foam Sheet

The dimensional stability specification of most foam sheet insulation is less than 2% in any direction. This is at 158 degrees Fahrenheit and 97% humidity for 24 hours. If you think about it, an 8 ft long sheet of foam insulation is permitted to change up to 1.92 in., 96 in. times .02 = 1.92 in. That is a pretty good change in dimension and is the reason that many insulation installers lay down 4 × 4 ft sheets rather than 4 × 8 ft sheets. A 4 × 4 sheet has a maximum change of .96 in. The open joint seams between sheets is then .96 in. total. In practice, the insulation is laid with 1/4 (.25) in. joints or just touching.

The dimensional stability of sheet foam insulations is, in reality, less than 1% and is more likely to be 0.5%. At 0.5%, a 1/4 in. gap between 4 × 4 ft sheets will be sufficient. This 1/4 in. wide joint is another good reason for using two or more thin layers of insulation rather than one thick layer. The air gaps of the first sheet are covered with the second sheet.

Layering Foam Insulation Sheets

When layering two or more foam insulation sheets, you should stagger the joints between sheets and layers so that no two joints lie on top of one another. The second sheet should be perpendicular to the first sheet. The second sheet can be mechanically fastened to the first or asphalt bonded to the first sheet. If asphalt is used, you must pay close attention to the temperature of the asphalt. The temperature is not to be so hot as to melt the surface of the insulation. Mopping is to be solid and applied at a rate of 25 + 5 lb per roofing square.

Bonding Foam Directly to a Roof Deck

Solid bonding to the roof deck is generally not recommended. The insulation should float on the deck so that each can expand and contract at their own rates. Some manufacturers do allow direct adhesion of their insulation sheet to corrugated steel decking. If done, then the adhesive is to run the full length and width of the deck ribs. When using asphalt or other hot applied adhesive, be sure it does not cool below its bonding temperature before the insulation is installed.

Most manufacturers recommend a felt underlayment or other resin type building paper be installed directly to the sheathing, before the foam insulation is laid. This underlayment is nailed to the sheathing and the insulation is solid mopped to the underlayment using 25 + 5 lb of asphalt per roofing square.

The asphalt used should be Type III or Type IV asphalt at an applied temperature of 400 degrees Fahrenheit. If the insulation board is polystyrene, it will be necessary to lower the asphalt temperature to 230 degrees Fahrenheit. Check specifications sheet for the foam you are using, it may recommend otherwise.

Insulating Foam Sheet at Walls

Wherever the foam insulation sheets butt to a vertical wall or curb, a 1/4 in. gap should be left for expansion. This gap will be covered with felt plies during the roofing. Be sure the metal flashing over the edges of the felt plies is firmly attached to the wall structure. There have been instances where the metal was poorly attached and blew off. In doing so, it exposed the felt ply edges and the felt lifted. This then lifted the insulation and the entire roof section blew off with the next good wind gust.

Bonding Foam Insulation for Fire Resistance

To minimize fire danger from within the building, a fire-resistant barrier should be installed between the sheathing and the insulation. Some insulation is laminated with perlite or other fire-resistant material and can be installed directly on the decking. Other insulation sheet may not have this provision. It is permissible to install a layer of gypsum board between the decking and the insulation sheet. The gypsum board is mechanically fastened to the decking and the insulation sheet is either solid mopped or mechanically fastened to the gypsum.

Foam Insulation Over Poured Concrete or Gypsum

Installing foam insulation over newly poured concrete or gypsum is not recommended. The concrete or the gypsum will contain a large amount of water and this water will seep into the insulation, ruining it.

To insulate a new concrete deck, use one of the several lightweight insulating concretes that are on the market today. These not only provide us with a smooth solid roof deck but also provide insulation value to the building. This concrete has very small granules, usually glass or minerals such as perlite or vermiculite, mixed through it. The granules provide air space that acts as the insulation.

Note: Do not mix perlite with vermiculite, the two are not compatible when used as a filler in lightweight concrete. Perlite is obsidian, a volcanic glass, and vermiculite is a hydrous silicate mineral. (Hydrous = water; silicate = acid) (acid + glass = no glass remaining).

Perlite and vermiculite are expanded insulations. They have been expanded in size by 4 to 20 times with the application of 1,400 to 2,000 degree Fahrenheit heat. Both soak up water like a sponge.

Tapered Foam for Roof Drainage

There are several different tapered foam systems in use. I will describe one system. The foam insulation is sold as wedge-shaped panels and is used to form a roof drainage slope on an otherwise dead flat roof surface. Each wedge is 2 ft by 4 ft and has a flat bottom, flat vertical front and rear. The slopes are 1/8 in., 1/4 in., 1/2 in., and 1 in. from front to back. There are 8 standard sizes ranging from 1/4:1/2 in. (front edge:back edge) to 2-1/4:2-1/2 in. The B standard sizes, when placed together, form a 16 ft long by 4 ft wide ramp. By combining ramps into various patterns, hips, valleys, and crickets can be formed.

8

Flashing and Vents

INTRODUCTION

This chapter discusses roofing felt, flashings, and vents and how they work. Included are the installation methods for most types of flashing you will encounter. The chapter is essential reading as it is referenced throughout this manual. Flashing prevents water entry into a building at all junctions of the roofing to a wall, skylight, chimney, parapet, dormer, vent, or other roof penetrations. Without flashing, most roofs would leak like a sieve. The venting portion of this chapter describes the types of vents available and their usage.

Included in the chapter is a small section on rain gutters and their installation.

ROOFING FELT

Roofing felt can and should be used where there is a hip, ridge, valley, sharp corner, or overhang covered with shingles. Roofing felt is not waterproof. It is resistant to water for short durations.

See Fig. 8–1. Roofing felt has directional stability and should be laid with its long direction along the length of the roof. By directional stability I am referring to its ability to expand and contract without splitting (see Fig. 8–2). Installed along the length of the building, it can better expand and contract with the building. Installation is easier and the felt's ability to shed water is better.

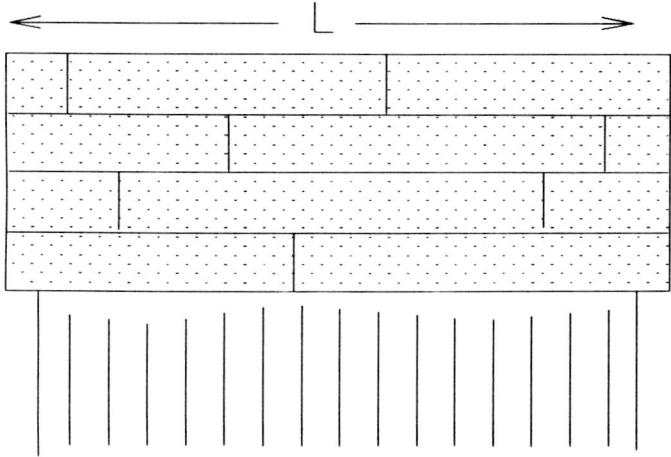

FOR BEST RESULTS RUN FELT
LENGTHWISE WITH ROOF

Figure 8–1 Felt installation on roof sheathing

FELT STRENGTH

MAXIMUM STRENGTH

MINIMUM STRENGTH

Figure 8–2 Felt strength is lengthwise

Felt used to seal the roof's ridge should run along the length of the ridge (see Fig. 8–3). It should overlap the field felts on both sides of the building. Do not apply felt over the ridge using short pieces or the felt will split where shown.

Felt used as flashing or as an underlay to valley flashing should run lengthwise with the valley (see Fig. 8–4). Do not install as lengthwise strips across the valley, it will split. The exception is where you bring the roof field felts into the valley. These should cross the valley on to the next roof in an interlaced pattern. You then add additional felt, lengthwise with the valley, over these crossing felts.

In dry desert areas, it may be permissible to use a single sheet of #15 felt, 36 in. wide for the valley under full or half-lace shingles (see Fig. 8–5). Most localities will, though, require a metal valley or a valley formed with a #90 cap sheet installed over the felt. The #15 felt protects the valley metal or the cap sheet from moisture escaping from inside the building.

When attaching felt to metal flashing at the rake or eaves, do not apply the mastic in solid strips (see Fig. 8–6). Apply as small, 1 to 2 in. strips, spaced 4 to 6 in. apart. This allows any trapped moisture to escape from under the felt. It also allows for expansion and contraction at very different rates between the metal flashing and the felts.

METAL EXPANSION AND CONTRACTION

Metal expands and contracts with changes of temperature, as do all materials (see Fig. 8–7). If you nail two items tightly together and both items expand and contract at the same rate, there is no problem. But, if the items expand and contract at highly different rates, you have

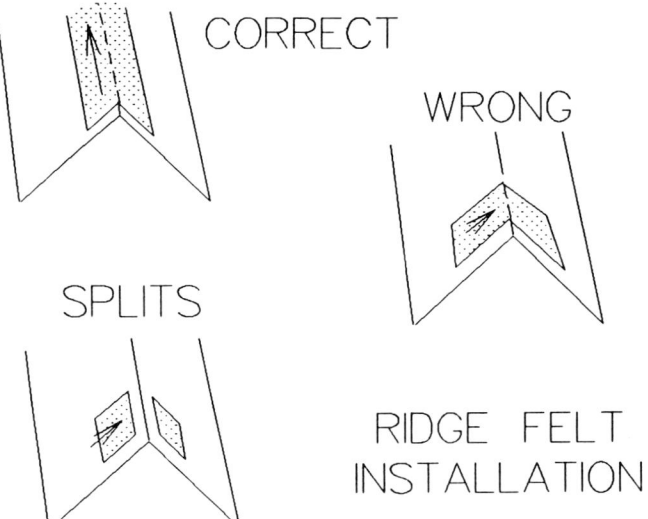

Figure 8–3 Installing felt along ridge

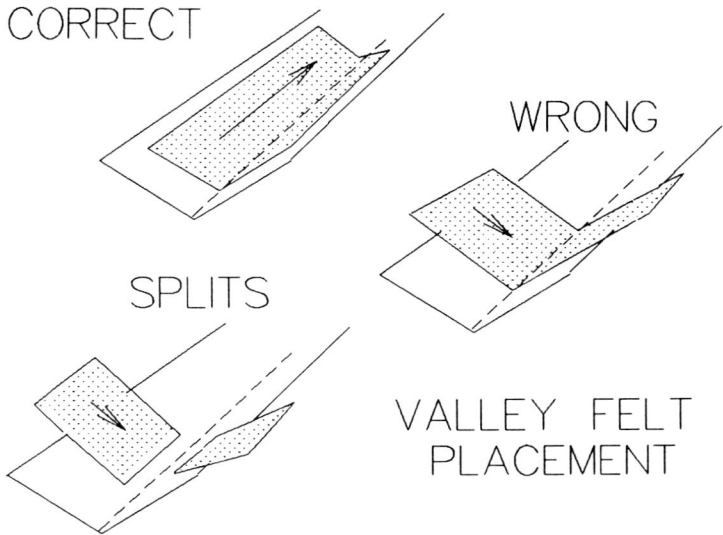

CORRECT

WRONG

SPLITS

VALLEY FELT
PLACEMENT

Figure 8–4 Installing felt in a valley

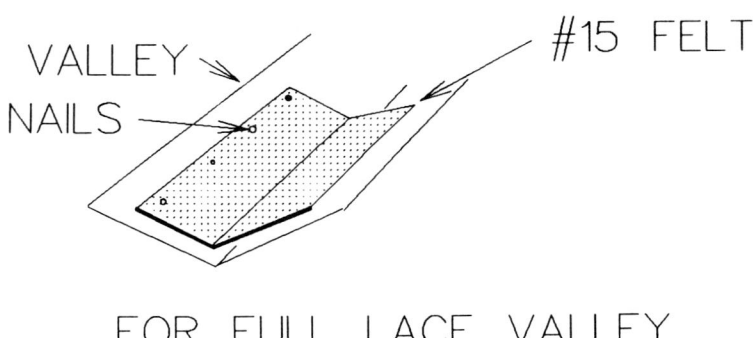

VALLEY

NAILS

#15 FELT

FOR FULL LACE VALLEY
CHECK LOCAL CODES

NAIL 1" FROM EDGE ON 12" CTR.

Figure 8–5 One type of valley

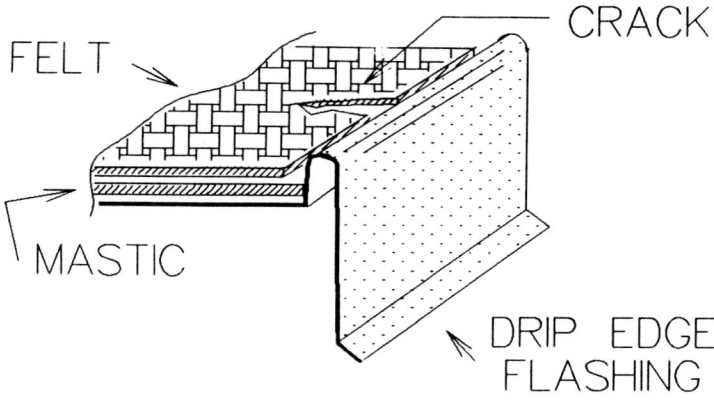

MASTIC APPLIED SOLID WILL LEAD TO FELT
CRACKING DUE TO METAL EXPANSION

Figure 8–6   Expansion and contraction of dissimilar materials can cause problems

major problems. This is the case when metal is nailed tightly to wood. The metal expands
and contracts at a much greater rate than the wood. Shown is a metal drip edge nailed tight
to the wood sheathing. The drip edge buckles and with it the overlying shingles buckle.
Keep it loose. When nailing drip edging or any metal flashing, the nail heads should just
contact the metal, not dent it. If possible, purchase material that is preslotted at the nailing

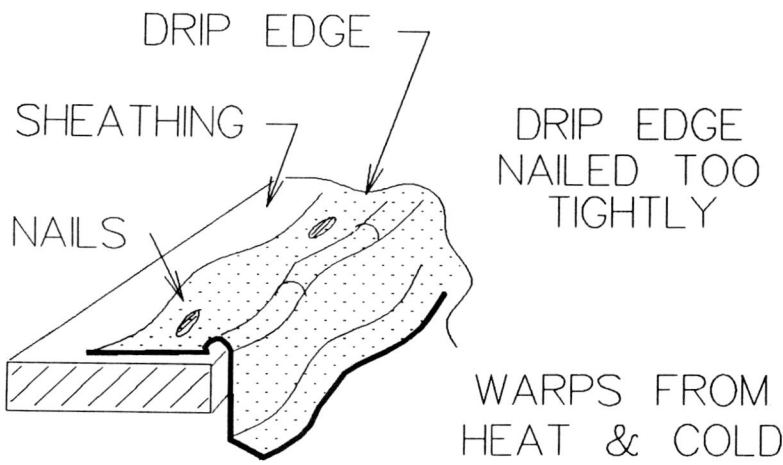

Figure 8–7   Keep it loose

points. If using metal thicker than #28 gauge, predrill the nailing holes with a drill bit that is 10% larger than the diameter of the nail shank.

DRIP EDGE FLASHING

Figure 8–8 is an end view or cross-sectional view of a drip edge. Water will cling to items via surface tension. Water also seeks its own level, or the lowest level it can. Try this: dip a pencil into a pan of water, now hold the pencil flat, horizontal. You will see the droplets of water clinging to the underside. If you rock the pencil back and forth, the droplets will move back and forth but will stay adhered to the surface. That is an example of surface tension. Now tilt the pencil so that the point is straight down. The water will fall off at the tip. The reason? The tip is sharp and narrow and provides a very small surface for the water to cling to. You have broken the surface tension.

This is the theory behind a drip edge. We provide a very small, sharp, surface at point A. The water loses its hold and falls into the rain gutter or to the ground. Drip edging should not contact the fascia board. It should be 1/4 in. or more away, as shown in Fig. 8–9. The reason is that a tight connection between the drip edge and the fascia can result in trapped moisture and dirt. It can lead to surface tension and capillary action of any water running off the drip edge. Surface tension and capillary action will suck water into the fascia and eventually rot it. As an added precaution, the fascia board should be painted with an oil base primer and then top coated with a latex enamel before the drip edge is installed. If you are using stain, then use an oil base stain. This will help waterproof the fascia board and it will last years longer. I highly recommend using treated lumber for all fascia board.

Figure 8–10 shows two acceptable methods of installing a drip edge at the rakes. For best appearance and maximum coverage of the edge of the sheathing, the drip edge is installed before the shingles (Fig. 8–10B). For maximum protection from wind uplift, the drip edge is installed over the shingles as shown in Fig. 8–10A.

See Fig. 8–11. This is not the recommended method of nailing shingles to a drip edge. In this instance, it was deemed necessary by someone. Spot the reason? This drip edge is on the upper side of the roof. This is the last course of shingles installed.

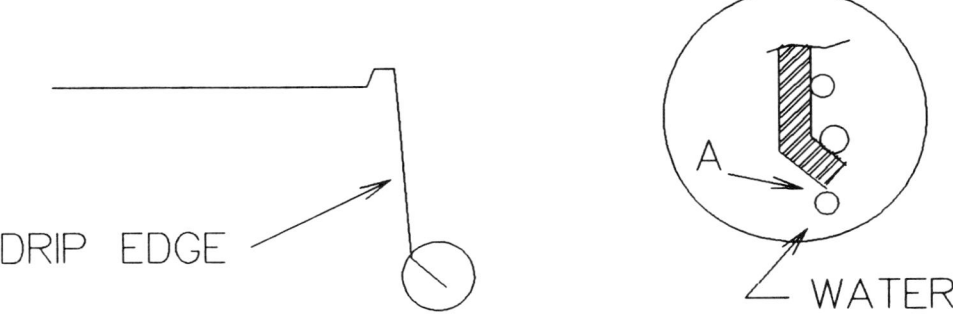

Figure 8–8 How a drip edge works

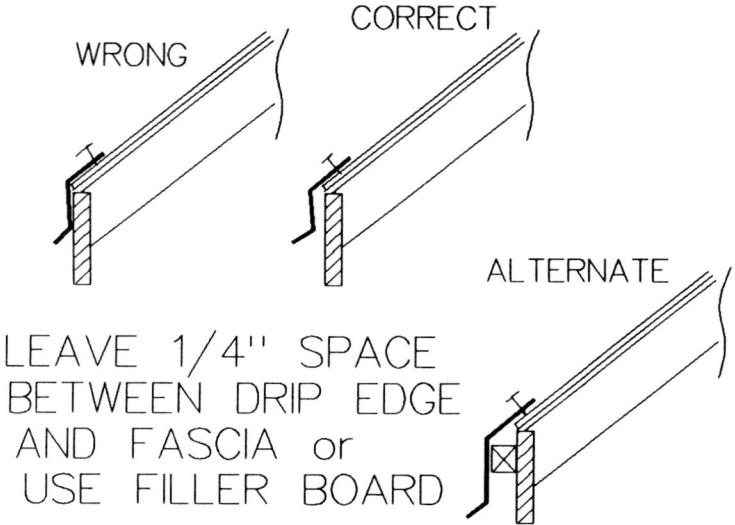

WRONG   CORRECT

ALTERNATE

LEAVE 1/4" SPACE
BETWEEN DRIP EDGE
AND FASCIA or
USE FILLER BOARD

Figure 8–9   Correct installation of a drip edge

DRIP EDGE POSITIONS

USE A FOR MAX. WIND/SNOW RESISTANCE

USE B FOR APPEARANCE

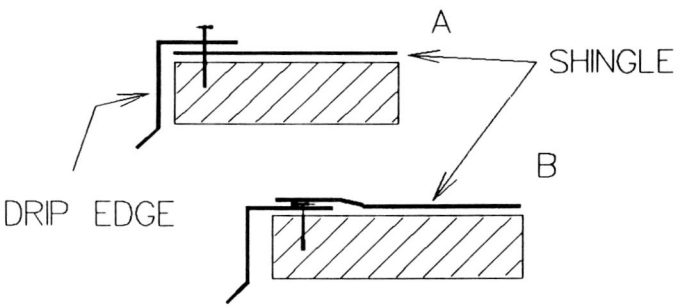

A

SHINGLE

B

DRIP EDGE

Figure 8–10   Drip edge installed for multiple use

Figure 8–11   Improperly installed drip edge

What should have happened is that the shingles should have been applied first with the edging over the top of them. As it is now, water will run under the top shingle. In either instance, the nails should have been grommeted or, at the very least, covered with mastic.

Figure 8–12 shows the proper installation technique used to secure the metal drip edge on the rake of a corrugated roof. A nailer is installed along the rake, and the drip edge nailed to it. Grommet nails are required on the top side of the drip edge.

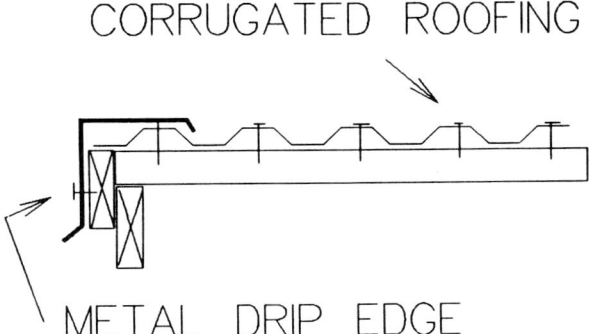

Figure 8–12   Drip edge installed on rake of corrugated roof

Nosing

Nosing is very much like drip edging, except it does not have the drip edge; nosing is L-shaped only. It is used to finish off rakes and other areas where a drip edge is not required.

BIRD STOP FLASHING

See Fig. 8–13. This is someone's idea of roof flashing. After the birds moved out, the bees moved in. This is NOT bird stop flashing. Bird stop flashing is used to prevent birds and other small animals from nesting up under shingles. Bird stop is sold in strips and is formed to fit the type and brand metal sheeting or tile being installed. It is required under the leading edge of corrugated metal roofing and '~' shape clay tile.

GRAVEL STOPS

Figure 8–14 shows the typical installation method used to install a gravel stop on a flat roof. The gravel stop prevents loose gravel and stone from being washed over the roof edge

Figure 8–13   Roofer created a home for birds and bees

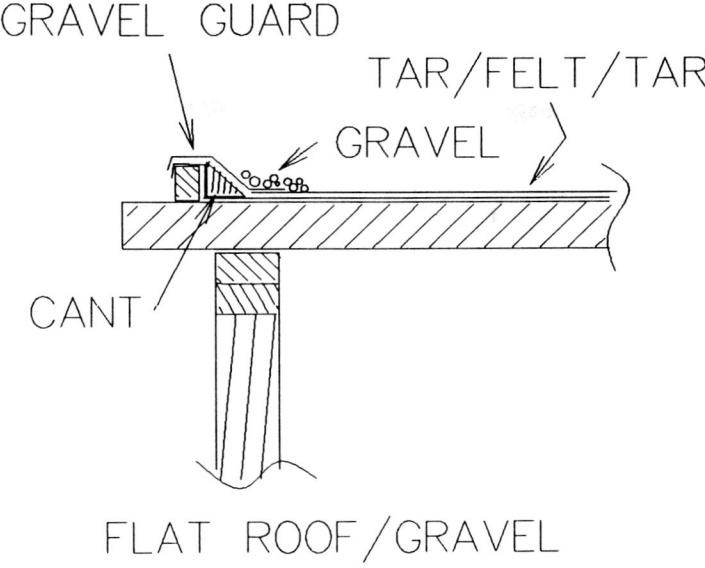

GRAVEL GUARD

TAR/FELT/TAR

GRAVEL

CANT

FLAT ROOF/GRAVEL

Figure 8–14 Gravel guard

during heavy rains. Continuous application of mastic is required between all layers of the roofing and the gravel stop to prevent water entry into and under the roofing. Flush mounted roof drains should be provided to prevent standing water.

RIDGE CAP FLASHING

Ridge flashing is to be installed over, not under, the field shingles adjoining it. There are several types of ridge flashing available. Some are for wood and shake shingles, they are '∧' shape and may or may not have an expansion bulb on top. The expansion bulb allows the metal to expand and contract without tearing up the roofing. It looks like an inverted U. The sides of the '∧' are smooth and should be nailed to the roof with gasketed nails.

Ridge flashing for metal roofs is either flat sided or corrugated. The flat-sided flashing is used on low-profile metal tile or noncorrugated metal sheet roofs. The corrugated side flashing is used to cap off corrugated metal. You must purchase corrugated flashing to match the corrugated roofing; the flutes of each must fit tightly together for best results.

See Fig. 8–15. If the metal ridge cap is not corrugated, and does not interlock with the corrugated roofing material, then you must install 1 × 2 nailers along the length of the ridge. These nailers are omitted if the ridge cap is corrugated and matches the roofing material. Nail through the corrugated cap and roofing and into the ridge beam. Always use gasketed nails for this application.

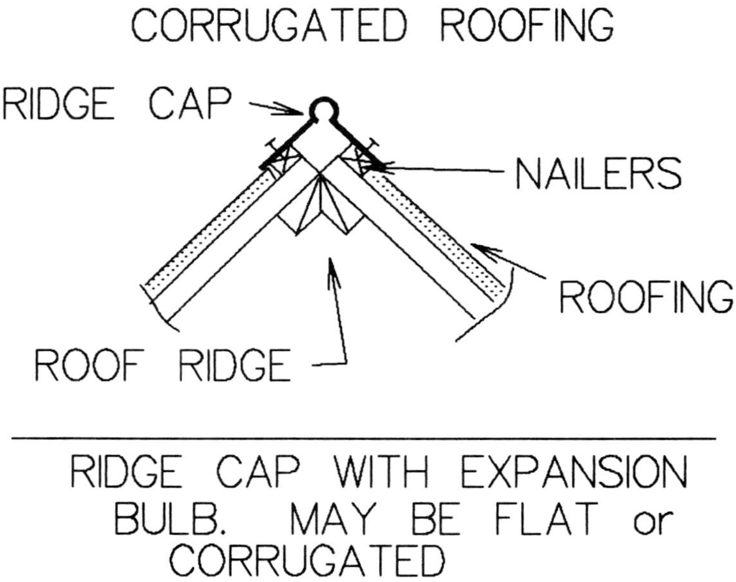

CORRUGATED ROOFING

RIDGE CAP

NAILERS

ROOFING

ROOF RIDGE

RIDGE CAP WITH EXPANSION BULB. MAY BE FLAT or CORRUGATED

Figure 8–15   Ridge cap for corrugated roofing

CHIMNEY FLASHING

Chimneys are a great source of water leaks. Most chimneys are constructed of brick and mortar and do not provide a smooth sealing surface to the junction of the chimney and the roof shingles. Chimneys not installed at the ridge of a roof become dams for any water or ice on the roof above the chimney. Chimneys can change temperature from ice cold to roaring hot in a matter of minutes. There is a considerable amount of soot expelled from a chimney, and this soot, mixed with rain or moisture, becomes an acid. The acid settles on roof shingles and metal flashing and eventually dissolves them. There are sparks being generated by the fireplace below, and these sparks, if not contained, can set the roof on fire. This is especially true if you have a tile or wood roof.

The mason or the homeowner should have taken care of the spark problem. There are spark arrestors and wire screens designed to contain sparks. The soot and acid can be taken care of with a scrubber. A scrubber is a chamber designed to clean the air. It costs into the hundreds of dollars and most homeowners will not spend the money for one. The alternative to this is to apply a coating of roofing cement or mastic to all metal flashing that surrounds the chimney. Roofing cement is asphalt and is acid resistant.

The solution to the ice or water damming is to install a cricket on the upper side of the chimney. The cricket directs water around the chimney. The cold to hot problem is cured with proper flashing design and with the use of counterflashing. This allows for expansion and contraction between the various components.

See Fig. 8–16. When you install flashing around a chimney, skylight, or other roof protrusion, you should follow this rule. Start at the bottom or lower section and work up,

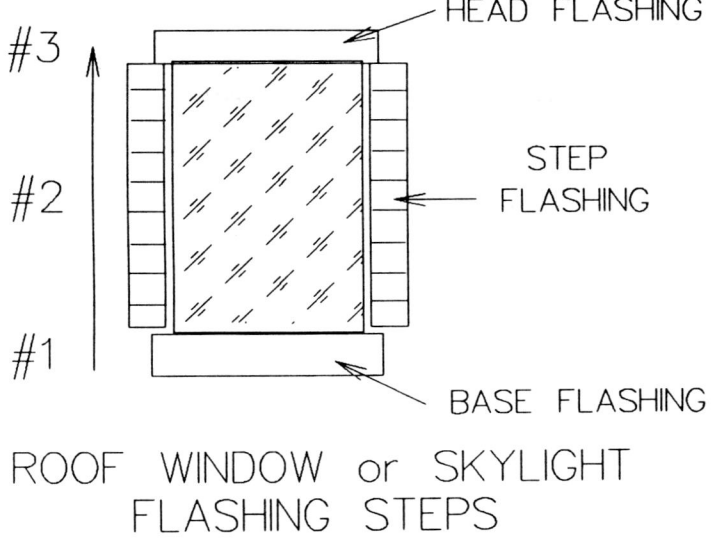

Figure 8–16   Flashing around a dormer or a chimney

covering each prior section of flashing and roofing material with the next. The bottom flashing or base flashing is installed first, then the right and left side step flashings, and finally the top or head flashing. By doing this, any water that runs onto the flashing at any point will be harmlessly carried away to the lower roofing. Flashing installed from top down will always result in a leak.

See Fig. 8–17. Occasionally it becomes necessary to get two different trades coordinated, the roofer and the mason. Chimney flashing should be grouted into the chimney for best leak-proof results. But what often happens is that the roof is installed first and the mason leaves off the flashing or installs the flashing incorrectly. Sometimes during new construction, and usually during a reroofing, the roofer who installs the flashing has to chisel out the chimney grouting. If he or she does a poor job, the mason gets the blame.

The roof has to go on first and then the chimney with the roof flashing. If not working simultaneously, the roofer has to make a second trip. If the chimney goes in first, then the flashing can only be built to the brick layer where the flashing is inserted. The roofer then puts on his roof and flashing to that brick course. Then the mason has to make another trip to complete the chimney. Add to this a sheet metal worker, who may or may not be under separate contract to install the metal, and you have potential problems. The solution to this is counterflashing.

COUNTERFLASHING

Counterflashing is referred to by some as cap flashing. Counterflashing is generally installed by the mason. It now does not matter who goes first. If the roofer is first, he or she

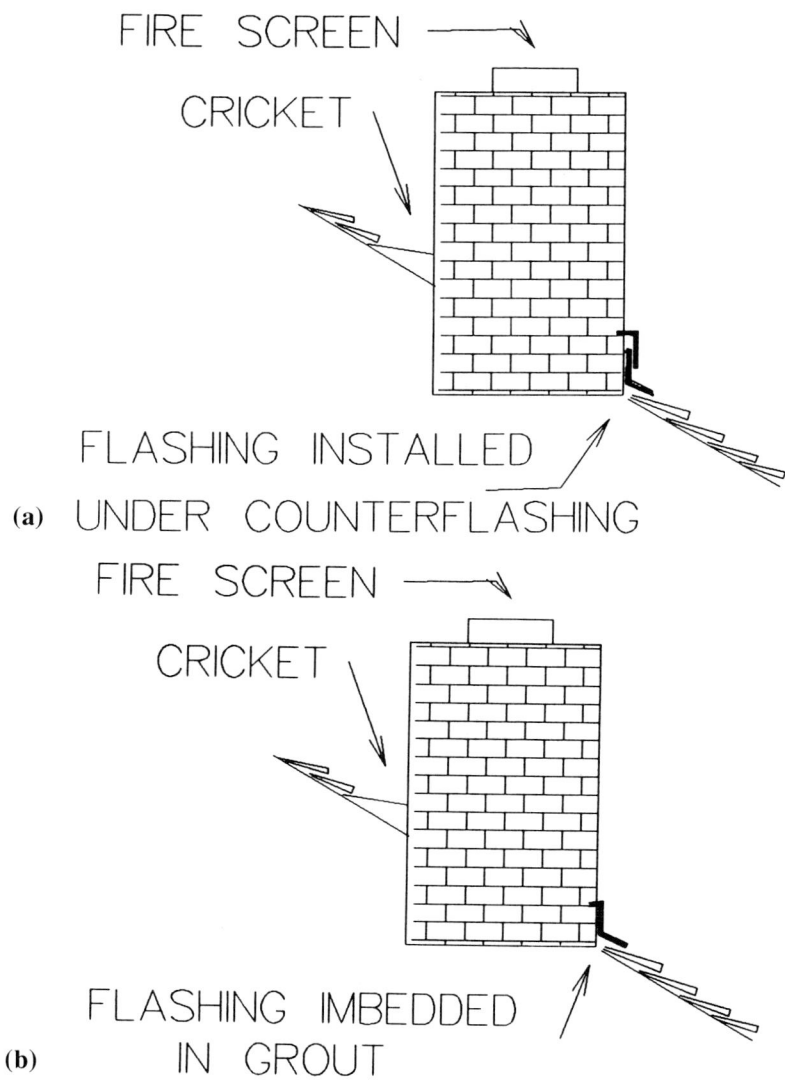

Figure 8–17   a) Base flashing on chimney, not recommended method; b) Base flashing on chimney, recommended method

puts on the roof and the chimney flashing. This flashing attaches to the roof sheathing and ends in the air where the chimney should eventually be. The masons then build their chimney and install the counterflashing over the top of the roofer's flashing. If the masons built their chimney first, they install counterflashing with the bricks. The roofer can then roof the building by slipping the chimney flashing up under the counterflashing. Each person has made only one trip. The chimney is properly flashed and the customer gets a better job.

Figure 8–18 illustrates what happens if the chimney flashing is installed after both the roof and the chimney are installed. Water running down the bricks gets behind and under the roof flashing. Also, the only spot for nailing on the flashing is through it, right into the shingles. Each exposed nail now becomes a leak source.

Counterflashing at a chimney serves a dual purpose, that of preventing water entry into the junction and isolating the junction. Even if the counterflashing is seated tightly to the roof flashing, there will be some air space. This air space helps to dissipate the heat buildup in the bricks. The large metal surface also acts like a heat radiator and dissipates heat. This keeps the roof flashing cool and it does not expand and contract away from the shingles or sheathing.

CHIMNEY FLASHING

Some chimney step flashing may be formed differently than wall to roof flashing. Wall to roof step flashing is normally shaped like an L. In well-made chimneys, the counterflashing and the step flashing are L-shaped. The short leg of the counterflashing L is inserted into a mortar joint with the long leg pointed down. The roof flashing is installed with the short leg of its L pointed up and under the counterflashing, the long leg of the L under the shingles (see Fig. 8–19).

The practice of not using counterflashing is common. The flashing used is a Z-shape flashing. The top of the Z is embedded in a mortar joint and the bottom end of the Z is under the shingles. This is acceptable but will cause problems with expansion and contraction and when it comes time to reroof (see Figs. 8–20 and 8–21).

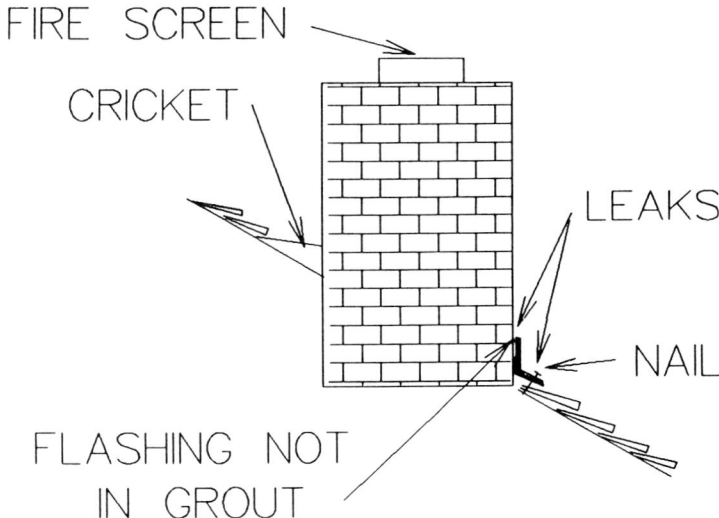

FIRE SCREEN

CRICKET

LEAKS

NAIL

FLASHING NOT IN GROUT

Figure 8–18　Improperly installed base flashing

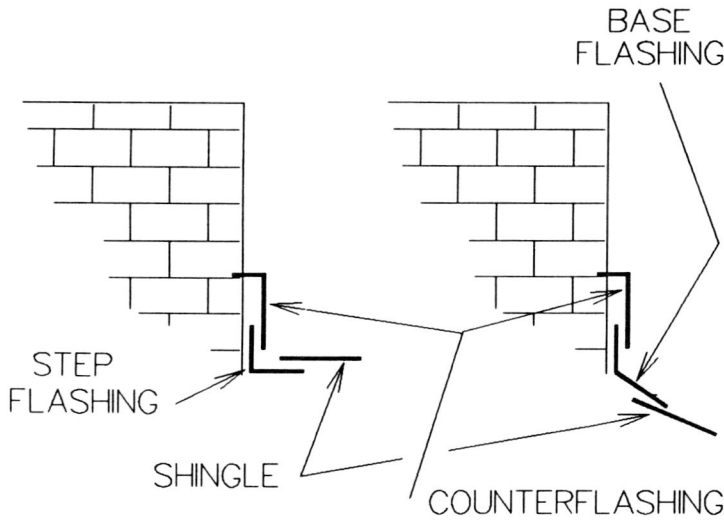

Figure 8–19   Base flashing should go under the counterflashing, over the shingles

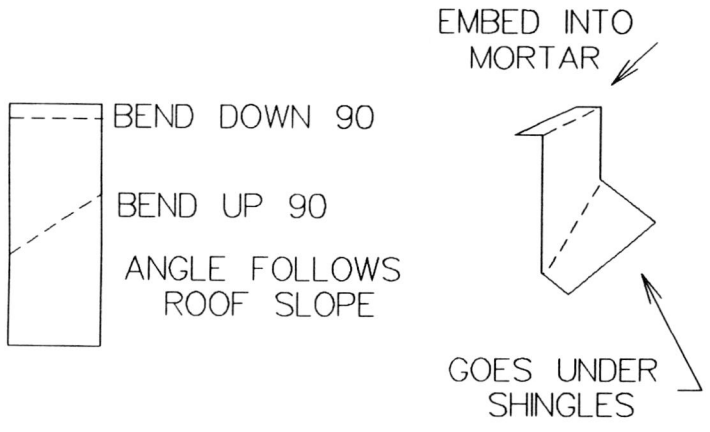

Figure 8–20   Step flashing, not recommended method

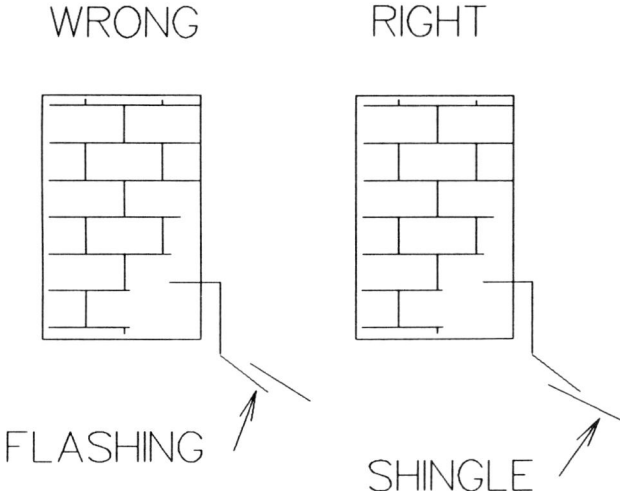

Figure 8–21   Z-type flashing

CRICKET FLASHING

Figure 8–22 is the sheet metal design for a cricket. A cricket is used in place of the head flashing, above the chimney, as a water diverter. You will require this piece and a few small pieces of flat sheet metal. You may require some 1/4 in. exterior plywood as a backer to the triangular portions of the cricket.

The plywood is installed as a ramp from the roof to the chimney. On each corner, install corner flashing where the chimney wraps around to the plywood. The cricket is installed over the plywood and the corner flashing. Cover the two splits in the cricket with flat pieces of flashing. Over the top of the cricket install counterflashing where the cricket contacts the chimney.

The width of the cricket is the width of the chimney. The length is the distance from a point 6 in. up the chimney to a spot on the roof that is just slightly higher. You want to have at least 1/8 in. per linear foot of slope from the roof to the chimney. The cricket metal should be painted to match the roof shingle color or covered with mastic to prevent corrosion from fumes.

VALLEY FLASHING

Valleys are subject to staining more often than other areas of a roof (see Fig. 8–23). They receive the bulk of the runoff water and are a collector of leaves and other wind-blown trash. Valleys should always be kept clean.

Valleys are a common source of water leaks into a building. They should not be, if designed and constructed correctly. Most leaking valleys can be traced back to a roofer who

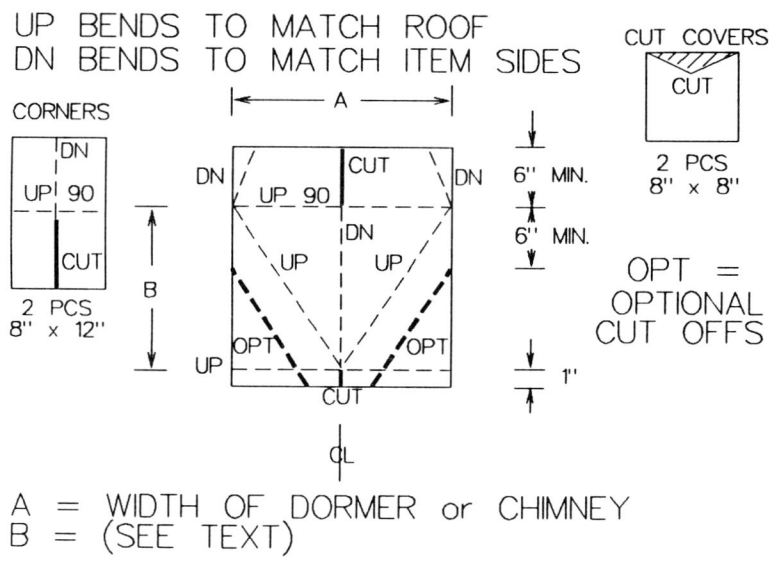

UP BENDS TO MATCH ROOF
DN BENDS TO MATCH ITEM SIDES

A = WIDTH OF DORMER or CHIMNEY
B = (SEE TEXT)

Figure 8–22   Cricket metal layout

Figure 8–23   Valleys collect stains and dirt

made a simple, and sometimes silly, mistake. Here are a few acceptable valleys and a review of mistakes made.

Smooth Cap Sheet Valley, Style 1

This is one of the simpler valleys to make (see Fig. 8–24). Do check your local building code before fabricating it. Some localities no longer permit this open-faced valley design. The design can allow water running off one roof to cross the valley and go under the shingles of the intersecting roof. This design may be used if it is used under a half- or full-laced shingled valley.

Construction is to apply a 36 in. wide, 65 lb felt the full length of the valley. The felt is to be nailed 1 in. from its edges every 12 in. on both sides of the valley. To this, cement a sheet of 20 in. wide, 26 gauge galvanized or other sheet metal. Do not nail the sheet metal. Bring shingles 4 in. onto the sheet metal from both sides. The shingles are not nailed to the sheet metal. Shingles should be cemented in place with a minimum of 1 in. wide strips of roofing cement.

Smooth Cap Sheet Valley, Style 2

Figure 8–25 illustrates valley flashing made from #90 mineral faced roll roofing or #90 cap sheet. First line the valley with a 12 in. wide strip of roofing, mineral side facing the sheathing. Apply a 24 in. or wider sheet face up over the first sheet. The valley can now be shingled using a full- or half-lace shingle pattern or, sometimes, used as is. NOTE: Some building inspectors will require a sheet of flat metal flashing be used on top of the #90 roofing and under the woven roof felts.

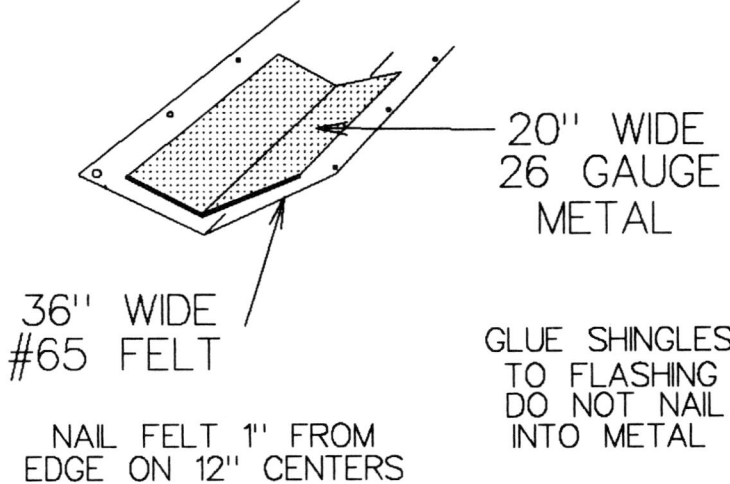

20" WIDE 26 GAUGE METAL

36" WIDE #65 FELT

NAIL FELT 1" FROM EDGE ON 12" CENTERS

GLUE SHINGLES TO FLASHING DO NOT NAIL INTO METAL

Figure 8–24 Open sheet metal valley, may not be code approved

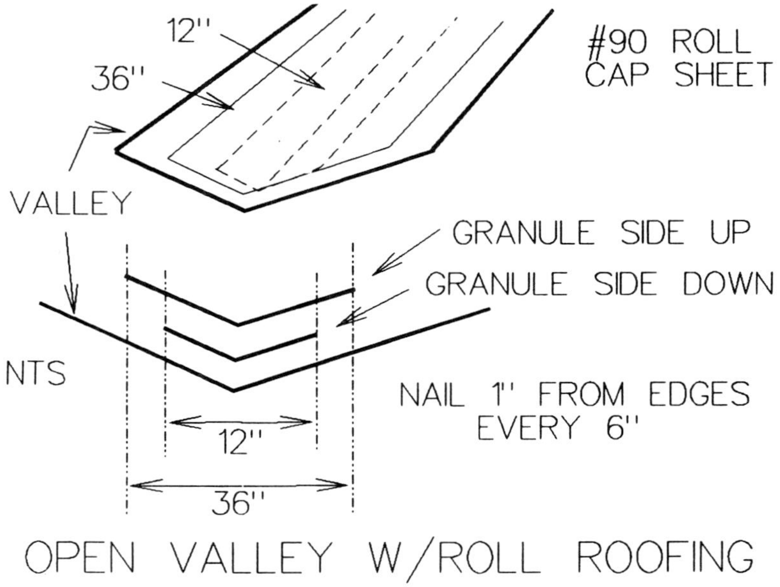

12"

36"

#90 ROLL
CAP SHEET

VALLEY

GRANULE SIDE UP

GRANULE SIDE DOWN

NTS

NAIL 1" FROM EDGES
EVERY 6"

12"

36"

OPEN VALLEY W/ROLL ROOFING

Figure 8–25   Open valley using roll roofing

Recommended Valley Flashing

Most current building codes require that a valley be constructed of metal. Code also requires that the valley contain a splash diverter.

In exposed valleys, shingles not interlaced, one should use flashing (see Fig. 8–26). Shown is W type valley flashing. This flashing is generally made from 28 gauge galvanized steel. The shape is somewhat like a W, therefore the name. The center rib is the splash diverter. This diverter is 3/4 to 1 in. in height. Its purpose is to break up and direct the flow of water entering the valley. If it were not there, water from one side could conceivably cross to the other side and wash up under the other side's shingles. The water runoff channel of a valley should be 5 to 6 in. wide on each side of the splash diverter. The water runoff channel is the clear distance between the roofing and the splash diverter.

A layer of #30 felt should be installed under the metal valley flashing. This keeps moisture, escaping from the building's interior, from rusting the flashing's underside.

If more than one strip of felt or one strip of flashing is required to cover the length of

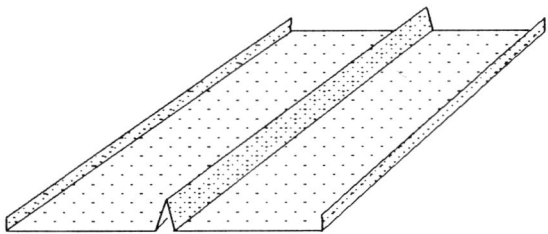

Figure 8–26   W-flashing for valley construction, code approved

the valley, then install the lower strips first. The upper strips must be installed so that they lap 4 to 6 in. over the lower strip. If this is not done, the water will run off the upper flashing and under the lower flashing.

When bringing shingles into a valley, one must take care to properly cut the shingles at the correct angle and placement in the valley. A mistake made by beginners is not finding the center of the valley. Thus, they cut one side very close to the valley center, the other side way up the slope. The shingles should be cut 5 to 6 in. on either side of water center for best leak-free results. To test, pour a cup of water in the valley at the ridge and note the runoff.

The corners of the shingles entering a valley must be dubbed. This prevents water entering under the shingles from flowing along the top edge of the shingles and under the roof field shingles. This step is omitted by more beginning roofers than you can count.

Valley metal should be nailed with a minimum of nails to help prevent buckling and leaks. Nails should be placed near the flashing edges, coated with mastic, and then covered with the shingles. If possible, do not use nails of a different material than the flashing material. The use of different metals can lead to galvanic action. This forms a corrosion that can eat at the metals and rust them away, leaving a hole into which water may enter. My recommendation is to use aluminum flashing and aluminum nails.

EXISTING VALLEYS

On buildings that have metal flashed valleys, check for the type of flashing used. In the above instances, we had unequal volumes and forces of water coming off each roof surface. We found that the water coming from the higher or greater sloped roof ran up onto the lower roof. Many installers have used smooth metal flashing in the valleys of roofs such as this. The metal flashing that should have been used is W flashing.

While you are checking the valley flashing, look for signs of rust and weak spots. Push in on the metal with your fingers the full length of the valley. Any soft or weak areas could mean that the flashing is rusting out from the underside due to having no felt between it and the roof sheathing. Or, there is a leak in the shingles somewhere.

Check the seal at the edge of the flashing by gently pulling upward with your fingers; a good seal will not come loose. On older roofs, 12 years of age and up, figure that most of the sealer and flashing must be replaced during a reroofing job. This, by the way, can add a full day to your estimate.

Sealer should be applied smooth and if exposed, should then be painted with an aluminum paint. Sealer should be somewhat pliable to the touch and not be brittle or cracked. It should not be oozing out from under shingles or flashing.

DORMER FLASHING

Dormers break up an otherwise dull-looking roof line. Some are functional in that they are part of an occupied room and provide air and light. Others are purely decorative and do nothing. Either way they are used, they can be a source of water leaks. Again, the leaks can usually be traced back to a roofer who did not understand the principles behind proper flashing.

The area where a dormer meets a lower roof is a potential source of water leakage (see Fig. 8–27). This area is often improperly constructed. Shown is a situation where flashing was not installed. Water dripping off the siding will enter under the shingles and get trapped. Wind-blown dirt and leaves will frequently get trapped, decompose, and settle under the shingles. This will eventually lead to moisture under the shingles and dry rot. The dry rot will eat through the roofing felt and into the sheathing. In freezing weather, the trapped water will turn to ice and form an ice dam under the shingles. This, in turn, will not only tear up the shingles but possibly the siding as well. It is therefore a good idea to add a piece of flashing.

See Figure 8–28. Well, we tried to prevent water from entering under the shingles. We put on the flashing, right over the outside of the siding. Water will run down the siding, climb right up over the top edge of the flashing, and fall harmlessly on to the shingles, right? No! Water can and will find the smallest of cracks to enter. In this instance, it will enter behind the flashing and under the shingles as it did in Fig. 8–27.

If you put mastic behind the metal flashing, that would probably keep out the water for a short period. There are three dissimilar materials involved: the siding, the shingles, and the roofing; the mastic gives a fourth variable. Each of these materials has a different rate of expansion and contraction. It will not be long before the bond between the mastic and the flashing, or the mastic and the siding, fails. If wood shake or wood shingles are used, the problem is further complicated. Wood absorbs water and mastic will not stick to water or wet wood. We shall try again.

Figure 8–29 presents a better solution. Here the flashing was placed under the siding. We also ran a sheet of #30 felt under both the siding and the roofing before installing the flashing. The felt is there as a backup to the flashing. Now any water running down the siding empties onto the top of the shingles and is carried harmlessly off the roof. Note that the #30 felt is lapped onto the second course of shingles. Even if the flashing leaks, the felt will carry the water off the roof.

See Fig. 8–30. A small roof helps keep rain and snow out of a garage. Shown is flashing properly installed under the siding and over the shingles.

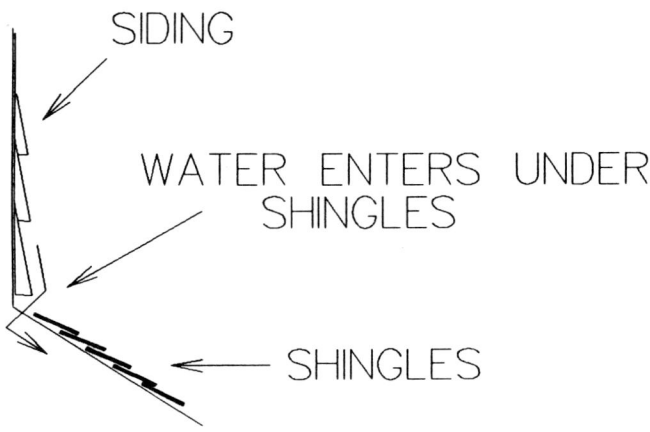

SIDING

WATER ENTERS UNDER
SHINGLES

SHINGLES

Figure 8–27   Lack of flashing will result in leaks

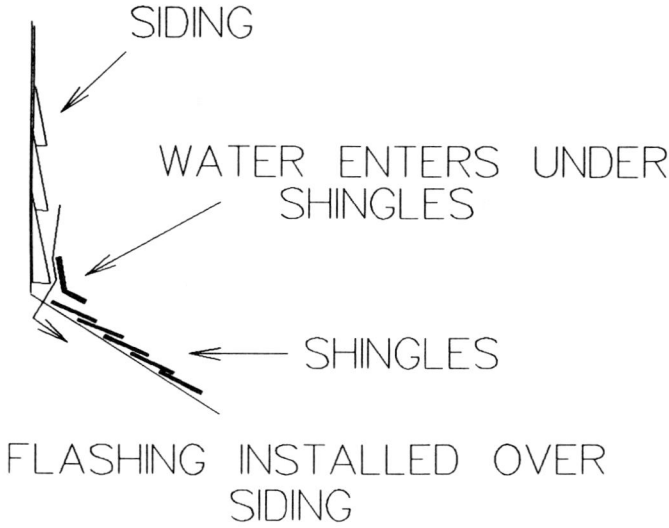

SIDING

WATER ENTERS UNDER
SHINGLES

SHINGLES

FLASHING INSTALLED OVER
SIDING

Figure 8–28 Flashing improperly installed over siding

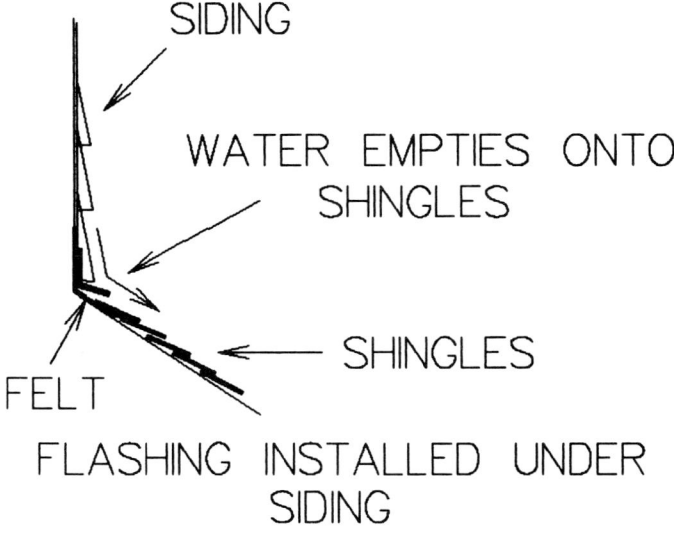

SIDING

WATER EMPTIES ONTO
SHINGLES

SHINGLES

FELT

FLASHING INSTALLED UNDER
SIDING

Figure 8–29 Flashing properly installed

Figure 8–30   Typical siding to roof intersection

In Fig. 8–31, we have a construction problem, that of getting two different contractors coordinated to do the job correctly. It is generally the roofer's job to install flashing. The roofing is usually put on before the siding, but not always. Sometimes, especially when adding a second story or a dormer, the roofer is called last. By then the siding is on and the proper flashing is not. Now the roofer has to tear off the bottom row or two of siding with its starter strip before installing the roof flashing. The correct order of construction is:

1. Roofing
2. Felt
3. Last row of roofing
4. Flashing
5. Siding starter strip
6. Siding

Note that all nails are covered using this method.

BASE AND HEAD FLASHING

The lower horizontal flashing is called *base flashing* and it should wrap around the edge of the dormer and part way up the dormer's sides. The step flashing installed on the sides of the dormer will lap over these side laps. Base flashing is used at the bottom of dormers,

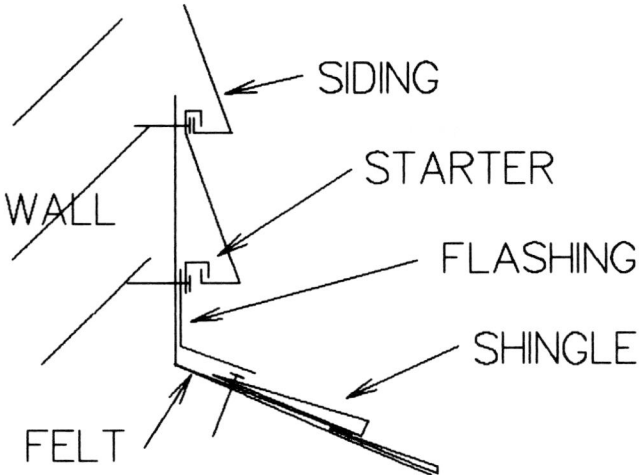

SIDING

STARTER

FLASHING

WALL

SHINGLE

FELT

Figure 8–31 Proper installation hides the nails

skylights, and chimneys. Be sure to lap the first step flashing onto the base flashing, not under it.

For appearance sake, it is permissible to shingle onto the base flashing and up to the siding or item. This is providing the lower edge of the flashing exits on top of a shingle that is one or two courses below the siding or item.

If you are shingling with wood or shake, do not cover the exposed flashing with shingle; prime coat and then paint the flashing instead. All exposed nails should be gasketed nails or covered with a layer of mastic or roof cement.

Head flashing is used on the top horizontal intersection of a roof and a skylight or a chimney. The head flashing is the last flashing to be installed on these items. It wraps around and over the step flashing. Do not install it before the step flashing. Since dormers, skylights, and chimneys vary in size, most base and head flashing must be fabricated on site. Figure 8–30 shows the sheet metal layout. The exception to on-site fabrication is for skylights. Most skylight manufacturers sell or provide flashing kits with their product.

STEP FLASHING

Step flashing is the recommended flashing for shingling next to a wall that transverses up a roof slope. For example, firewalls, dormers, and second story additions generally transverse up the roof.

Figure 8–32 illustrates step flashing sheet metal design for flashing used alongside of dormers. Step flashing is installed during the shingling process from the bottom up toward the ridge—shingle, flashing, shingle, flashing, etc.

I recently looked at a condominium complex that had fire walls extending above the roof line in several areas. The roofers did use step flashing to join the shingles at the fire walls. But they made one mistake. There is a dashed line of silvery metal showing along the

BASE FLASHING   28 GAUGE SHEET METAL

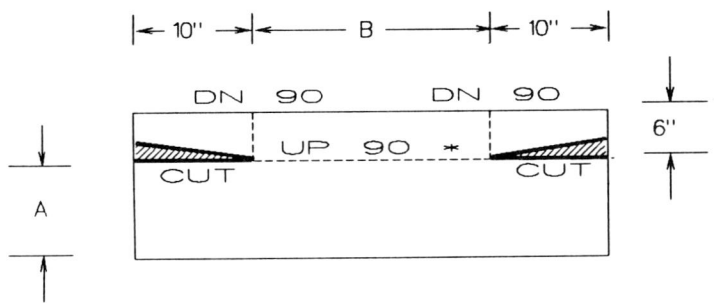

* BEND TO ROOF SLOPE
    CUT ANGLE TO FIT

    A = SHINGLE WEATHER EXPOSURE + 1"
    B = WIDTH OF DORMER or CHIMNEY

STEP FLASHING

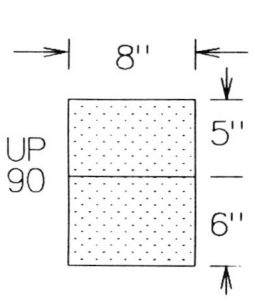

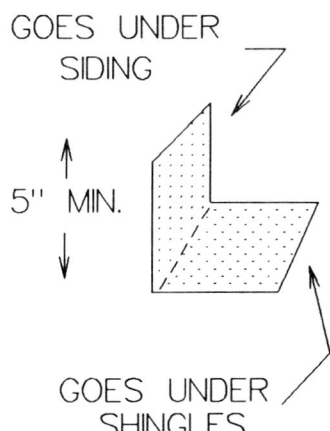

USE FOR DORMER & SIDING

Figure 8–32   Sheet metal designs for base and step flashing

intersection. Closer examination revealed the reason. The shingles were butted to the flashing and the flashing was showing through the shingle end cutouts. I did not climb on the roof to get a closer look, but here are two possible causes of the dotted line.

1. The roofers started their shingling from the step flashing and worked toward the other end of the roof. This created a situation where every other course of shingles was a whole shingle and still had its end cutout in place. The flashing was showing through this cutout. If this is the case, then it would have been better if the roofers started with 2/3 shingles cut flush.

2. The roofers started at the other end of the roof and worked across to the fire wall. The roof rake to fire wall measurement was a multiple of 36-1/16 in., the length of a three tab shingle with expansion space. The roofers then ended with full shingles, on every other course, butting up against the fire wall. If this was the case they should have started at the firewall and followed the procedure given in 1. An alternative would have been to paint the step flashing to match the shingle color or at least apply a coating of mastic to the flashing. A dark color, such as the black color of roofing mastic, would not have been so obvious.

SHED DORMER FLASHING

Figure 8–33 shows the flashing used on a shed dormer. Flashing is used at all points where the dormer intersects the roof plane. The top edge can be flashed with #30 roofing paper. The roof shingles then overlap on top of the dormer shingles. The side wall flashing must be step flashing (Fig. 8–32). The base flashing is according to Fig. 8–30.

GABLE DORMER FLASHING

See Fig. 8–34. The gable dormer is a little more difficult since the upper connection to the roof plane is an inverted V. Valley flashing is used here, the upper or roof side placed under the shingles, the lower or gable side placed over the shingles. You can do full-lace valleys at this intersection, but you may find it difficult. The shingles may not line up properly. It is recommended that a standard W-flashing valley be use. The side walls are step flashing (Fig. 8–32) and the bottom is base flashing (Fig. 8–30).

PAINTING THE FLASHING METAL

Most steel flashing metal is coated with a protective layer of galvanizing. Galvanized metal will last several years in normal service, but will eventually begin to rust. Figure 8–35 shows a good reason for not depending on the galvanized coatings used to protect roof flashing. I recommend that you take the extra time and paint all flashing metal; it goes a long way in preventing roof stains.

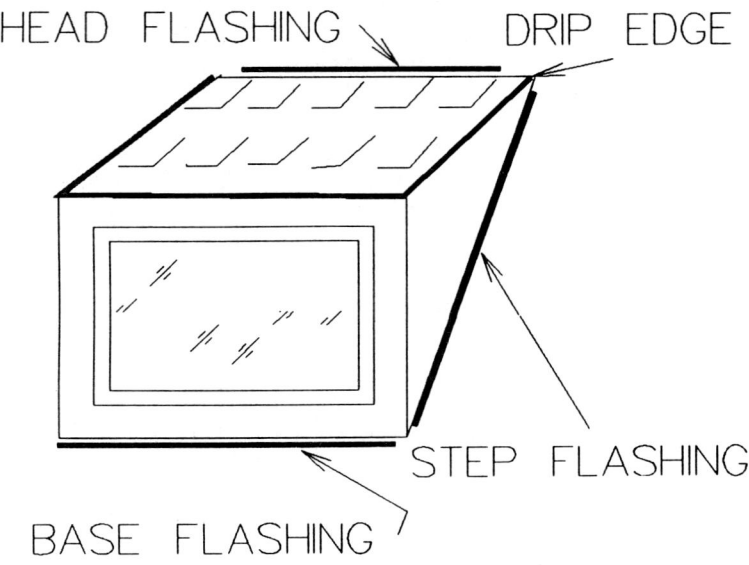

HEAD FLASHING

DRIP EDGE

STEP FLASHING

BASE FLASHING

Figure 8–33   Flashing used around a shed dormer

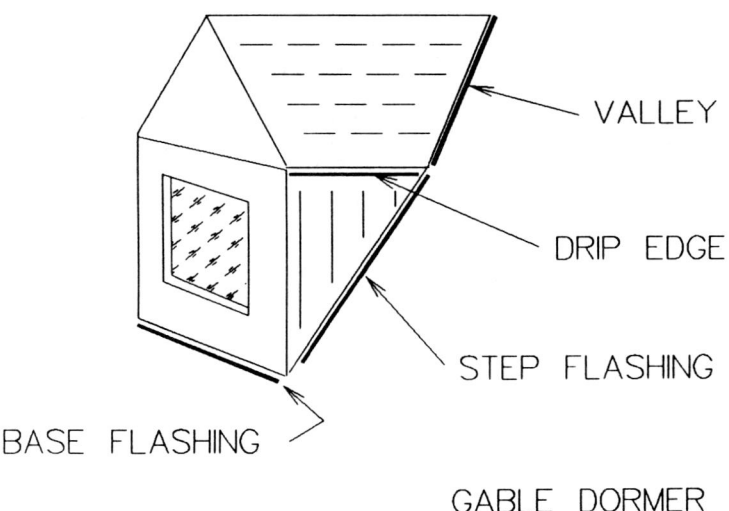

VALLEY

DRIP EDGE

STEP FLASHING

BASE FLASHING

GABLE DORMER

Figure 8–34   Flashing used around a gable dormer

Figure 8–35   Roof stain due to unkept flashing

Figure 8–36 gives another reason for painting the flashings—the lower roof to siding junction looks ugly due to the exposed flashing. A little paint, the color of the siding, would make this junction disappear.

For appearance sake, I recommend that even aluminum flashing be painted. Figure 8–37 shows a capped stove pipe flashing primed with a zinc metal primer. All flashing should be properly prime coated before it is installed or top coated with another paint. The large plastic dome item is the power air vent cover. It too was painted to match the roof and the other flashing. Figure 8–38 shows the result of painting the vent flashings. There are nine different vents, of which five are shown in the picture. Color and placement of the vents hide them from view.

Figure 8–36   Rusted flashing stains siding

Figure 8–37   Prepainted flashing

VENTS AND VENTING

You will find all sort of vents on a roof. There are vents for the sewer system, the attic area, the heating appliances, and the furnace. Vents, penetrating a roof, are a major source of leaks. Here are a few problems you should avoid.

Figure 8–38   Painted vents blend into roof

Vent Too Low to the Roof

When a roof vent is placed too low to the roof line, the vented fumes do not immediately escape into the atmosphere. Figure 8–39 shows a section of this roof that has been ruined by fumes. What happens is that the surface tension of the roofing material pulls the air flow to the shingles. Figure 8–40 shows the air contacting the roof on the left side. This air flow then follows the roof up the left side, over the ridge, and down the right side. As it does, it intermixes with the fumes and allows them to contaminate the roofing material.

Vent Fumes

Figure 8–41 illustrates a good place for birds to nest. The void or buckle is caused by a combination of airborne chemicals and smog coupled with cold weather contraction. The wood stove vent (Fig. 8–39, top left center), was too low to the roof. Shown here it is fixed by extending it 2 ft higher. Figure 8–39 showed the roof area affected. Figure 8–40 showed why.

In Figure 8–39 we saw the stove vent at the top left center. This vent has already been extended upward into the air stream. It had been some 24 in. shorter. Figure 8–42 now shows that the air at the roof is clean air. The fumes are now mixing with the higher air flow and are being carried away from the building. These fumes will then mix with more air, dilute, and become somewhat harmless.

This surface drag of fumes happens to all vents, stove, furnace, and drain/waste

Figure 8–39   Fume damage to asphalt shingles

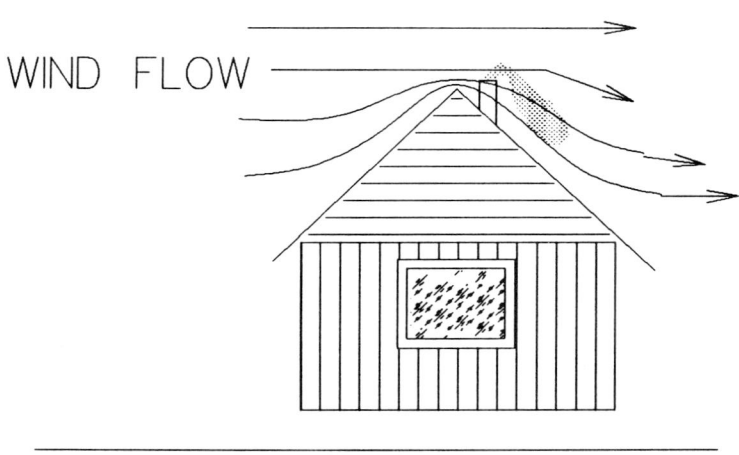

WIND FLOW

VENT NOT HIGH ENOUGH

Figure 8–40   Wind flow over a roof

(sewer) venting. Most building codes require that the vent pipes extend above the roof or roof-protruding fire wall by a minimum of 6 in. This is a very unclear specification. Does it mean above the roof surface or above the roof peak? Many plumbers end the vent pipes 6 in. above the roof surface. I was told by one building inspector that I had to extend the pipes to a minimum of 18 in. Another told me I needed to have the vent pipes higher than the ridge by 6 in. Interpretation of the U.B.C. is not always concise or easy. It is to your and your customer's advantage to extend these pipes if they do not meet current code, or if you are seeing fume-related roof problems.

Just a little more about the code and vent pipes. Any vent pipe that is within 10 ft. of

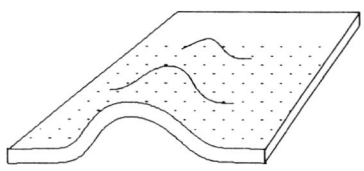

EFFECT of SMOG & CHEMICALS
ON ASPHALT SHINGLES

Figure 8–41

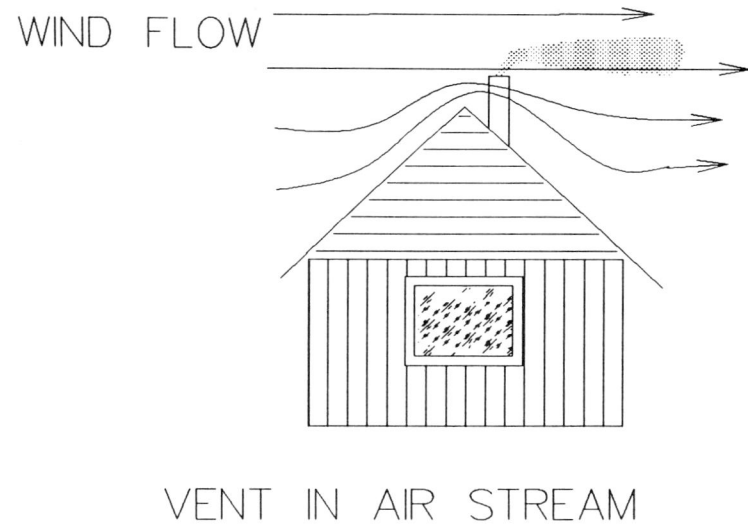

WIND FLOW

VENT IN AIR STREAM

Figure 8–42 Vent is high on roof, fumes are properly removed

a window, door, air intake, vent shaft, or opening shall be extended 7 ft. above the roof and securely braced.

The reasons for the code rules? First is that fumes from stoves contain carbon monoxide, hydrocarbons, and other chemicals that can attack the shingles. If the shingles are wet, the problem is compounded as many fumes, when mixed with water, form acids, which are harmful to the roofing, the gutters, and the plants near the building.

The fumes from DWV (Drain/Waste/Vents) may contain bacteria and oxides. These bacteria and the oxides can settle into and react with the roofing material and decompose it. Plus, each of the mentioned fumes are a health threat to persons or animals who breathe them and should be vented high into the airstream.

Mastic Used as Vent Pipe Flashing

Figure 8–43 is someone's idea of vent pipe flashing; as you can see the mastic is drying and pulling away from the roof shingles. This is a leak situation and should be corrected. The proper flashing for vent pipes is with the use of commercially available metal or plastic sleeves and boots.

Vent pipe flashing is installed as follows:

- Step 1. When installing vent flashing. shingle up to the vent. Slip the flashing over the vent pipe and onto the shingle (Fig. 8–44).

- Step 2 (Fig. 8–45) is to shingle over the lower edges of the flashing. Leave a water runoff channel, exposed flashing, below the vent pipe.

Figure 8–43   Mastic by itself does not make for good flashing

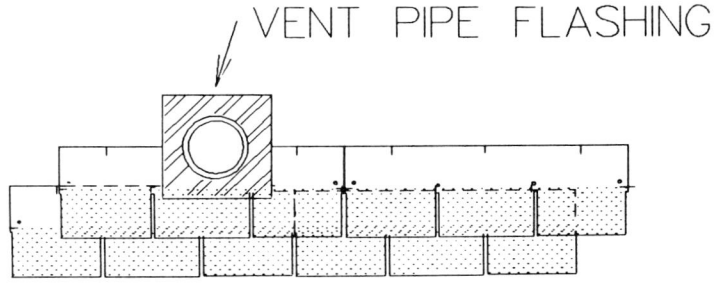

STEP #1

Figure 8–44   Step #1 of a properly flashed vent

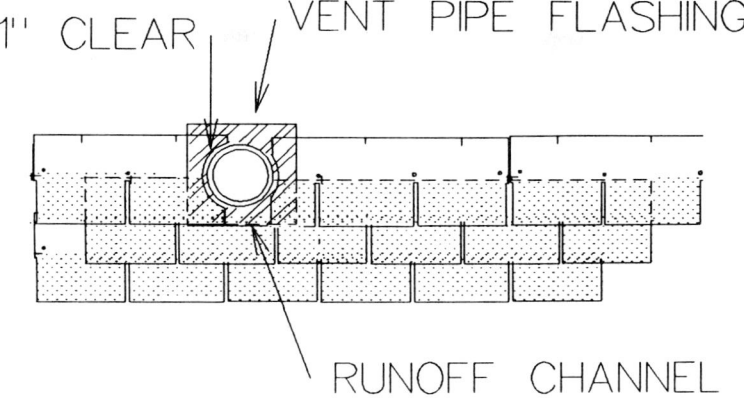

STEP #2

Figure 8–45   Step #2 of a properly flashed vent

• Step 3 (Fig. 8–46) is to shingle above the vent pipe and over the second course of shingles. Cut this shingle to conform. The cutout should be 1 in. greater in diameter than the diameter of the vent pipe.

• Step 4 is to slip the rubber flashing boot down over the pipe and onto the flashing flange (Fig. 8–47).

Roofing cement, mastic, may be applied to the upper side of the cutout. It also may be applied under the shingle tabs to glue them to the flashing. Do not apply mastic in continuous strips, apply in spots only. A continuous strip of mastic will hold and trap any water that gets under the shingles or flashing, creating an undesirable condition, a leak.

Do not side butt two shingles directly above and next to the vent (Fig. 8–48). Instead, always use a full-size shingle above the vent. This may entail cutting tabs off one or two shingles (Fig. 8–49). Do not center a tab cutout above the vent pipe (Fig. 8–50).

A rain bonnet is not required since these vent pipes feed directly into the sewer system. A screen may be required if you are in an area that contains mice or rats, or in an area where young adults have access to the roof, to prevent things climbing or being thrown into the vents and clogging up the system.

Stove Pipe Vent Flashing

Use the same flashing method as shown in vent pipe flashing. Be sure to install a rain bonnet over the inlet.

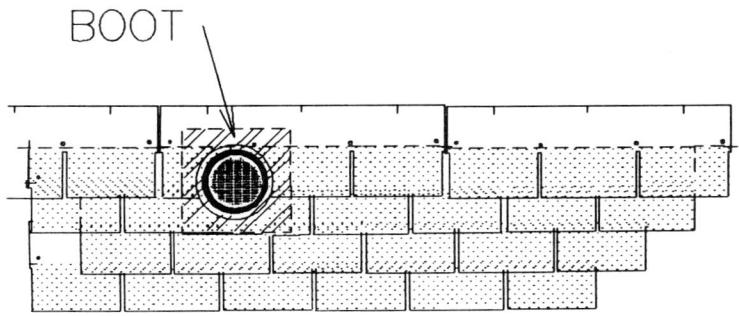

STEP #3

Figure 8–46 Step #3 of a properly flashed vent

Water Heater Vent Flashing

Same as the stove vent flashing. A rain bonnet is required.

Furnace Vent Flashing

Same as the stove vent flashing. A rain bonnet is required.

Attic Venting

It bothers me that a building contractor will design a building to sell at over $200,000. He will use the finest materials available, and then vent the attic with one 12 in. square vent in

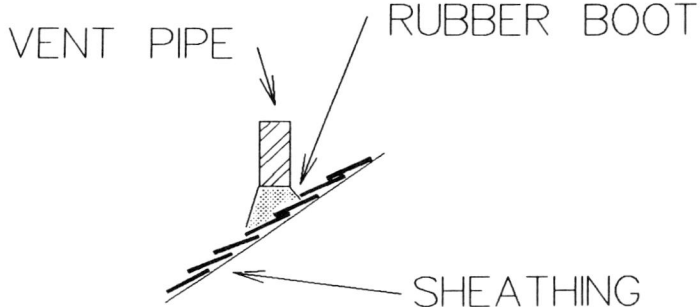

Figure 8–47 Step #4 of a properly flashed vent

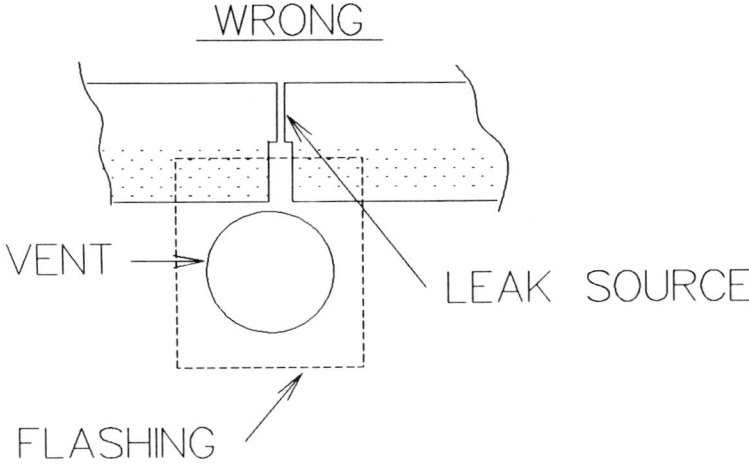

Figure 8–48 Do not join shingles above a vent

the gable. Venting of an attic space is important and the more air flow you have, the better things will be. The rule is that you *must* have at least two opposing vents sized as such. For every 300 sq ft of ceiling area directly below the attic area, you need a vent with a net open area of 1 sq ft. A Hip roof building needs 1 sq ft of vent for each 900 sq ft of ceiling area directly under the attic.

The vents on a gable house should be located as close to the roof ridge as possible and

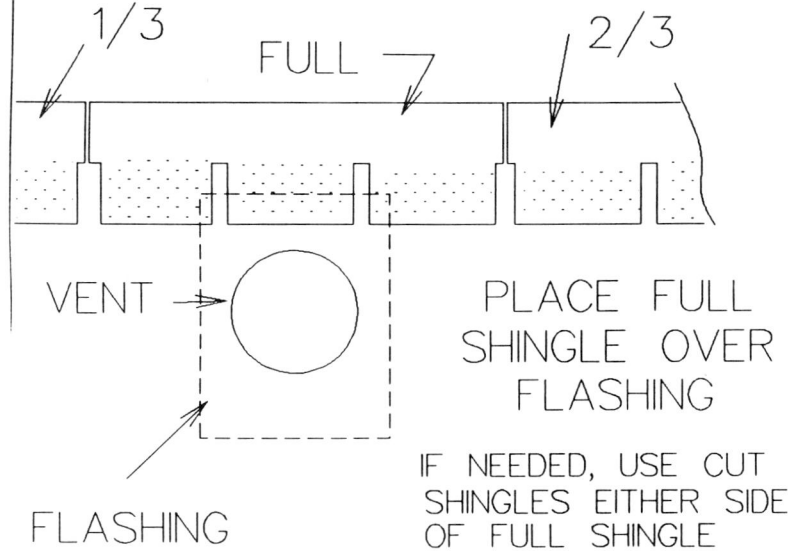

Figure 8–49 Always use a full shingle above a vent

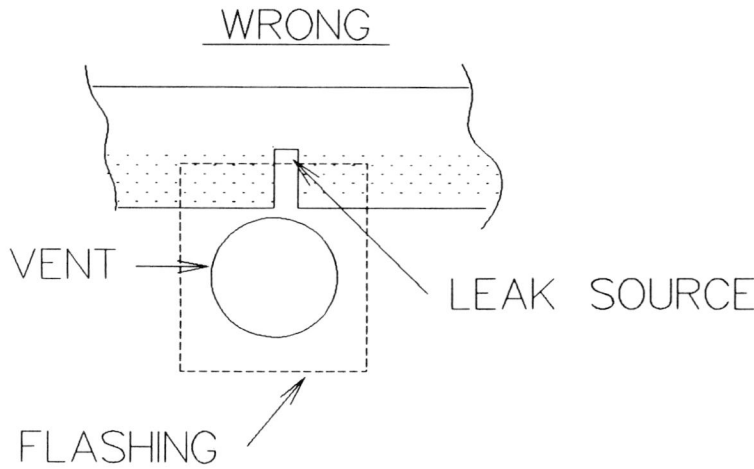

Figure 8–50   Watch where the cutouts fall

in opposite gables. On a Hip type roof, one vent should be in the soffit area and the other at or near the ridge. There are ridge vents that install along the ridge and these are recommended. When installing ridge and soffit vents, be sure to balance the system properly. The ridge vent must have the same or slightly more free opening than the total of the soffit vents feeding the enclosed vented area. Be sure that each section of the building is vented. The extended sections of T and L shaped buildings are sometimes forgotten by many contractors.

The Indiana Vent

Figure 8–51 illustrates the Indiana vent. I was somewhat amazed at seeing this while driving through Indiana. Most all the homes in several areas had multiple roof vents. Spacing was about 6 ft on center across the entire roof, just below the roof peak. Then I made the connection. The area I was in is known for building many of the finest RVs and campers in the country. The roof vents used are remarkably similar to those used on RVs. I suspect that the cost is very low. What I do not like about the venting system is that with so many vents, there is an increased likelihood of water leaks.

Ridge Vents

See Fig. 8–52. The ridge vent is becoming popular in many sections of the country, but mainly east of the Mississippi. This vent is installed where one would normally install the ridge cap shingles. If painted to match the shingle color, it blends in very well. The one thing I don't like about ridge vents is that they are frequently used alone. To be fully effective, there should be full-length eave vents installed on both sides of the building.

There are some ridge vents on the market that can be shingled to match the roof. The

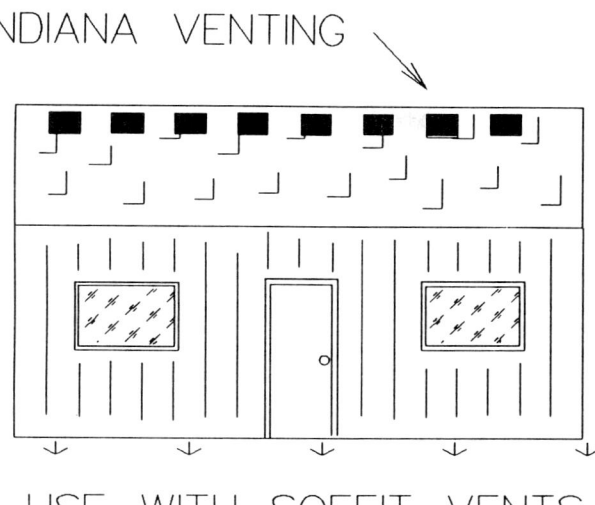

INDIANA VENTING

USE WITH SOFFIT VENTS

Figure 8–51 Indiana vents

vents are nailed through the top and not on the lower edges. These vents are designed to be nailed in this location. Some roof peak vents are designed with rain gutter troughs and are nailed to the roof through the trough. Use gasketed nails for these vents.

Turbine Vents

See Fig. 8–53. The turbine vent is primarily a West coast, dry area vent. Rising heat inside the attic, coupled with a slight breeze outside, makes this vent's vanes turn. The vanes then

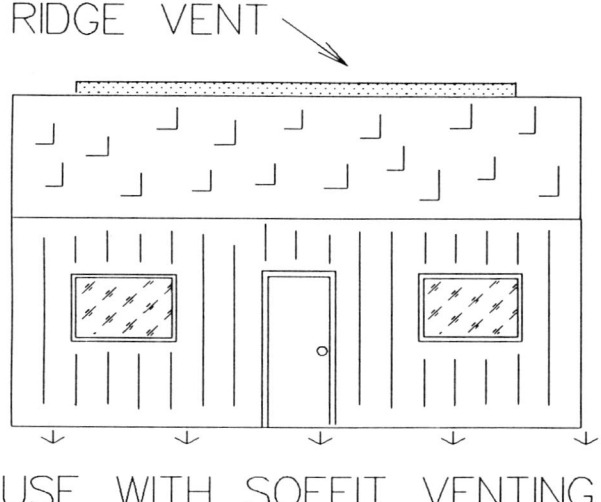

RIDGE VENT

USE WITH SOFFIT VENTING

Figure 8–52 Ridge vent

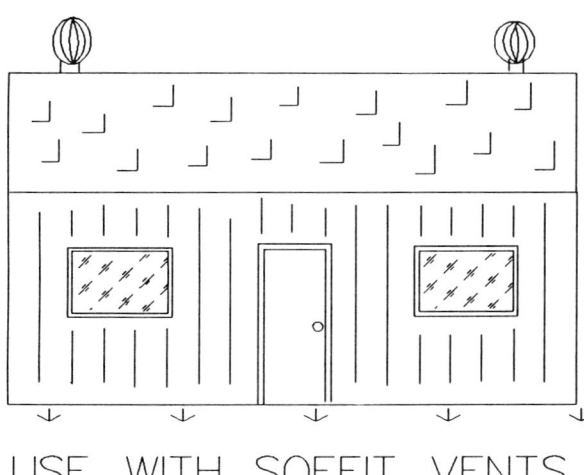

TURBINE VENTS

USE WITH SOFFIT VENTS   **Figure 8–53** Turbine vents

draw out the excess heat and moisture. Eave vents should be used, but frequently are not. Although the design is good, there is a chance of water entry into the building during heavy rain storms.

Power Vents

Figure 8–54 shows two power vents mounted on the roof. There are opposing gable vents on the other side of the building used for makeup air.

Power vents are somewhat new to the market. They consist of an electrically pow-

Figure 8–54   Power vents

ered, motorized fan and thermal detection device. Gable or eave vents should be provided since a large amount of air is expelled in a very short time when the fans are running.

Power vents pull hot air out of the attic area on demand. The unit has a plastic or metal housing, a motor with a fan, a thermocouple, and a controller. The thermocouple detects the temperature of the attic air and compares this with the dial setting temperature on the controller.

Figure 8–55 is a diagram of the system's components. The controller is wired to 117 VAC. This can be hard wired, installed in a 4 in. junction box, or soft wired, plugged into a 117 VAC socket (check local electrical codes). The controller has a relay with normally open contacts. The relay coil is connected to a temperature sensing bridge network. When the sensor senses air temperature equal to the dial setting temperature, the relay activates and its contacts close, connecting the motor to the 117 VAC line voltage. The fan spins and draws the hot air from the attic. When the air temperature drops below the dial setting, the relay de-energizes and turns off the motor. There is usually an on/off switch on the controller or AC line so that the unit can be serviced safely. A plastic or metal hood covers the unit and prevents water entry into the building.

I find that the units work very well in eliminating hot attic air. I find that one must provide a good solid base for the units to mount on. Power vents, with age, generate vibrations, which can be heard throughout various rooms or the entire building. Since power vents operate on air temperature, they usually only turn on during the late morning and afternoon hours when most people are awake or at work. Thus, the vibration is not very noticeable. To a day sleeper, the vibration can be disturbing. I did like the idea that my air conditioner very seldom ran, the power vents tended to keep the ceilings cool and, there-

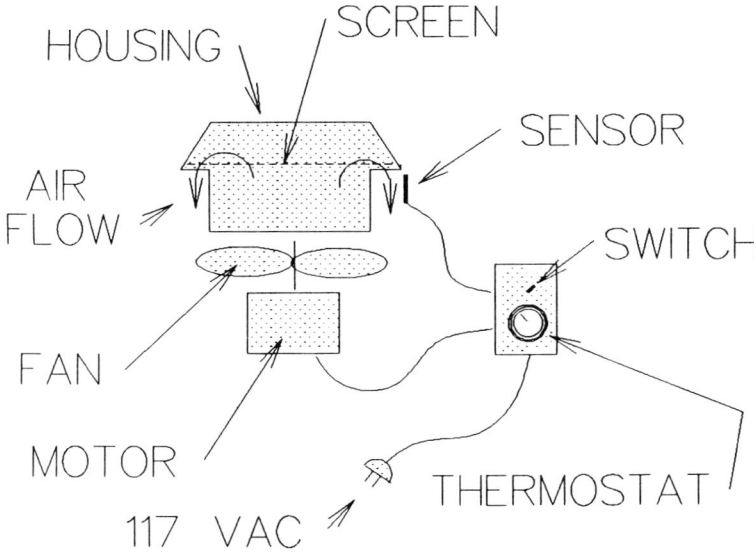

Figure 8–55   Power vent system

fore, the building cool. Power consumption of the fans is moderately low, with the cost being less than a dollar per day to operate.

Triangular Gable Vents

Figure 8–56 illustrates the old standby in an updated condition. These triangular shaped gable vents have been around for many years. They work well. The newer ones are made adjustable. That is, the vent shape can be very quickly changed at the job site to accommodate various roof pitches. The old method was to design and fabricate on the job site. These adjustable vents are readily available on the East coast, and usually have to be special ordered on the West coast. Cost is moderate, under $30.00.

Standard Gable Vents

See Fig. 8–57. Here is the old, old standby. Rectangular or square gable vents are probably the most used of any of the vents shown. They do a good job, the cost is low, and availability is coast to coast.

Soffit Venting

Soffit only venting will work if the attic air space is small. In other words, soffit vents work on low pitched roofs. If the attic space, roof pitch is more than 4/12, then I recommend the addition of ridge or gable vents (Fig. 8–58).

See Fig. 8–59. Vented soffit panels were used to solve the problem in Fig. 8–60. The

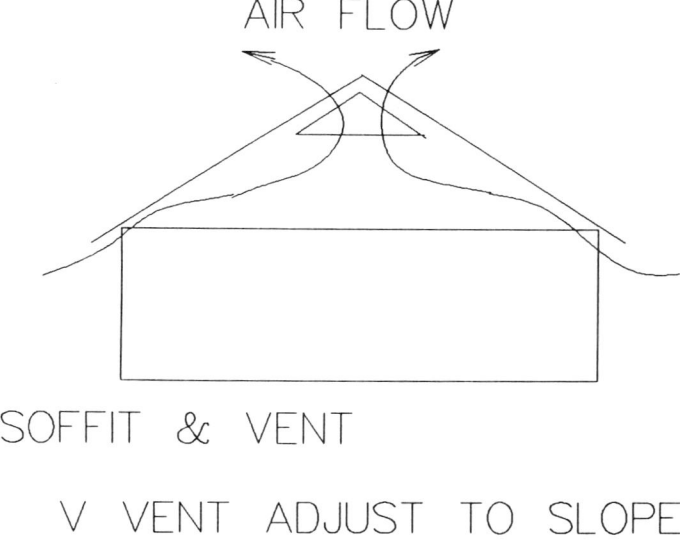

Figure 8–56   Adjustable gable vent

HOT AIR OUT

RECTANGULAR GABLE VENT   **Figure 8–57** The old standby

lack of air venting on this porch roof caused frequent paint peeling. Moisture would get trapped between the soffit and the roof sheathing. Being unable to vent itself, this trapped moisture attacked the sheathing, the soffit, the paint, and the roofing.

A low-cost soffit vent can save hundreds of dollars. Figure 8–61 shows an individual vent. As a note of interest, these individual vents were installed for a reason. The building was built without any attic venting. It did rain inside the attic/roof, it did ruin the insulation, and it did ruin the ceiling drywall. The builder didn't know any better some 20 years ago. The $5,000 plus lesson smartened him up real quick.

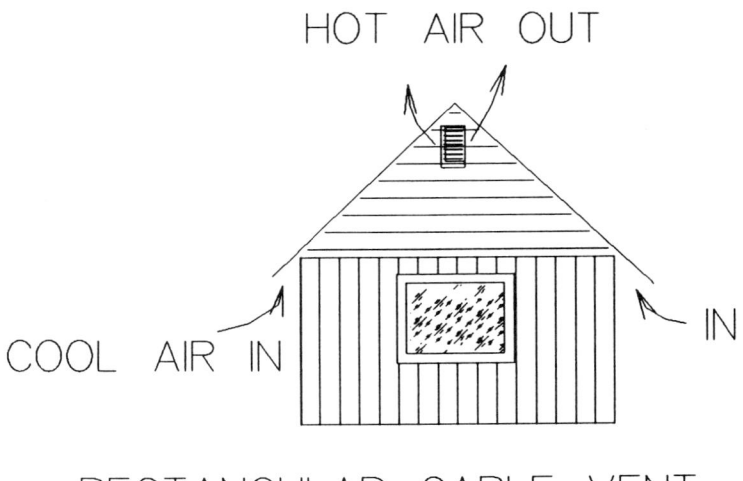

HOT AIR OUT

COOL AIR IN   IN

RECTANGULAR GABLE VENT

Figure 8–58 Use soffit vents or blocking with screened holes

Figure 8–59 Paint peeling due to moisture being trapped between soffit and roof sheathing

Figure 8–60 Vented metal soffit prevents problem in Fig. 8–59

Figure 8–61 Low-cost soffit vent

Installing Soffit and Soffit Vent Panels

Figure 8–62 shows a simple method of installing a continuous eave vent system. A wire screen is stapled to the tail rafter where the soffit should be. This screen prevents insects from entering the building. Two aluminum or plastic channels are then stapled, or nailed, to the rafters as shown. The rear channel faces off the top edge of the siding. The front channel

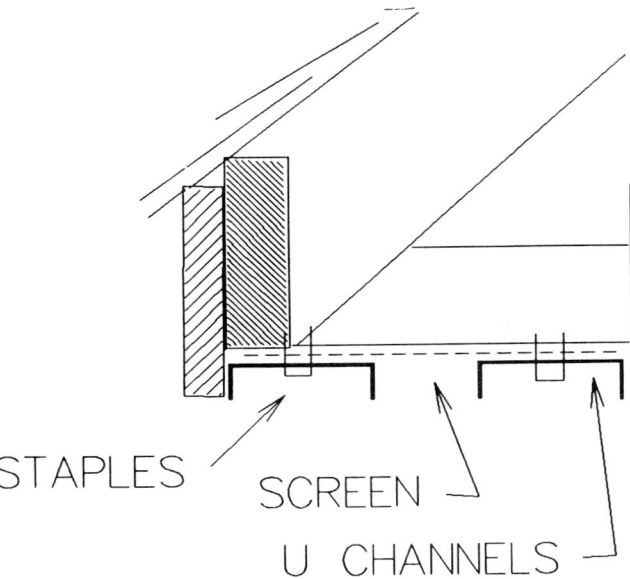

STAPLES

SCREEN

U CHANNELS

Figure 8–62 Inexpensive soffit vent design

completes the back edge of the fascia. Both channels have a down facing edge that acts as a drip edge to condensate. The entire assembly, the two channels and the screen, forms the soffit and the vent. This system is simple, functional, and good looking. It uses low-cost materials and installs in minutes.

Figure 8–63 illustrates F channels used for attachment of soffit. These extruded F channels are made from vinyls or other plastic and from aluminums. They are face nailed, upside down, to the fascia and the building wall. The soffit is then slipped into the channels. You also can use a combination of a J channel and an F channel (Fig. 8–64).

J and F channels are available from companies that supply vinyl or aluminum siding. The fascia and soffit on most buildings that are vinyl or aluminum sided is supplied and installed by the siding contractor. Channel color is usually white, although brown and other colors are sometimes available in stock or by special order.

See Fig. 8–65. Aluminum and vinyl fascia is available from most siding companies. The soffit is installed using two J or one J and one F channel. An undersill trim piece is installed from rafter to rafter just under the shingle overhang. The fascia strip is then locked into the J channel and the undersill trim. Face nailing of the fascia is not required. The fascia may be removed and replaced without damage for maintenance or painting.

VAPOR BARRIERS AND ROOFING

Vapor barriers are required in all walls and ceilings that are heated or contain moisture generating equipment. This means the baths, kitchens, laundry rooms, sauna rooms, utility

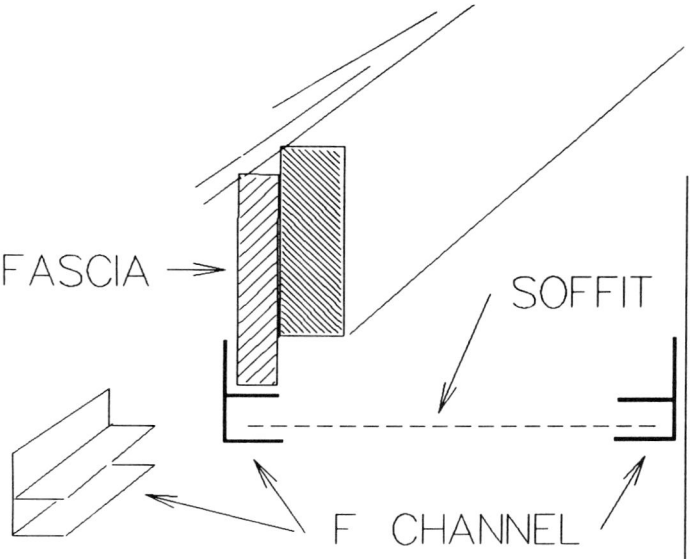

Figure 8–63   Use of "F" channels for holding soffit in place

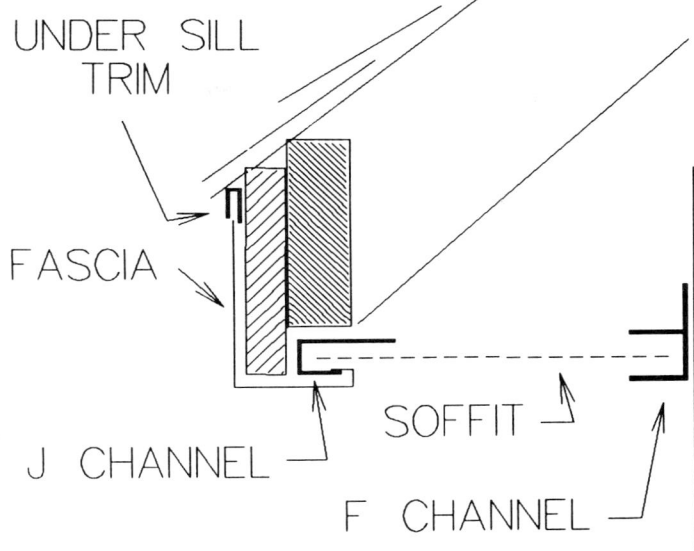

Figure 8–64   Use of "F" and "J" channels for holding soffit

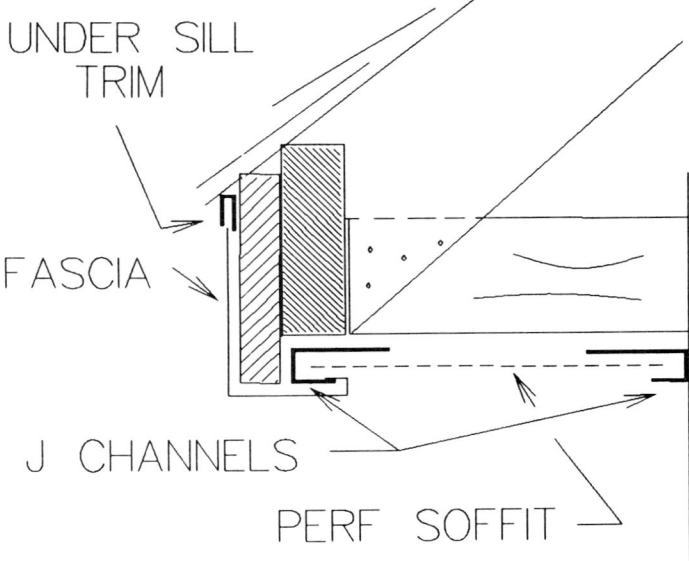

Figure 8–65   Use of two "J" channels for holding soffit

rooms, and the water heater area. The vapor barriers prevent moisture from exiting through the walls or ceiling and into the insulation. Moist insulation does not insulate.

See Fig. 8–66. Here is one effect of not having a proper or properly installed vapor barrier in the ceiling of a building. It is also the effect of not having proper attic air venting. The moisture collected in the attic settled into the ceiling drywall. This then softens the drywall to the point where it sags between the ceiling joists. Eventually the drywall will soften to the point that the nails holding it to the joist will pull through and the ceiling will collapse into the room.

If no other vapor barrier exists, a vapor barrier can be painted on. Use a sealer prime such as B.I.N.™. Paint the ceiling and walls of the nonvapor barrier room(s). This product is one of several white pigmented shellacs used to seal drywall and plaster.

INSULATION AND ROOFING

Fig. 8–67 illustrates another killer of an otherwise good roofing job. Weeks after you installed your roof and are now on another job, the insulation contractor shows up and does his job. He pops in the ceiling insulation using no vapor barrier, or he improperly installs it facing to the attic. A few months later you get the call, "My roof leaks, I have water all over the kitchen and bathrooms. The ceiling is falling apart." What happened? The water vapor generated by washing dishes, taking showers, etc., went through the ceiling and into the attic. Here it rises to the underside of the roof sheathing or to the underside of the upside down insulation vapor barrier. It then cools off and condenses into rain. This rain, water droplets, falls back to the ceiling and either warps the ceiling or comes through it into the

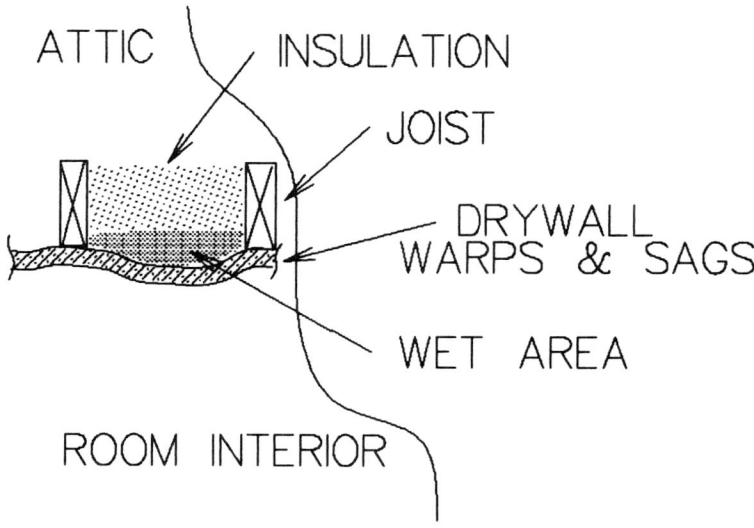

Figure 8–66 Lack of vapor barrier

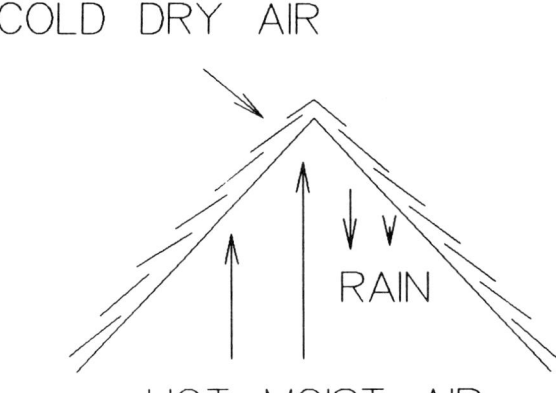

COLD DRY AIR

RAIN

HOT MOIST AIR

Figure 8–67 Lack of vapor barrier

building. What you now have is ruined insulation, ruined ceiling drywall, and possibly ruined furniture and carpets. And you will get the blame, because your roof leaks.

What I am saying is *look*. Look to see that the roof vent system is installed correctly. Look to see that the insulation contractor did his job correctly, that of placing the vapor barrier toward the room side of the ceiling as in Fig. 8–68. If you don't look, it becomes your problem. You may spend more time trying to convince your attorney, your insurance company, your boss, and the court judge that it was another contractor's poor quality work that caused the problem, that it was not the roof that you installed.

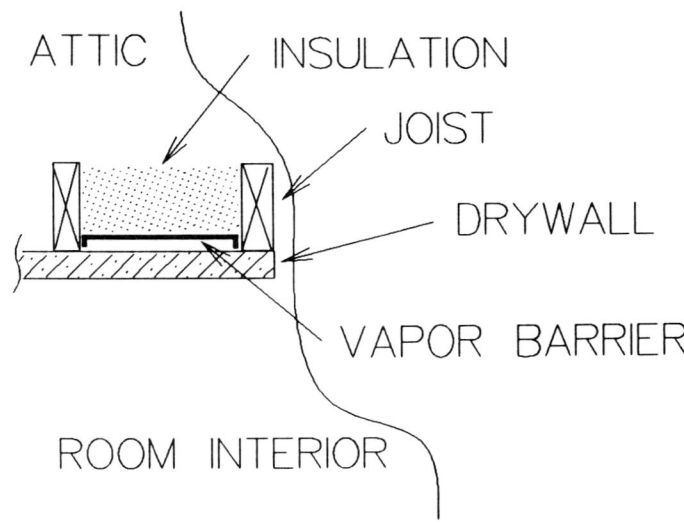

ATTIC   INSULATION

JOIST

DRYWALL

VAPOR BARRIER

ROOM INTERIOR

Figure 8–68 A small air channel between vapor barrier and dry wall will keep everything dry

RAIN GUTTERS

See Fig. 8–69. Unless you live in the northern portion of the country, you probably have never seen wooden drain gutters. They do exist and have been used for generations. The use of lightweight, formed on the job site, aluminum gutters is new to the building trade and is taking over. New in that homes have been built for thousands of years, aluminum gutters for less than fifty. Formed at the site and molded vinyl plastic gutters have been around for even less time.

There are four basic materials in use for building drain gutters. These are wood, painted or galvanized steel, aluminum, and plastic. These materials are suitable in that they are low cost, can be formed into gutters, and are weather resistant. Each will impart a feel or style of its own to the building it is installed on. Each will require preventive maintenance if it is to do its intended job for any length of time.

Don't indiscriminately tear off old wooden gutters and replace them with plastic or aluminum. The building's architecture, its feel and style, may be changed to something out of place. If the wooden gutters are not destroyed from lack of maintenance, then I recommend leaving them in place. Fix the dry rot, recoat the inner gutter with waterproofing, and paint the outside. These wooden gutters have done their job for 50 or more years, and with a little TLC they will continue to do so for another fifty.

The roofing business gets slow from time to time. Fortunately, you have the equipment, the insurance, and the skills to do roof drain gutter inspections and repair. Get your-

Figure 8–69   Wooden gutters are still around

self a supplier and some inspection forms. Actively advertise for and solicit drain gutter work. This will help you get your foot in the door. You do a good gutter replacement job during a cold day in January and you will get the reroofing job in July. It may be too cold in winter to do a reroofing job that takes weeks. But, even in the dead of winter, a few good hours can be found for gutter replacement. Many customers discover that they have gutter and roof problems during the winter months. They will be more receptive to your suggestions and help when the rain starts pouring.

See Fig. 8–70. Second-story rain gutters should be connected to the first-story gutters or run separately to the ground. Second-story gutters should never empty on to the first-story roof. The volume and force of the water exiting the drain will destroy the roofing by abrasion. Runoff water will be forced up under shingles if the angle of water discharge is low. Rust and paint oxidation dissolved in the runoff water will stain the lower roof.

Gutter Hangers

New products are introduced to the marketplace every day. Using these products correctly and for the right application is a problem. There are gutter strap hangers popping up on roofs in every locality. Most of these hangers are being installed on old roofs and being installed improperly. Strap hangers are designed to be nailed under the shingles. Proper nailing requires that the shingles be installed after the hangers are in place.

I have seen case after case where the hangers are nailed to and through the shingles (Fig. 8–71). It looks bad, it is bad. Use spike and ferrule or fascia mount hangers if you are not going to reroof. Use strap hangers if you are reroofing. Be sure the straps are securely fastened to the sheathing.

Clean Those Gutters

See Fig. 8–72. Gutter systems become clogged with leaves and loose roof granules.

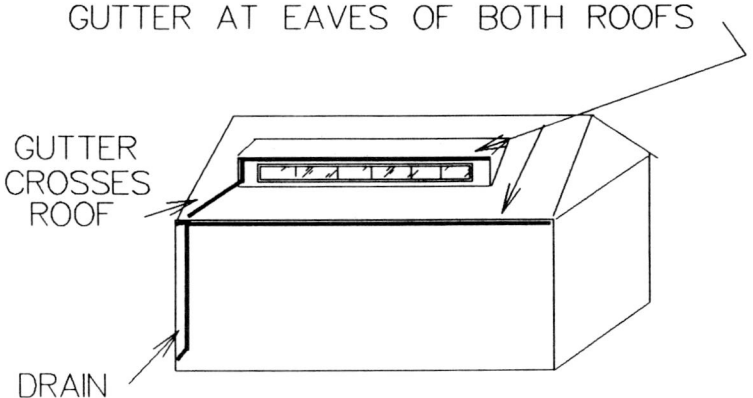

Figure 8–70   Gutter crossing roof is necessary for proper drainage

Figure 8–71   Improper installation of strap hangers. This eventually pulled out.

Figure 8–72   Clean the gutters at least once a year

Ice Dams in Gutters

The methods of preventing ice dams in gutters are as follows: The initial construction should be such that there are mesh wire leaf guards installed on all rain gutters. You can put these on afterward, but prevention of a problem is usually better than a cure.

Second is the fall/early winter preventive maintenance of cleaning out the gutters. You, as a roofer, can do this. During your P.M. (Preventive Maintenance), you should check for the gutter slope to be sure that it does drain to the down spouts. A 1/16 to 1/8 in. per linear foot slope is required. Check the down spouts for proper drainage, no blockage. See that they are draining out from the building and high enough off the ground so as not to get ice or snow clogged in the winter.

9

Caulk, Sealer, and Coatings

INTRODUCTION

This will be a short chapter introducing you to the various roofing cements, roof coatings, and their usage.

Roof coatings are cements, sealers, paints, and mastics. Most coatings are used to seal joints, to provide waterproofing, and to provide color to a roof. Some are used to cement roofing plies together, others to coat a roof in one or two quick applications. Some are used to reflect the sun's heat and U/V radiation, keeping the building's interior cooler. Others are used to repair leaks or fill cracks and holes. Some coatings will fireproof a roof, some will lower the fire rating of a roof.

SURFACE PREPARATION

In order for any glue or binder type material to stick to another material, the surfaces of the materials being bonded must be clean. There should be no oil, grease, dirt, dust, or any other loose material present. This may take a good brooming or vacuuming and a washing to accomplish. Bonding takes place at the molecular level. The molecules of one material combine electrostatically with the other or the molecules combine by an interweaving with those of the second material. If there is any foreign matter blocking the molecules, then the materials will not bond.

Patch It Up

All voids, splits, cracks, blisters, out of level places, etc., must be fixed before attempting to apply a coating. All oxides, efflorescence, loose paints, rust, or stains must be removed before coatings are applied.

Application Tools

Most sealers and caulks come in tubes and can be dispensed with the use of a caulking gun. Inexpensive caulking guns are available at most hardware stores. For fast production type work, I recommend a pneumatic powered caulking gun. The initial cost is more and you need an air compressor, but the work will go faster and be neater.

Paints can be applied with brushes, rollers, or with airless spray guns. The airless is the best tool for most applications. If there is a big danger of getting paint on surfaces that should not be painted, then a roller is best. Save the brush for the trim and detail work.

Liquid roof coatings can be applied with specially tipped airless spray guns, rollers, or with a roofer's mop and squeegee. Tight, difficult to access spots must be brushed.

The heavier coatings must be applied with a trowel. These are the plastic cements and cold asphalt cements.

Cautions

Most roofing asphalts and plastics are safe to use, some are safe to use only if you follow directions. Most will burn if they contact an open flame. Some give off vapors that will explode violently if ignited. Read the labels.

Cleanup

Here are some ingredients used for cleaning unwanted roofing cements off hands, tools, and other surfaces.

- Asphalt-based coatings will clean up with mineral spirits or paint thinners.
- Neoprene-based coatings will clean up with xylol or xylene, available at most hardware stores.
- Latex paints and caulks will clean up with soap and water.
- Silicon-based products will usually clean up with paint thinners.
- Oil and alkyd-based coatings and paints will clean up with paint thinners.
- Epoxy coatings will clean up with diacetone alcohol if the epoxy has not set up. After the epoxy has set hard, you must use a sander or grinding wheel or discard the tool.

- Polyester resin coatings will clean up with acetone or nail polish remover. (This also works on super glues.)
- Polyurethane can be cleaned with paint thinner or acetone.

Surface cleaning

- Old oil base paint can be removed with commercial paint strippers that contain methyl chloride.
- Latex can be removed with methyl chloride or acetone.
- Concrete can be cleaned with muriatic acid, available at most paint or hardware stores.
- Ceramic tile and glass will require etching with hydrofluoric acid.
- PVC, (polyvinyl chloride), polyethylene, and polypropylene should be cleaned with methyl ethyl ketone (MEK).
- Use 20% solution of nitric acid to clean stainless steel.
- Rust can be removed with phosphoric acid, navel jelly™, or oxalic acid.
- Clean galvanized metal with denatured alcohol and soap and water.
- Aluminum can be cleaned with tri-sodium phosphate (T.S.P.) or Soilex™.
- Muriatic acid or white vinegar is used to remove efflorescence.
- Mud, dirt, and clays can be removed with soap and water.
- Use mineral spirits or toluene to remove grease and oil. Liquid sandpaper will do a good job in a minimum of time.

Heavy cleaning.   If the surface is very dirty, rusted, or flaked, then I recommend the use of a wire brush and a little elbow grease. Use a brass or stainless steel bristled brush, do not use a steel bristled one.

Steel brushes leave small particles of steel on the surface. These particles will rust under the new coating. Hand or machine sand wood and metal where the surface is very rough or deteriorated. Use aluminum oxide or silicon carbide paper for best results. Start with a low number grit, 60, 80, or 100 grit for fast removal of the defects. Then change to the next higher number and sand again. Repeat this until you have a smooth clean surface. Always use a rubber backer under your sanding pad and always wear protective eyeglasses.

Primers

Metals will require a first coat of a primer. Use a primer to clean and etch the old surface so that the new coating will bond. Use balsam oil or wood primer on new wood to be painted. Use fish oil, tung tree oil, or linseed oil primer on new or bare metal. Zinc primer can be used on metal flashings. Asphalt cement primer should be used on all bare concrete. NOTE: Many paint stores will inform you that thinned latex paint is used as a primer for latex paint. This is the wrong advice. Latex paint does not have the proper ingredients or the acid etchants that make a primer a primer.

Weathering

Most new surfaces should be allowed to weather before being coated. Brick, block, and concrete should weather for up to 12 months, wood for a maximum of 10–20 days. Galvanized metals must weather 30 to 60 days. Weathering is not always possible when you need to complete the job quickly. You can use artificial weathering to speed up the process. Clean the surface as recommended, then sprinkle the surface with a water mist two to four times a day. Allow surface to dry between each watering. Artificially weather for one to two weeks, then clean with the recommended cleaner and clear water rinse. Allow surface to dry for one to two days, prime, and top coat.

Weathering does two things: it opens up the pores in the material and it cleans them. The open pores provide spaces for the new coating(s) to adhere to.

Storage

Read the labels. Most coatings are very specific about not being frozen or overheated. These coatings are chemicals, which can and will change their properties if stored incorrectly or for too long a time. Many oil-base products will smell rancid when bad. Do not use the product if it is a dated product and the expiration date has past. You will only make more work for yourself, because the product will fail and have to be removed.

Application Temperatures

Most coatings must be applied in their liquid or semiliquid state. When the materials get too cold, they tend to become very stiff or solid and application becomes difficult. In ambient temperatures of under 45 degrees Fahrenheit, I recommend that you heat the materials or not use them. There are a few exceptional products that can be applied down to 0 degrees, but not many.

Too hot a temperature is also not good. The material thins out too much and you will not get the required coating thickness for best protection. Do not use if the ambient temperature is over 95 degrees Fahrenheit, or the surface temperature of the item being coated is over 110 degrees Fahrenheit. Wait for a cooler day.

Mornings and Evenings

Early morning application of coatings is not recommended in most localities. The overnight dew has made everything wet or damp. Most coatings will not adhere well to a wet surface. Give the surface(s) a few hours to dry before commencing work.

Late afternoon application of coatings is not recommended. The temperature from day to night drops very quickly in most localities. This sudden temperature drop sends many coating materials into shock. They wrinkle, buckle, blister, and eventually crack and peel. Allow at least 2 to 3 hours of drying time before drastic temperature changes.

The morning wetness and the afternoon temperature changes can be duplicated during the day by a rainstorm. If you know that rain is coming, plan to work on another day.

JOINT SEALERS AND CAULKS

Sealers and caulks do just that, they seal against intrusion of air and water. Since they are generally used to fill cracks, they are elastic. Any two items that expand and contract will crack at their intersection. The crack may be 1/64 in. wide one day and 1/4 in. wide the next. The caulk must adhere to and stretch with both materials. There are 11 main types of caulks and sealers, as follows:

- Butyl rubber sealant
- Epoxies
- Silicon RTV
- Synthetic rubber
- Hypalon™
- Neoprene™ sealant
- Oil-base caulk
- Polysulfide sealant
- Polyurethane sealant
- Vinyls
- Latex-base caulk

Many sealants can be obtained as liquid pouring grade sealants that are self-leveling. Some can be obtained as solids as rope, tubes, beads, ribbons, etc. These are squeezed into place to form the seal. The caulks are in paste form and designed to hold without sagging or running.

The method of setting up varies in three main groups:

1. Hot melt thermoplastics are rubber-asphalts, asphalts, and rubber-tars. The chemical reaction thermosetting compounds are the polysulfides, polyurethanes, and silicones. The solvent release compounds are the neoprene, hypalon, and butadiene styrenes. Epoxies are thermosetting but do not fall into the above groups.

2. The polysulphides are two component elastomers. The polybutenes and polyisobutylenes should be used in concealed areas. These sealants remain soft and tacky and will pick up dust and dirt if left exposed.

3. Good caulk can stretch up to 100% or more without loosening or splitting. The caulk should last, in service, for 15 to 20 years or more. Most say they do, but most do not. I recommend inspection and possible replacement every 5 years at the minimum. Here is a short description of a chalking or elastomer joint.

Elastomer Joints

See Fig. 9–1. An elastomer joint must expand and contract without the elastomer detaching from the joined or sealed materials. Figure 9–1 (top) shows a joint as made. Figure 9–1

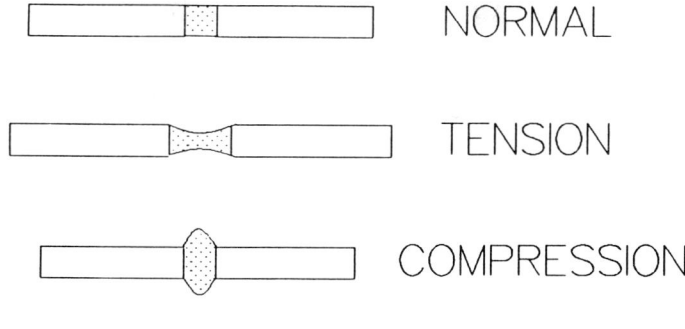

FORCES ON AN ELASTOMER JOINT

Figure 9–1 Elastomer joint under tension and compression

(middle) is the same joint when the items joined have contracted away from the joint. In Fig. 9–1 (bottom), the sealed items have expanded into the joint. This is how things are supposed to work, and here is what happens if the elastomer is improperly applied.

See Fig. 9–2. This elastomer joint not only bonded to the two materials, but also to a third bottom material. During expansion and contraction, the elastomer is pulled and pushed unevenly. This will result in the elastomer to material, bond failure. The watertight seal will be broken. If a bottom or third surface is encountered, then you should use a slip sheet. This slip sheet is nothing more than a piece of waxed paper or polyethylene plastic applied to the third surface. The elastomer sealant, when cured, will not stick to the slip sheet, and you will have a properly sealed joint.

The following are general descriptions of the products you may eventually use.

Elastomeric Mastic

This is a specially formulated asphalt used to seal around vent pipes, gutters, metal edges, etc. It works best when applied 1/8 in. thick, covered with fabric reinforcement, and recoated. Plies can be built up to 1 in. thick. Henry #209 Elastomer Mastic™ is recommended for use in repairing Henry Ruftac™ self-adhering membranes.

Reflective Coatings

These coatings are applied to the entire roof surface. They are the bright white and aluminum-filled coatings. The bright white coating is usually a latex or water-base acrylic coating. The aluminum-filled coatings are asphalt-base coatings.

Reflective coatings are not long-lasting coatings. The useful life expectancy of most does not exceed 3 years. The coating film may not fail, but the reflective qualities of the coating do.

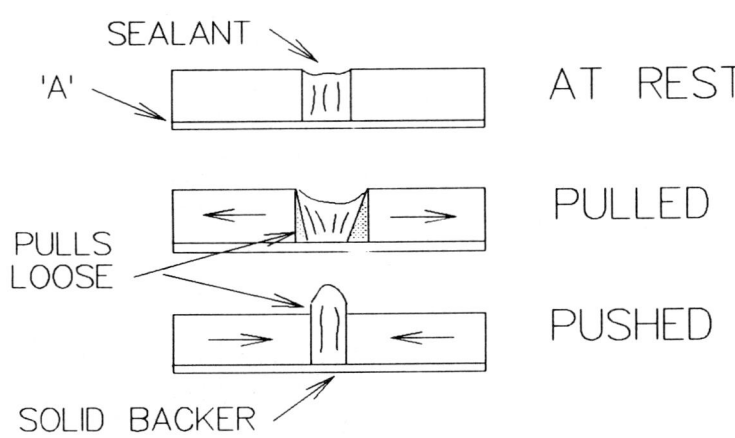

Figure 9–2   Unequal forces on an elastomer joint

Reflective coatings are not to be considered waterproofing coatings. Apply reflective coatings in a thin coat of approximately 13 to 20 mils thick. A mil is a thousandths of an inch (1/1000). As you can see, this is not a very thick coating.

Reflective coatings are not to be considered abrasion-proof coatings. Most are soft and will damage easily. If possible, do not pile things on them or walk on them.

The acrylic coatings are considered fire-resistant and usually will meet or exceed Class II fire ratings. Be careful of your FM, Factory Mutual, and local building code requirements. A roof or building takes on the fire rating of the lowest classification material contained within it. Putting a Class II coating on a Class I roof may lower the roof's rating to a Class II.

Reflective coatings are designed to reflect sunlight, its U/V radiation, and its heat. This in turn keeps the building's interior cooler. This saves energy since the air conditioning does not have to work as hard removing heat from the building. Roof temperature can be as much as 30 degrees cooler when using a reflective coating instead of a black asphalt coating. Plus, the reflection of the U/V radiation will help keep the underlying materials from deteriorating.

Reflective coatings are designed to be used just as they come out of the can. Thinning is generally not recommended. Application of the coatings is by brush, roller, or airless spray.

Reflection qualities of reflective coatings come from the metal particles suspended in the coating base. These particles are either aluminum flakes, zinc oxides, or other metals. As the coating dries, the particles float to the top surface skin and form the thin reflective layer. This reflective layer is easily damaged, so use care when working on the roof.

Reflective coatings are designed to be used over a waterproofing coating. They are to be used over smooth-surfaced roofs, i.e., BUR.

Some reflective coatings contain additives that help prevent fungal and algal growth.

Some contain plastic or glass fibers for added strength and dimensional stability. Most will remain flexible in cold weather.

Bright white coatings should be applied in two or more layers, each perpendicular to the other. The heavier aluminum coatings are applied in a single coat with a brush.

Both coatings can be applied to metal or asphalt. Both are recommended as a top coating on trailers and mobile homes. This can become a career specialty in itself, especially if you reside in the South or Southwestern United States where the sun always shines and trailer parks abound.

Coverage of reflective coatings is 1/2 to 3/4 gal per 100 sq ft of smooth roofing surface. Coverage should be from 3/4 to 1-1/2 gal per 100 sq ft on rough surfaces. Drying times vary from 1 to 4 hr.

Standing water is detrimental to the coatings and, therefore, the roof drainage system should be kept in good working order.

Bright White Coating for Urethane Foam Roofs

Some recreational vehicle manufacturers use urethane foam for their roofing material. The foam provides insulation value and weather resistance. Use aluminum or bright white coatings formulated for use over urethane. Henry #280™ and Henry #281™ are formulated for this purpose.

Aluminum Coating for Early Morning Dew Application

Use an aluminum coating formulated for application over damp surfaces. Henry #229™ is formulated for this purpose.

Aluminum Coating for Application over Rust

Use Henry #222 Alumi-Liner Cooler Coating™.

Plastic Cements

Plastic cements are asphalts that are modified with the use of mineral stabilizers and fiberglass or polyester fibers. Plastic cements are used to bond and coat flashings.

Application is with a trowel and the coating is applied 1/8 in. thick for best results. If a second coating is required, then it is recommended that a reinforcing tape be used between layers. Apply plastic cement to clean, dry surfaces only. Do not attempt to bond it with wood or shake shingles. The oils in the shingles, plus the moisture absorption of the shingles, will eventually break the bonds.

It takes about 75 lb of plastic cement to cover 100 sq ft of surface to a depth of 1/8 in. A gallon of material should cover about 9 sq ft. It will take 11 gal to cover 100 sq ft.

There are especially modified plastic cements that are called all-weather cements. These all-weather plastic cements can be used during extremes of ambient temperature and on wet surfaces. They are formulated to replace water and can be applied under water. GAF Jetblak™ is one such plastic cement, Henry #208™ is another.

Plastic cements skin over, but do not dry hard. They stay pliable and flex with the item(s) on which they are adhered. This makes the plastic cements an excellent choice for metal to wood or metal to asphalt seals.

Plastic cements are resistant to damage from acids and alkalis and, therefore, can be used on concrete and bricks. Some manufacturers specify specific plastic cement use for their warranties to be effective. GAF is one such manufacturer and specifies GAF Jetblak Flashtite™ for their built-up roofing guarantees.

Plastic cements should be used as they come from the container. If thinning is necessary, use just enough mineral spirits to obtain the consistency desired. Do not over thin.

For quick and easy shingle tab glue down, use a plastic cement in a caulking gun tube. Shingles at the rakes and eaves of buildings in high-wind areas should have a drop of cement under each tab. This holds true for all shingles installed on slopes of 3:12 or under, and all shingles installed on slopes greater than 9:12. Special adhesive cements are made for the purpose.

Adhesive Cements

Adhesive cements are asphalt-base roofing cements modified for superior adhesive qualities. They are used to seal down shingles and for filling joints. GAF Shinglstik™ is one such cement.

The Gibson-Homans Company produces an excellent neoprene adhesive cement that can be used between metal and red cedar or redwood. It may, in time, break loose from the wood but being neoprene based, will prevent tannic acid staining. The product is recommended by the company for use as a sealant for gutters, downspouts, fasteners, asphalt roofs, and driveways. The product is Black Jack 1010™.

Neoprene is a rubber and requires a period of 4 to 7 days to cure solid. It is very flexible and can recover 100% from stretching over 250%. It is resistant to water, oil, and solvents and has very good resistance to sunlight. Cleanup is with mineral spirits.

Latex- Or Oil-Base Caulks

These are readily available in most lumber and hardware stores. Many are good, long-lasting products, some are not.

Caulks are formulated for specific use, color, and durability. Choose the product that is correct for your application. Some caulks may be painted after they cure, the silicon-base caulks generally cannot be. Silicon caulks are considered the best caulks and many are precolored. Colors available, at this writing, are white, brown, black, and bronze.

Clean up latex caulks with water, oil and silicon caulks with mineral spirits or paint thinner. Application is from a caulking gun and in 1/8 in. wide strips. Use a wet finger or a wet plastic paddle to smooth out the caulk and have it conform to the surface.

Asphalt Primer or Asphalt/Concrete Primer

When sealing concrete or cracks in concrete, you must prime the concrete for it to accept the heavier top coatings. Asphalt/concrete primers are used for this purpose.

These are adhesive-modified, thinned asphalts designed to seep deeply into concrete surfaces. These primers are recommended whenever a concrete deck or wall surface is to be covered with a built-up roof. Asphalt primers may be brushed or sprayed. Coverage is 150 to 300 sq ft per gallon.

Henry #105™ or GAF Asphalt/Concrete Primer™ are two such products. Cleanup is with mineral spirits or paint thinner.

Cold Application Cements

These are fiber reinforced, asphalt-base coatings designed to replace hot asphalt in built-up roofing systems. The cements will bond felt ply layers and gravel or granule toppings securely.

Henry #203 Cold Application Cement™ is used to bond Henry Rufon™ polyester fabric. Henry #203 should not be used over saturated felts or solvent-sensitive roof insulation. NOTE: Cold application cements contain a high percentage of solvents and are flammable. Keep them away from open flame.

Cold application cements should be applied as they come from the can; additional thinning is not recommended. Application is via roofing brush, rollers, or spray gun. They remain tacky for up to an hour, giving adequate time for cutting, positioning, and installing the roofing plies.

Application rate is 1 to 1-1/2 gal per 100 sq ft. Cleanup is with mineral spirits.

Gravel Binder Cold Process Cement

Use Henry #202 Gravel Binder™ as a cold process gravel binder.

Asphalt Emulsions

Emulsions are particles suspended in a liquid or a liquid suspended in a liquid. Examples are asphalt in a solvent or clay in an asphalt.

Asphalt emulsions are basic roof coatings and can be used as a final surface coating for built-up roofing, composition roofing, and metals. With the use of reinforcing fabric, they can be used on brick, block, or other masonry. Asphalt emulsions can be used for waterproofing below or above the ground. They are recommended for adding corrosion resistance to steel pipes, rails, tanks, flashings, and other structures above or below ground.

Asphalt emulsions are made from asphalt that has a low softening point. This enables the coating to remain flexible and be somewhat self-healing to small cracks and splits. The emulsions are very waterproof and have good resistance to sunlight.

Application is with a brush or roofer's brush. The 3 or 4 knot roofer's brushes will work fine. Application rate is 15 to 30 sq ft per gallon.

Asphalt emulsion for polyester fabric.   Polyester fabric will dissolve in some solvents. It is necessary to use an emulsion formulated for use with polyester. Henry #106 is one such product, it contains a special clay type asphalt.

Lap Cements

Lap cements are designed to seal the seams of roll roofing during installation. Lap cement is to be applied to the roll roofing joints at a rate of 270 to 290 ft per gal in strips 2 in. wide. Do not use Lap cements on flat roofs of less than 1:12 slope; use Cold Application Cement instead.

Paints for Wood

Wood trim used for fascia, soffit, and moldings should be painted to protect them from the weather. The biggest mistake a roofer makes is not allowing the new wood time to weather. Wood contains natural oils, solvents, and a high content of water. Paint will not adhere to these items and a weathering of 10 to 15 days is recommended.

After the weathering time is up, wash the wood with clear water and allow 2 to 3 days drying time. Prime, using a good wood primer, and paint with a good quality latex paint. I do not recommend oil-base paint, because although it gives superior film resistance, it will eventually flake, chip, and peel. Latex paints tend to eliminate these problems. Most latex paints and stains are non-air polluting and meet or exceed EPA (Environmental Protection Agency) requirements. Cleanup of latex is with soap and water.

On rough surface lumber, or in high sunlight areas, use latex stain. Stain is a dye and will adhere to the wood far better than paint. The main disadvantage of stain is that it fades and must be recoated every 2 to 3 years. Paint is a surface coating and will eventually crack or peel, but paint should last from 7 to 12 years between coatings. Use nonchalking paints only. Chalking or self-cleaning paints will stain roofing materials and shingles if used on trim above the roof line.

Paints for Metals

Metal trim and flashings should be painted for best long-lasting results. Painting the flashing the same color as the roofing makes for a professional looking, finished job. Galvanized metals should be cleaned with soap and water, dried, and then recleaned with denatured alcohol. This cleaning process removes the dirt and oils left over from the manufacturing and handling processes. When dry, prime and top coat with a latex paint. This cleaning process replaces the 30- to 60-day recommended weathering process.

Paint may be applied with a brush, roller, or spray gun. Spray is fast and easy providing you do it before you install the metal. After installation, spraying is difficult if not impossible. The flashing strips are too narrow and you will have paint everywhere. This is OK if the siding and roofing have not yet been installed.

I do not recommend the use of oil-base paint on metal flashing. Oil-base paint will not expand and contract with the metal and will peel.

Application rate of latex paint is 400 sq ft per gallon, 350 sq ft per gallon if sprayed. Do not apply in heavy thick coats, two thin coats will last longer. Cleanup is with soap and water.

Concrete and Asbestos Tile Sealer

Thoroclear #777™ is one product used for sealing and waterproofing brick, block, masonry, concrete tile, and asbestos tiles or shingles. The sealer is clear and invisible to the eye. It will last for 8 to 10 years of normal weathering. It is not affected by temperature change, sunlight, or abrasion.

Application by low-pressure spray is recommended—use a pump sprayer. The surface should be flood coated, allowed to dry for 24 hours, and then reflood coated for best results.

The product must not be applied if rain is expected within 24 hours or if it has rained in the past 96 hours. Coverage is 150 to 200 sq ft per gallon.

Thoroclear #777 is NOT to be used on limestone, glass, aluminum, or porcelain. Doing so will damage these items. Thoroclear #777 is an exterior use product only.

If there are cracks in the concrete, these must be repaired before sealing or top coating. Specially formulated one component patching materials are available. One such product is Thorocrete™.

Concrete Patching Cements

THOROCRETE is a one-component, polymer modified, cement-base patching material for concrete and masonry. It is sold as a powder and must be mixed with water. It is nonflammable, vermin-proof, and not affected by exposure to water.

Application is with a trowel to fill voids up to 1 in. For voids over 1 in. wide or deep, allow the first layer to cure for 24 hours and apply a second layer. Finish with a thin brush-on coating to smooth with existing concrete. Coverage is 20 sq ft, 1/4 in. thick, for each 40 lb bag. Cleanup is with soap and water. Acryl 60™ a liquid plastic additive, and can be added to increase strength.

For small patches, holes, and mounting of equipment mounting bolts, use a quick set hydraulic cement such as THOROGRIP™. THOROGRIP is a pourable, waterproof, non-shrinking, quick setting, hydraulic compound. It can be used to secure aluminum or steel mounting bolts to concrete without negative effects. The product sets hard in 15 min. so only mix as much as can be used in 15 min. Full-cure, maximum pullout resistance is obtained in 60 min. A 1 lb package will fill 14 cubic in. Cleanup is with soap and water.

For areas subject to water seeping through a crack or hole, use a product such as WATERPLUG™. This hydraulic cement can be used to seal cracks and grout holes that pipes, wires, or hoses pass through. Being a hydraulic cement, it expands as it cures and fills in all minute voids. Setup is very fast, about 3 min., so do not mix more than can be used in 3 min. WATERPLUG will stop water flow and can be applied to a surface that is under water. It bonds to concrete, masonry, aluminum, and steel. Coverage is 17 cu in. per pound of material. Cleanup is with soap and water.

Fire-Retardant Coatings

The aluminum reflective coatings are somewhat fire-retardant in that the metal floats to the surface film and forms a metal layer. This layer is resistant to catching fire from sparks or open flame. Aluminum reflective coatings are not fireproof since they are asphalt-base coatings.

There are spray-on coatings that are used on the interior sides of roofs to make them fire-resistant. These coatings contain chemicals, usually nitrogen compounds, that are self-extinguishing. The coatings are sprayed into place to a thickness of 1/2 to 2 in. They not only do not burn, but insulate as well. This insulation value keeps the building's interior warm or cool and keeps the roofing from overheating and releasing flammable gases during a fire. A Class II metal decked building can be made Class I with the use of the proper coating, installed properly.

Additional Coating Additives

There are additives that can be used with or built into roofing materials to give specific qualities to the materials.

- Tributyltin is a mildewcide used in shingles sold in high-moisture areas. Tributyltin can be used to control algae growth.
- Pentachlorophenol is a wood preservative and a bleaching agent used to preserve wood or shake shingles. It usually waterproofs and may lighten the wood's color back to natural. Pentachlorophenol is used in a variety of clear coat, wood finishing products.
- Chlorpyrifos is an insecticide that is added to paints to keep insects under control. Other insecticides are Cinern, Paris Green, and Dieldrin. Paris Green is copper acetate and arsenic trioxide.
- Copper napthenate is a wood preservative used to prevent dry rot. It is effective on wood used in contact with water. It has wide marine use due to its ability to ward off mussels, barnacles, and other sea life.
- CCA, chromated copper arsenate, is a wood preservative used to extend the life of lumber for up to 40 years or more. It is the main chemical used for making treated lumber.
- Wax is used by some manufacturers of Cant Strips as a preservative.

These additives are all poisons. Use care when using these products or using products containing these chemicals. Wear a respirator when cutting or spraying these products. Wear gloves and wash hands before smoking or eating after using products containing these chemicals.

PRACTICAL APPLICATIONS OF COATINGS AND SEALERS

Please read the chapters on flashings and built-up roofing. These give further details of the construction methods and coatings used by the professionals. Here I will show a few common problem areas.

Mastic as a Chimney Flashing

See Fig. 9–3. Can one use roof cement or mastic for chimney flashing? It can be done. It will last for a few years, but it will require periodic maintenance, removal, and redoing. The biggest mistake some make when using mastic as flashing is that they apply it too thick. A thick coating, shown in Fig. 9–3, will dry out, crack, split, pull loose, and otherwise leak. It is better to use a thin 1/4 in. coating as in Fig. 9–4. The thinner coating stays more elastic and will more closely follow the expansion and contraction of the brick and roofing. Start the mastic in a grout joint. This helps lock the mastic in place and helps prevent a mastic to brick bond leak. The insertion of fiberglass reinforcement tape will help prevent the mastic from cracking.

Leaky metal flashing can be repaired with a combination of layers of mastic, fiberglass tape, mastic, and a final coat of aluminum fiber mastic. Repairs will last from 5 to 10 years. This system works well on newly installed flashing since it protects the metal from the weather, adding years to the roofing job.

Vent Pipe Flashing Using Mastic

On many older buildings, the vent pipes do not have metal flashing. The leak seal between the pipe and the roof was formed from roofing tar. Frequently, this tar was applied very thick (Fig. 9–5). Figure 9–6 shows the results of this poor application a few years down the line. The roof tar or mastic has shrunk and pulled away from the pipe and the roof, creating

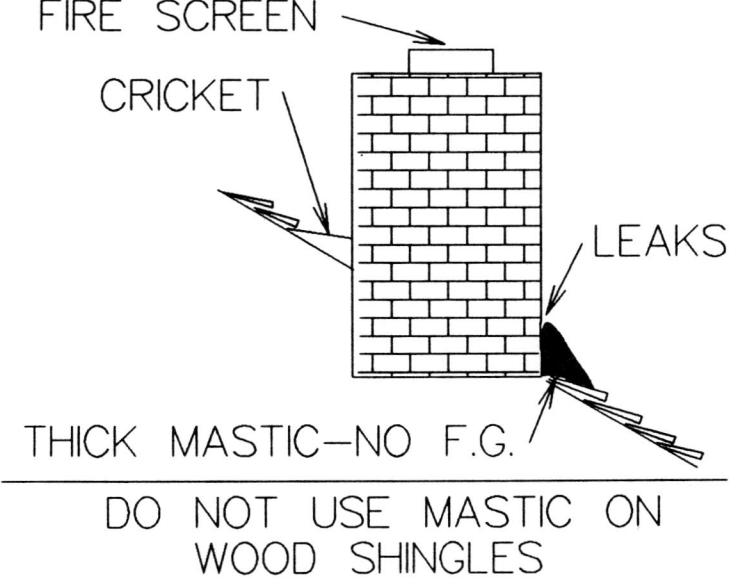

Figure 9–3 Mastic applied in a thick coating leads to leaks

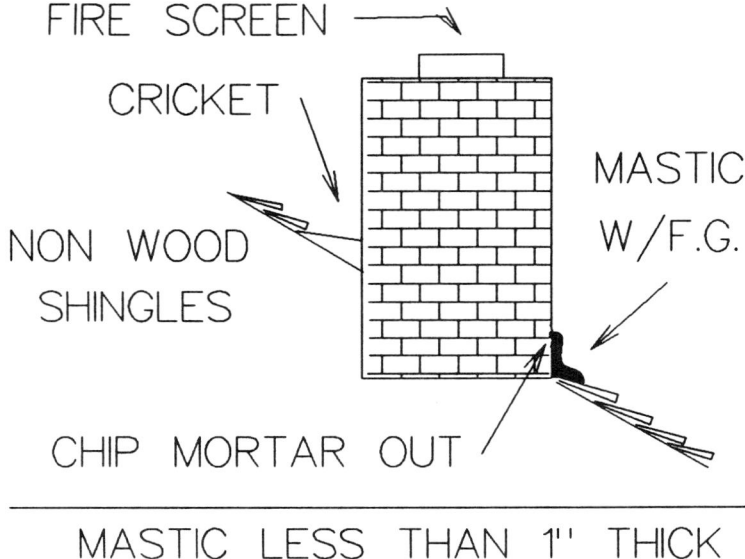

FIRE SCREEN

CRICKET

MASTIC
W/F.G.

NON WOOD
SHINGLES

CHIP MORTAR OUT

MASTIC LESS THAN 1" THICK

Figure 9–4 Mastic applied in a thin coating will last longer

leak sources. To make a short-term repair, remove all the old mastic. Clean the pipe with mineral spirits and scuff sand with #100 grit sandpaper. If the pipe is metal, paint it with metal primer topped with a good quality exterior paint. When the paint has dried 2 to 3 days, apply a 1/8 in. coating of mastic to the pipe and the roof. Then apply fiberglass patching tape covered with 1/4 in. of mastic. To this, apply a coat of aluminum fiber roof mastic.

If you are reroofing the building, then I highly recommend that the pipe be cleaned, primed, painted, and new vent pipe flashing be installed. The new flashings have a cone-

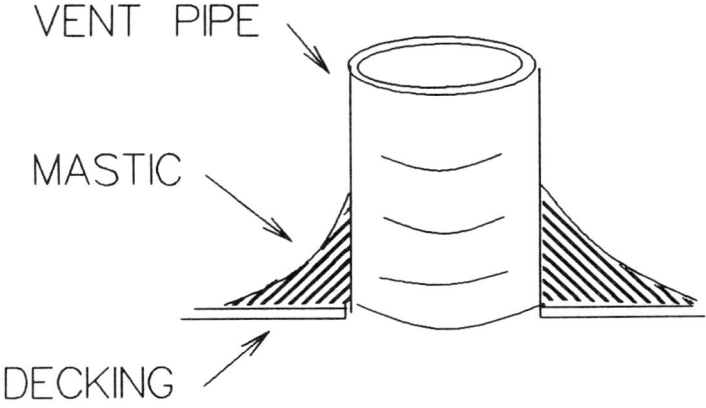

VENT PIPE

MASTIC

DECKING

Figure 9–5 As applied to a vent pipe

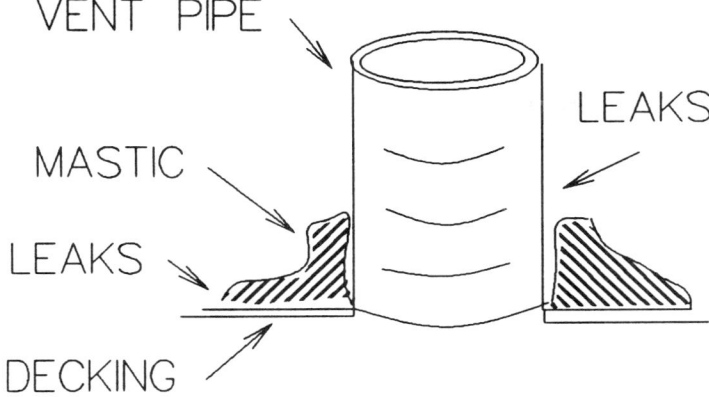

Figure 9–6   Figure 9–5 after a few months to years

Figure 9–7   A chimney cricket?

shaped base and a rubber sealer boot. They cost under $10.00 and will do an excellent job. I do recommend that they be prime coated and painted before being installed. Use a zinc oxide or fish oil metal primer and a good grade of exterior latex enamel. Paint color should match the applied roofing color.

The Chimney Cricket

Figure 9–7 illustrates an asphalt cricket. This is acceptable in this case because the roof slope is small and the rainfall is under 2 in. a year. There is chimney flashing present that should prevent any leaks.

10

Asphalt Shingles

INTRODUCTION

I would estimate that about eight out of ten residential buildings in the United States have asphalt shingled roofs. Asphalt shingles are good looking, long lasting, and easy to install. They are available from lumber yards and a variety of home improvement stores. A full range of colors and textures are available and many varieties cost under $10.00 a bundle.

Asphalt shingles are installed by everyone from the professional roofing contractor to the kid down the block. Most of the installations are correct, but a good many are not. Installation instructions are printed on each bundle of shingles. Here is a quotation from one roofer I talked to. "The instructions are good enough to allow you to shingle a shed roof without too many problems." This chapter will help you avoid installing an asphalt shingle roof that leaks, blows off, looks shoddy, or wears out prematurely.

SHINGLE CONSTRUCTION

Fiberglass shingles are another name for the common asphalt shingle. Most of today's asphalt shingles are constructed on a web or mat of fiberglass fibers. Some, still considered FG (fiberglass), use asbestos, or polypropylene, or other dimensionally stable plastic for a base.

The least expensive shingles use a web or net for a base. The more expensive use a mat (Fig. 10–1). The mat base is superior to the net base in that it is thicker, tougher, and less prone to splitting, expansion, and contraction. Unlike the net that can be distorted in length and width, the mat is unidirectional. The fibers lie in all directions and tend to prevent distortion in all directions. Mat shingles retain the asphalt better and minimize heat flow or melting. The mat shingles provide a superior nailing base, one that is resistant to pullthrough. Also, mat shingles are less prone to U/V radiation and heat destruction.

ALIGNMENT GUIDES

Manufacturers of asphalt shingles build alignment guides into their shingles. Across the top of the shingles you may find small slits, each about 3/4 in. long. These alignment slits guide the installer when he installs the shingles (see Fig. 10–2). Along the sides we have half a cutout slot that, when butted up to the next shingle, forms a full-cut slot. There is a horizontal alignment slit just above the cutout. Some shingles have shadow marks. These shadow marks, created by darker colored granules, should form a straight horizontal line from rake to rake (Fig. 10–3).

EXPANSION AND CONTRACTION

Shingles do expand and contract with heat and cold. This expansion and contraction is usually taken up by the shingle material's elasticity, but not always. Elasticity is the ability

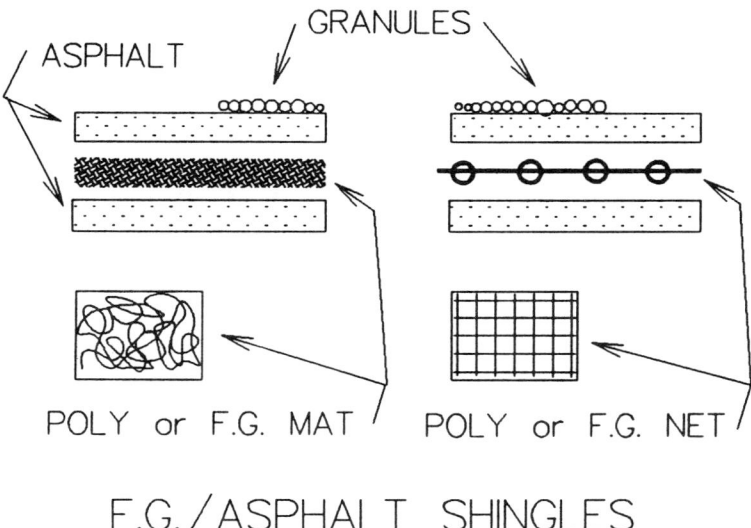

Figure 10–1   Mat versus web type shingles

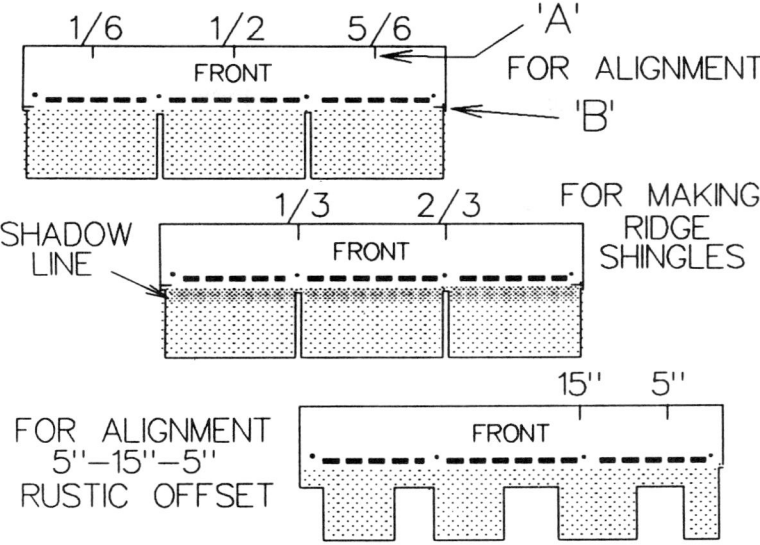

Figure 10–2   Tab type asphalt shingles

Figure 10–3   Shadow marks aid horizontal alignment

to be elastic, act like a rubber band. It is because of this expansion and contraction that one does not tightly butt shingles together during installation. Allow 1/16 in. side lap between shingles in warm weather, 1/8 in. side lap in cold—under 40 degrees Fahrenheit—weather.

SELF-SEALING TABS

See Fig. 10–4. On the face side of self-sealing asphalt shingles, you will find a row of black ovals. This is mastic that when heated will melt and seal the shingle to the shingle installed below it. Over this mastic is a thin plastic film. The film is there so that the shingles do not stick together in the bundle during storage. Remove it before installing the shingles. A full 80% of the roofers I have watched do not remove this protective plastic.

   The self-sealing properties of the shingle are lost if this plastic is not removed. By not removing this protective strip, the mastic melts under the strip and not to the next shingle. The seal is not made and the first good wind blows the roof away.

   You may wonder why the mastic is in several small spots, rather than in one continuous strip. The reason is that you want moisture entering under the upper edge of the shingle to escape. The nonmastic areas allow for this moisture to escape through the bottom edge of the shingle, the edge that is exposed to air flow.

SPECIAL-ORDERED SHINGLES

Figure 10–5 is one of the better challenges I've seen. Those are asphalt shingles (Fig. 10–6). See anything particular about the shingles? Check the coloration and pattern. I suspect that

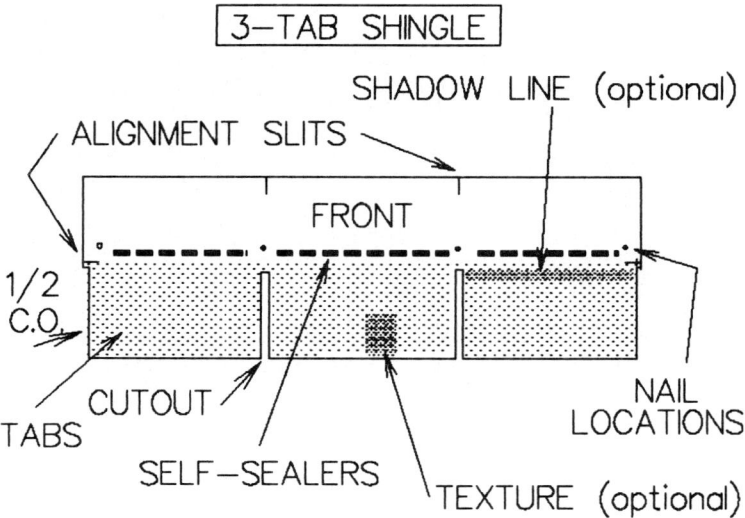

Figure 10–4   Makeup of a three-tab shingle

Figure 10–5   An interesting roofing challenge

Figure 10–6   Special-order shingle coloration

these shingles were custom manufactured for this job. The color is on one tab, not all the tabs.

The point is that you can purchase custom manufactured shingles, that is, if the order is large enough and you are willing to pay the price. Contact the roofing manufacturer's representative in your area for details.

FIRE RATINGS

Asphalt shingles are fire rated by classifications set down by the Unified Building Code and several testing agencies. See the chapter on specifications for testing methods.

The fire rating classifications are Class A, Class B, Class C, and unclassified. Use Class A for all residence shingling. Class B and Class C should *not* be used on residences as they may increase insurance cost. Unclassified shingles are not considered shingles, they are considered decorations. Unclassified shingles are used for covering vertical wall surfaces.

Most manufacturers of asphalt shingles produce only Class A shingles. But, there is a problem and that is in the installation of the shingles. The fire rating classification is only good if the shingles are properly installed. Improper installation, or using one roof component that is of a lower classification, will generally lower the classification of the entire roof.

WIND RATINGS

Asphalt shingles are wind rated as well as fire rated. For the most part, all shingles sold in your locality will be properly rated for use in your locality. The ratings are not so much on the material or construction of the shingles, but rather, on the installation techniques used. Underwriters Laboratories, Factory Mutual, manufacturers, and other agencies establish these requirements.

In the United States there are wind zones that dictate the use of roofing materials and installation methods. The worst zone is Dade County, Florida. Dade County therefore has very stringent regulations on the type and use of roofing. Toronto, Canada is also a high wind zone. Both Toronto and Dade County are subjected to winds of more than 100 miles per hour. Wind speed and wind uplift maps are available from most roofing manufacturers.

Both areas are located next to large bodies of water. Dade County is near the ocean and Toronto is near the Great Lakes. You will find high wind zones wherever there is a lack of obstruction to the wind. Your locality may be in a 40 mile per hour wind zone, but because of the immediate terrain, it may receive winds in excess of 80 miles per hour. It is for this reason that I suggest you use good roofing practice and follow all codes and the material manufacturer's instructions.

High Wind Areas

In high wind areas and areas subject to hurricanes and tornados, or on low slope roofs, it is advisable to add extra wind lift protection. Applying dabs of mastic under each shingle tab

(Fig. 10–7) will accomplish the desired results. The mastic can be troweled on or applied from a caulking gun. This extra little dab of mastic is required under shingles installed on roofs of over 21:12 slope.

COLORS AVAILABLE

There are a multitude of shingle colors to choose from, but most roofs are shingled in black, gray, or tan. Personally, I like the tans and grays, they blend with most siding colors and look good.

Solid colors, especially solid black, do not always produce an even coloration over the entire roof surface. I advise against using all solid color if possible.

Color is a personal preference item. I like the grays and tans, you might like the black or the whites. The main point is to avoid colors that will clash with the siding's color or that will clash with the other buildings in the neighborhood. There are bright yellow shingles available. If you shingle a building with them that is located in a neighborhood of tan roofs, you will soon be out of business. Public opinion will be against you. You have detracted from the neighborhood's look and feel and possibly lowered the neighborhood's overall resale value.

Asphalt shingles are supplied in basic black, off-white, brown tones, beige and tans, gray tones, red, yellow, and the light and dark blues and greens. The colors may be solids or blends of two or more colors. You will find shingles colored with shadow lines and ones colored to resemble shake shingles or slate tiles. Color is obtained from the top coating of granules. The granules used can be crushed ceramic, porcelain, rock, or minerals.

Residences will be roofed in the earth tones, while the bright reds, blues, and greens are normally reserved for commercial buildings. Commercial customers want you to see their establishments.

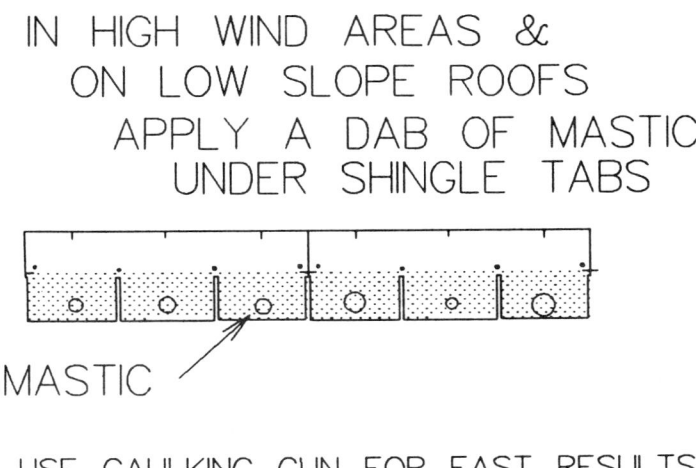

IN HIGH WIND AREAS &
ON LOW SLOPE ROOFS
APPLY A DAB OF MASTIC
UNDER SHINGLE TABS

MASTIC

USE CAULKING GUN FOR FAST RESULTS
PRESS SHINGLES INTO MASTIC TO SEAL

Figure 10–7   Glue down the tabs in high wind areas

Textures Available

Asphalt shingles need not be flat sheets of asphalt. New manufacturing techniques and new manufacturing equipment can produce asphalt shingles in three dimensions. The shingles can be cut with varying tab lengths and pressed to have varying tab depths. A close resemblance to wood or tile shingles is possible. The addition of varying color and color patterns can make for very realistic duplications of the real thing.

LOCKING SHINGLES

Locking shingles seem on their way out as the shingle of choice. Locking shingles are quickly being replaced with the standard three-tab shingle in most areas of the country. The only area that I know of that still stocks and sells locking shingles is the upper midwestern states. Most other areas will supply you with locking shingles, but by special order only.

Figure 10–8 shows several designs used for locking shingles. The tabs or ears insert into the next shingle and two nails are used to hold everything in place. The shingles are good, they hold tight in the strongest winds. The complaint about locking shingles is from the roof contractors. Locking shingles just take too much time to install. Three-tab shingles flop in place and can be power stapled in seconds. Locking tab shingles must be fitted in place, interlocked, and then nailed or stapled. But in high wind areas they do have superior holding qualities.

Accessory Shingles

To complete your near perfect roofing job, the manufacturers of asphalt roofing have developed special shingles for the hips and ridges. These shingles can be site made from standard

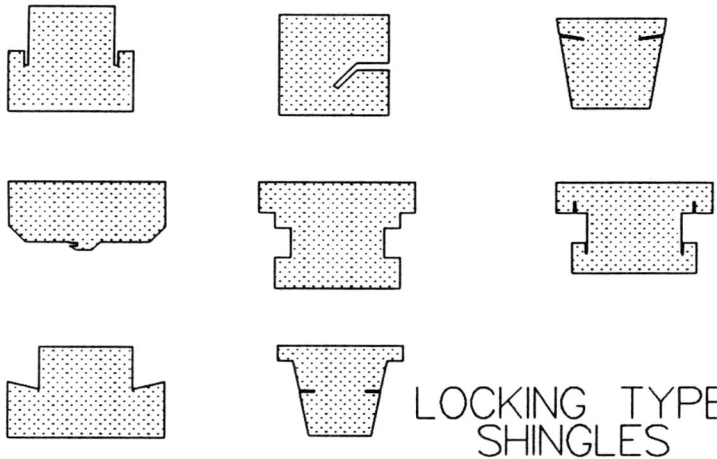

Figure 10–8   Examples of locking type shingles

shingles, but why bother? On-site fabrication takes time and manpower. The precut, pre-shaped shingles need only be nailed in place.

Some manufacturers are adding a fold to their preshaped ridge shingles. This fold is along the weather exposure edge and lifts the front edge of the shingles into the air 3/8 to 1/2 in. This gives the roof's ridge line a better look, according to some customers and installers.

Physical Sizes

At one time, all three-tab strip shingles were a standard 12 in. wide by 36 in. long. Today, this is no longer true. Various manufacturers have entered the market and with their entrance, changed the standards. We now have nonstandard sizes and metric sizes available. See Appendix B, table 1.

Weight Range of Asphalt Shingles

Standard weight was, at one time, 235 lb per roofing square. Today, we have 215, 225, 235, 240, 250, and 300 lb per square shingles. The 215 lb shingles are used on many factory-built homes, and the 300 lb shingle is being used for the high-end housing market.

Due to the weight and size of asphalt shingles, they are sold in bundles. It normally takes three bundles to make a roof square.

Temperature Range for Storage

Storage temperature of asphalt shingles is 40 to 80 degrees Fahrenheit. Maximum on-site storage is not to exceed 2 weeks. Shingles are to be raised above the ground and stored on pallets. They are to be covered with a waterproof canvas cover. Polyethylene tarpaulins are not to be used as protective covers.

The reasons for these storage rules are that shingles that are too cold are subject to damage by handling or moving, and shingles that are too hot will melt together into an unworkable mass. Polyethylene tarpaulins will hold in moisture and wet the shingles, and wet shingles will swell and distort. Waterproof canvas will keep water out and yet allow water vapor under the tarpaulin to be expelled into the air. The 2-week storage is because of the above reasons and the possibility of the shingles being damaged or stolen. The pallets hold the shingles off the moist ground and make transport of the shingles easy via a forklift.

Temperature Range for Installing

The installation temperature range recommended for asphalt shingles is 40 to 80 degrees Fahrenheit. Under 40 degrees, the shingles are brittle and difficult to handle. Over 80 degrees, they are soft and subject to handling damage. Walking on cold shingles will crack the edges of the shingles. Walking on hot shingles will remove or displace the protective granules.

Service Temperature of Asphalt Shingles

The service or installed temperature range is from −100 to +250 degrees Fahrenheit. Properly installed asphalt shingles are not subject to abuse or physical handling damage. The service temperature range is broad for this reason. The flow characteristics of the asphalt type used controls the upper temperature limit, the weather controls the lower useful limit. ONE CAUTION: Very cold weather will make the shingles very brittle and a strong wind can cause damage to the shingles.

Proper Slope Use of Asphalt Shingles

Asphalt shingles are not to be used on roof slopes of under 3:12 without the special permission of your local building inspector. They can be used on all other slopes up to vertical. On slopes over 21:12, they must be sealed with a dab of caulking under the leading edge of each tab.

   The reasons for these rules are that asphalt shingles are not watertight, nor are they fully wind resistant. Asphalt shingles guide water off the roof, they do not prevent water from seeping through to the roof. Use asphalt shingles on a low angle or low sloped roof and water will seep into the underlayment, sheathing, and the building. Shingles used on high angles or slopes will not self-seal. They will be blown off in moderate winds if not sealed under each tab.

   On rakes of buildings located in a high wind area, you should apply a continuous 1 in. wide strip of mastic under the shingles the full length of the rake. Each rake shingle should be cemented to the next from course to course. The alternative to cementing the shingles in place is to install nosing or drip edge over the shingles the full length of the rake.

Warranties and Longevity

Roofing warranties are only as good as the company and the installer who sold and installed the roofing. They may very well be out of business 15 to 40 years from now. Use the warranty as a guide to how long you should expect the shingles to last. Asphalt shingle warranties run from 15 to 40 years in 5-year increments.

   The thicker shingles weigh more and will last longer than thin or lightweight shingles. A 300 lb per roof square shingle should last 30 to 40 years. A 215 lb per square shingle will last about 15 years.

   Read the chapter on problems. Any incorrect installation of the shingles or any of the other problems listed can and will void a warranty.

Getting Materials to the Roof

Figure 10–9 shows a prepared roof being loaded with materials. The fence in front of the property may present a problem. How do you get the materials from the delivery truck to the roof? In this instance, by manual labor—everything is carried up the ladder.

   If it is a single-story building and if the delivery truck can get near the building, the

Figure 10–9 Loading the roof

materials can be handed from the truck to the roof. This will take at least two to three people, one on the truck and two on the roof. Do NOT offload all shingles to one spot on the roof, as the building's framing may not handle the weight. Distribute the bundles across the roof near the ridge if roofing felt has not yet been installed. Distribute the bundles across the roof in piles corresponding to where used if the felt is installed.

Many roofing material suppliers have specially equipped trucks. The trucks have a telescoping conveyer belt that can reach from the street, over fences and lawns, to the roof. This makes life much easier for all.

In tight quarters, you can use ladder lifts. These are crank up or motor driven lifts that attach to the rails on your ladder. Be sure that the lift can safely handle 70 to 100 lbs, the weight of a bundle of shingles. A lift with a minimum of a 300 lb rating is recommended. Forklifts, cranes, and booms can be used, but at a higher delivery cost.

STARTING THE JOB

Before you continue reading this chapter, I advise reading the chapters on flashing and vents, nailing, safety, and estimating. The job starts with good sheathing, proper flashings, good roofing felt and shingles. To skip any of these items will result in a roof that is of inferior construction. In this chapter, we will start our roofing with the felt underlayment.

Felt Underlayment

The recommended and code required weight of felt underlayment for asphalt shingles is 15 lb per roof square felt. You may use heavier felt but not lighter felt. NOTE: Check with your local building department, there are some new polyester base felts that are only 7.5 lb per square. These are authorized for use by some localities.

The reason we use the roofing felt is to protect the roof sheathing from moisture entering under the shingles. A second reason is to prevent moisture from the building getting under the shingles. Vapor barriers are supposed to do this, but they don't always work 100% efficiently.

A layer of felt must be installed at the eaves, rakes, and other flashing areas. There are several new products on the market that are now replacing the standard roofing felt in these areas. The new products are adhesive-backed plastic membranes. These peel-and-stick membranes are recommended in areas of high annual snowfall and rainfall.

What I do not agree with is the use of these membranes without additional roofing felt covering the roof field. Some sales flyers of these products show the membrane installed only at the eaves, rakes, valley and hip areas, but per code this is incorrect. And I have seen buildings in the Midwest roofed without felt applied to the roof field. It may be permissible by local code but it is not advisable according to good roofing practice.

Applying Felt Underlayment

See Fig. 10–10 for the proper overlaps of roof felt when installed under shingles and tile. Nails, staples, or roofing cement can be used for securement.

Application on Surface

The roof surface should be clean of dirt, bird droppings, leaves, and other foreign matter. The surface should be free of knots, cracks, and large voids. The surface must be dry and

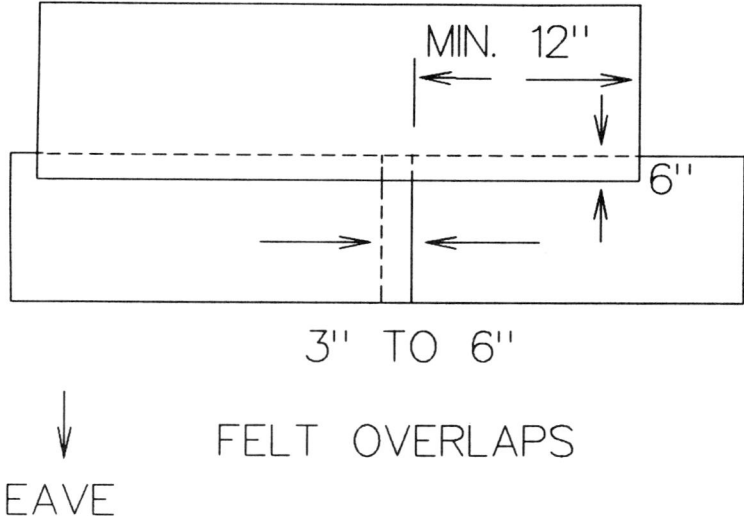

Figure 10–10 Proper felt overlapping

smooth. The roofing felt is to be rolled out horizontally from rake to rake starting at the eave line. The side to side overlap of roof felt sheets is to be a minimum of 4 in. The headlap of course to course is to be a minimum of 2 in. The felt is to lay flat to the roof with no bumps, wrinkles, or twist.

One layer of the proper weight felt is recommended instead of two layers of thinner felt. Using one layer lessens the chance of slipping and wrinkling felt plies. The exception is at the hips, ridges, and valleys. In these locations, a second layer is recommended and generally required.

The heavier weight felts will roll out smoother than the lighter weight felts. If the roof slope is steep, cut the felt into easily manageable lengths. Ten to twelve foot lengths will normally be easy to handle. Valleys should have a full length of felt from eave to ridge installed. Hips and ridges should have a full-length piece of felt if possible.

All felt is to overlap from the top down, over the last felt installed. Do not install two felts with the lower one overlapping the upper one. There are white guide lines printed on the felt; these are for side and head lap locations.

Shingle Coverage Patterns

Shingle coverage patterns vary and can be any pattern that produces a watertight roof. The more common patterns are the 3 in., 4 in., 5 in., and 6 in. patterns. Standard three-tab strip shingles are 36 in. long, with each tab being 12 in. wide. You may install these at a 6 in. course to course offset to obtain vertically aligned tabs in every other course.

The 3, 4, and 5 in. offsets will produce a stair-stepping pattern. The 3 in. offset pattern generates a more rustic look. Use the gauge on the roofing hatchet to aid in pattern generation. Many professional roofers use a 5 in. pattern, claiming it looks the best.

The weather exposure should be 5 in. maximum for all patterns. Smaller weather exposures may be used in high wind areas. The minimum economical weather exposure is 3 in.

The butt edge of the shingles should be in horizontal alignment from rake to rake except if using rustic shingles that have their butt edge trimmed at varying lengths. For best results, align all shingles using the side alignment slits. The hip and ridge shingle's offset and weather exposure patterns should correspond to the pattern used for laying the roof's field shingles.

Starter Strips

See Fig. 10–11. Starter strips are required at the eaves whenever you are shingling a roof with tab type shingles. The surface under the tab cutouts must be protected. You may purchase starter rolls for this purpose. The alternative, and the most common method used, is to use standard shingles laid with the tab cutouts facing *toward* the ridge.

The common mistake made is in using full-width shingles for this purpose. Figure 10–12 shows what happens. The thickness of two full size shingles, the starter and the first course, forms a void at point A. The second course of shingles, when warmed by the sun, settle into this void and their front edge lifts, leaving a gap for water entry. This is prevented

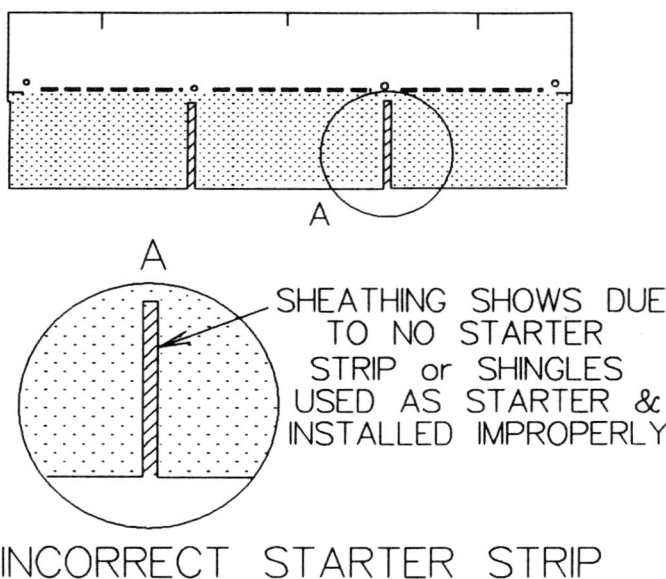

Figure 10–11   Starter strips must be used

by cutting the starter shingle tabs off just above or at the nailing line. This decreases the size of the void and conforms to the rest of the shingles laid.

See Fig. 10–13 for the recommended method of installing shingles over a drip edge. The drip edge is first nailed to the sheathing and then covered with a thin, 1/16 inch, layer of mastic. The starter strip is embedded in the mastic and nailed in place. A second layer of mastic is applied over the starter strip and its mounting nails. The first course of shingles is pressed into the mastic and over the drip edge. This method results in a water-resistant seal that is very resistant to wind lift and shingle blow-off. Apply the mastic as 2 in. long strips

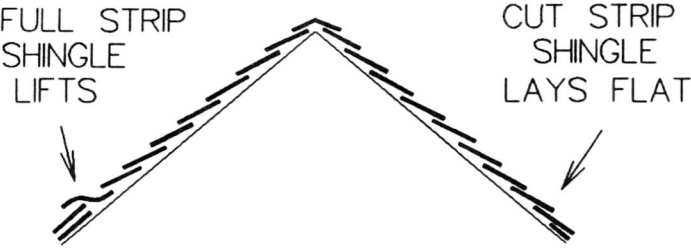

Figure 10–12   Trim shingles used for starter strips

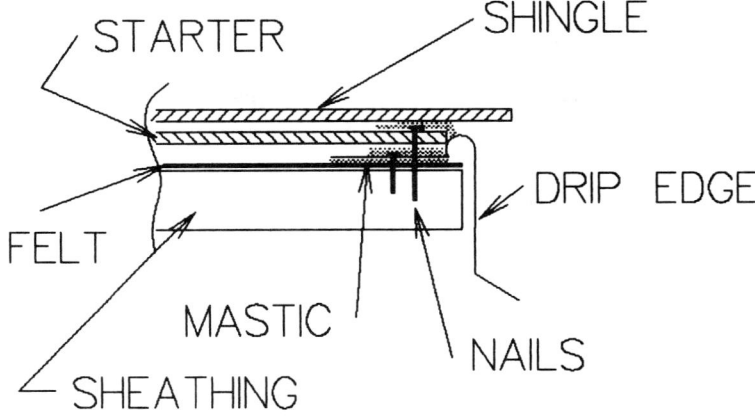

Figure 10–13 Use of a mastic at drip edge

with 1 in. of no mastic between spots. The open space allows moisture escape if any water does happen to get under the shingles.

Mastic Use at Drip Edge

See Fig. 10–14. On low slope roofs, the tray portion of the drip edge should be filled with roofing cement or mastic before the felt or roofing is installed. If not done, and though the roofing overhangs the drip edge, the surface tension between the roofing and water will allow the water to flow under the roofing. By using the mastic, the water lacks a place to enter and, therefore, must follow the surface of the drip edge, dropping harmlessly off the roof.

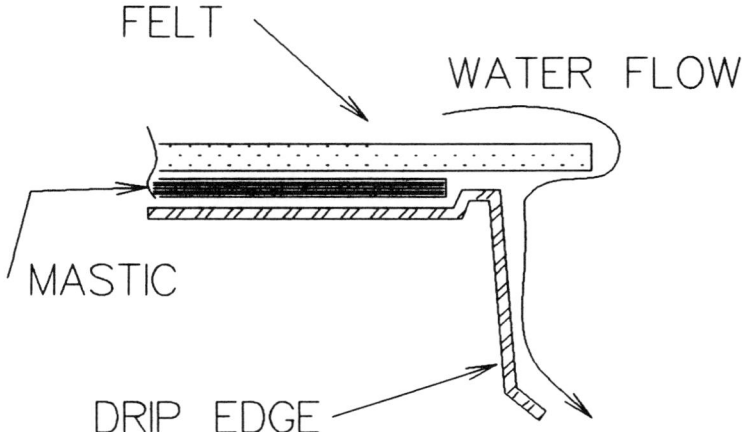

Figure 10–14 Mastic used to keep water from backing up under roofing

The mastic has a secondary benefit of cementing the felt and roofing to the nailed down drip edge. This helps prevent wind from entering under the roofing and lifting it off the roof.

The third benefit of applying mastic is the prevention of rust formation under the roofing when you use galvanized steel drip edging and nails. I highly recommend that all metal used on a roof be primed and painted before the roofing is applied. Yes, it is another step and does cost a few dollars, but the completed roof will last a lot longer.

Shingle Overhang

See Fig. 10–15. Keep the shingle overhang between 1/2 in. and 3/4 in. If the shingle's overhang is too long, the weight of the unsupported shingle material will cause it to sag. As it feels the effects of weather, wind, and sun drying, it will flex and crack. This is not only unsightly, but also produces debris that can clog the gutter drains.

See Fig.10–16. Too short of a shingle overhang is equally incorrect. Water, running off the shingles, contacts the cut edge of the sheathing and is absorbed into the sheathing (Fig. 10–17). This will lead to dry rot of the sheathing or delamination of the sheathing plies. In either instance, a major repair will follow.

Figure 10–18 illustrates the correct amount of shingle overhang on eaves and rakes. The recommended overhangs are 1/2 in. to 3/4 in. at the eaves and 1/2 in. at the rakes.

Correct Shingle Spacing

Figure 10–19 illustrates correctly butted asphalt shingles. In warm weather, above 60 degrees Fahrenheit, the gap between vertical seams should be about 1/16 in. If installed during cool weather, less than 50 degrees, the gap should be about 1/8 in. The reason for the

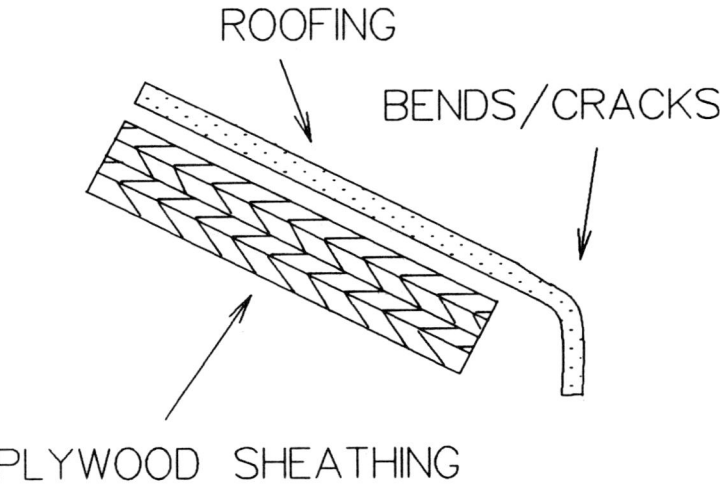

Figure 10–15 Improper length of shingle overhang

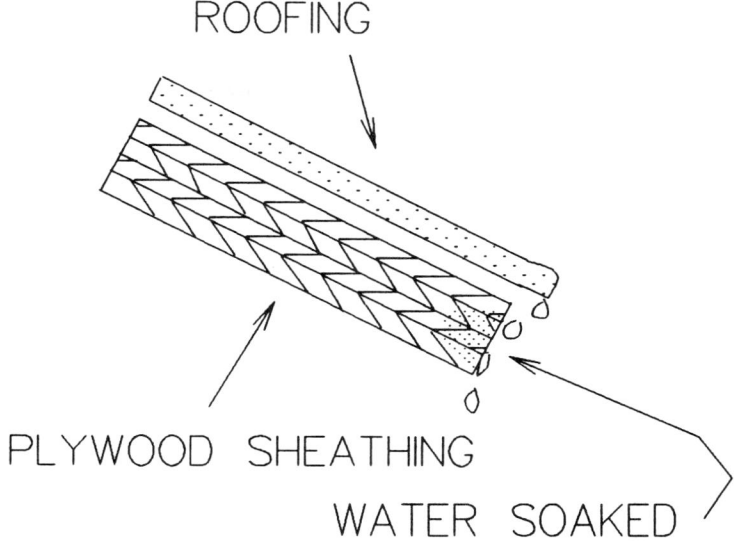

Figure 10–16 Improper length of shingle overhang

larger gap is that the shingles, when cold, are about 1/16 in. smaller than when warm. In very warm weather, above 90 degrees, the shingles will expand about 1/16 in. and the gap will fill to 0 inches. If the gap is not there, the shingles may buckle.

Shingle Layout and Alignment

Look at the rake edge in Fig. 10–20. This stresses the importance of proper planning and layout of shingles before nailing them into place. What went wrong? Was the rake sheathing not cut square? The shingles not aligned vertically? Distance between rake edges varied? Or all the above?

See Figs. 10–21 and 10–22. Shown here are two shingles with a side to side gap that is not even from top to bottom. In Fig. 10–21, the courses of shingles will begin to run

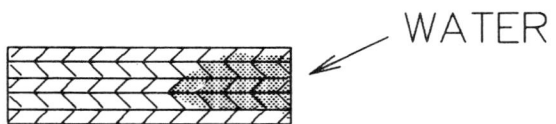

WATER SOAKS INTO EDGE OF
UNPROTECTED PLYWOOD

Figure 10–17 Water soaks into plywood and delaminates it

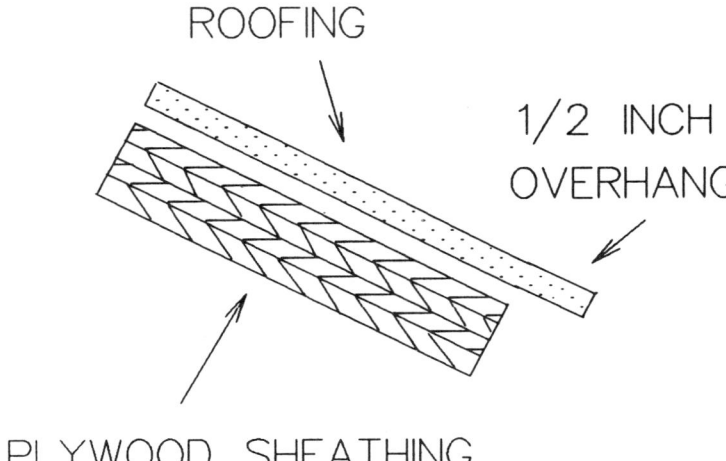

ROOFING

1/2 INCH
OVERHANG

PLYWOOD SHEATHING

Figure 10–18 Proper shingle overhang

uphill from this point. In Fig. 10–22, they will run downhill. Fig. 10–23 illustrates shingles that are not aligned horizontally. If you want random shingling, then purchase shingles made specifically for the purpose. The shake and rustic design asphalt shingles will do the job properly. Standard three- or five-tab shingles will leak if installed in a random fashion.

See Fig. 10–24. The alternate courses of asphalt shingles should line up vertically. If you have to, use a chalk line from the roof peak to the eave as a guideline. Make a visual check from the ground level to be sure.

Poor vertical alignment will not only look shoddy, but may result in leaks (see Fig. 10–25). The tab cutouts and side lap seams may line up too close to one another from one course to another.

See Fig. 10–26. There is an acceptable method of straight up shingling. This is not it. I have seen a few roofing jobs where the side to side seams were vertically aligned on each

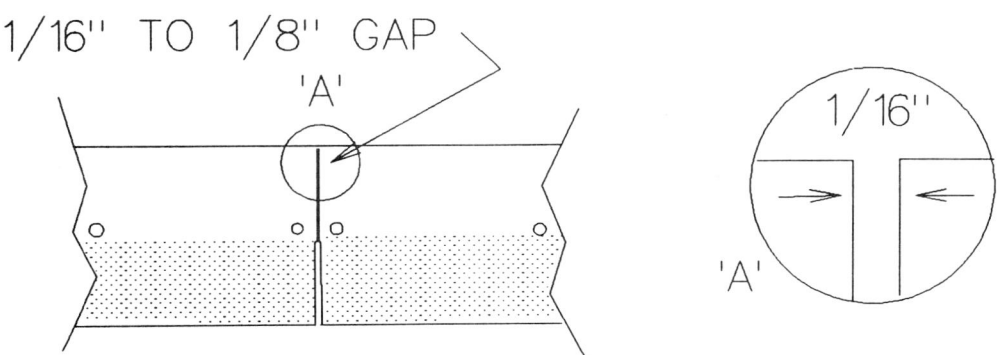

1/16" TO 1/8" GAP

'A'

1/16"

'A'

Figure 10–19 Proper gap between shingles

Figure 10–20 Rake edge shows lack of planning

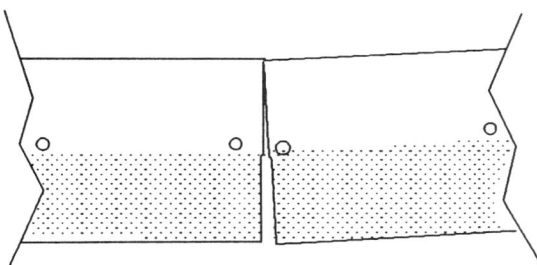

POOR ALIGNMENT

Figure 10–21 Poor side to side butting of shingles

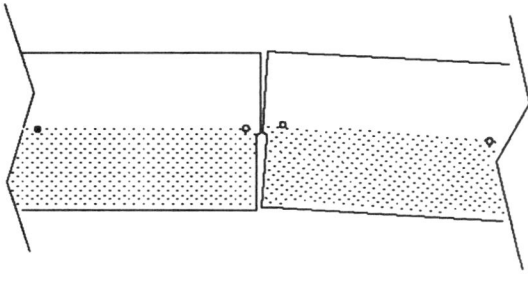

POOR ALIGNMENT

Figure 10–22 Shingles misaligned

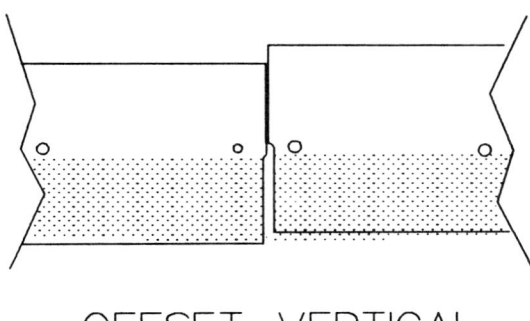

OFFSET VERTICAL

Figure 10–23   Shingles misaligned

course. Not only is this a leak-prone situation, but it looks bad. During cold weather, the shingles contract and leave large gaps from the eaves to the ridge, gaps that allow water and snow to take their toll.

Vertical misalignment happens when the installer does not use vertical chalk lines. It can happen when the vertical chalk lines are not properly plumbed. Most often, it happens when the installer started at the left rake. The rake is angled and not square to the roof's eaves and ridge.

See Fig. 10–27. Keep your chalk lines tight. String has weight, it will sag in the middle. Eyeball your chalk lines to make sure they are straight, that is, go to one end of the chalk line and at roof level, sight down it to the other end. Small variations from straight can then be seen if they exist. If you have a bow, then redo your chalk line before installing the shingles.

No chalk line was used in Fig. 10–28. The bow of the shingles is upside down—what

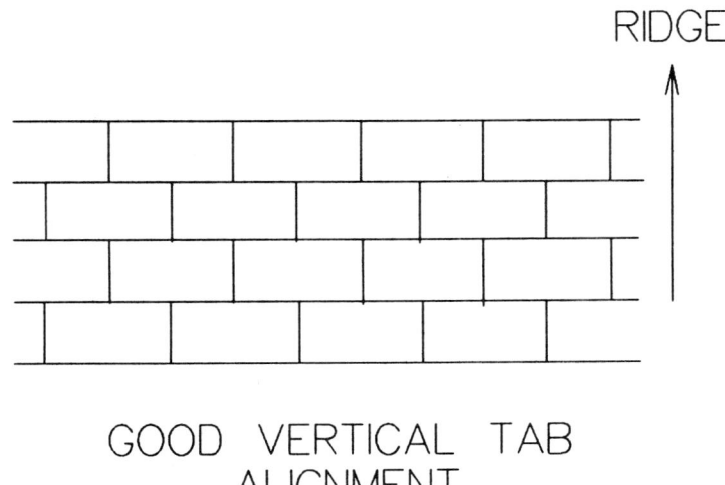

RIDGE

GOOD VERTICAL TAB
ALIGNMENT

Figure 10–24   Good vertical tab alignment

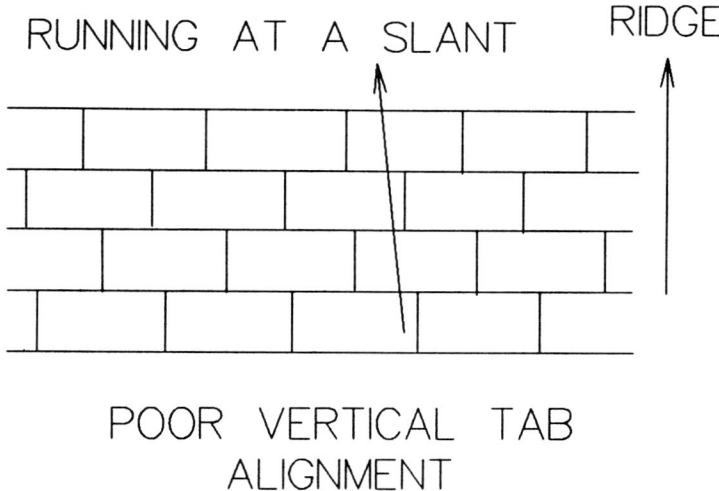

Figure 10–25 Shingles running at a slant

happened? This is what can happen when you use the straight up method and start in the middle of the roof working outwardly to the rakes. Use the chalk line.

See Fig. 10–29. There are a few things that can cause this problem of upward or downward sloping shingles. The first is that installer is not making properly spaced side laps. The second is that the installer did not make proper measurements at the rakes when doing his/her chalk lines. The third, less common problem, is that the roof trussing and

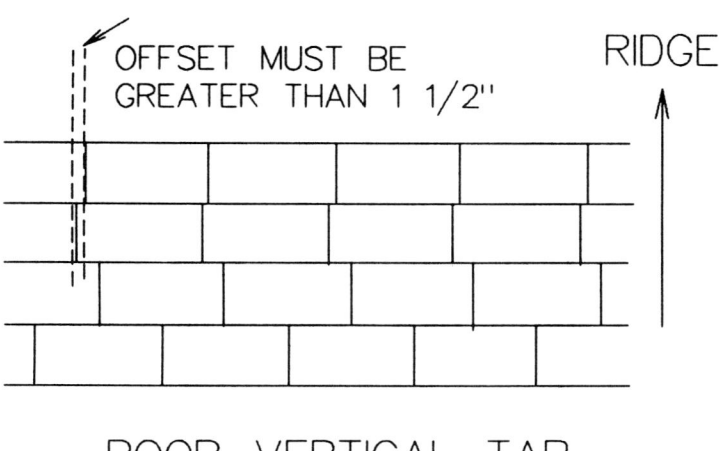

Figure 10–26 Very poor installation

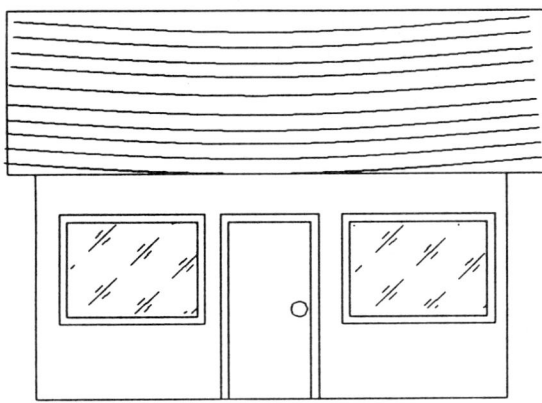

Figure 10–27 Chalk line not tight

sheathing are installed incorrectly. The distance from the eave to the peak is different at each rake. Thus, the starter course of shingles, although straight with the sheathing, is not parallel with the building's lines.

See Fig. 10–30. Good horizontal alignment produces a clean, crisp-looking roof. It is one indication that the installer was a professional.

Ridge Cap Shingles

Ridge cap shingles can be purchased as such or made from standard shingles. Figure 10–31 shows two methods of cutting three-tab shingles into ridge cap shingles. Both methods are correct. The V cut shingle gives a better look and is the suggested method.

When installing the ridge cap shingles, you can start at one end of the building and work across to the opposite end. This is acceptable practice. I personally like to start at the rakes and work in toward the middle. I think it gives a better look to the building.

On dormers or ridges, you apply shingles by starting on one end of the building and continuing across to the other end. This is fine except somewhere you should reverse the

Figure 10–28 Center to rake shingling

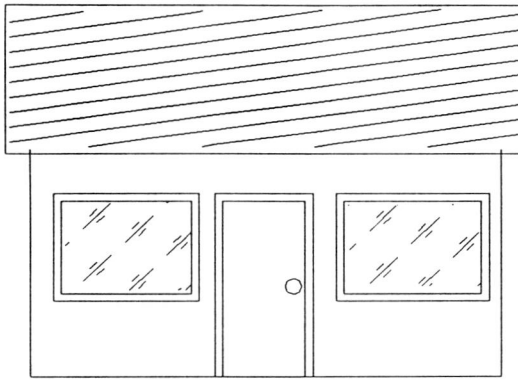

Figure 10–29 Nonuse of chalk line and proper measuring

direction of the shingles in order to end properly. The point where the direction changes must have cap shingle that covers the exposed nails of the shingles under it. Any time you have an exposed nail through a shingle, you have a place for a leak. Apply mastic over these nails. I prefer to cement this cap or joint shingle in place.

The ridge cap shingles should overlap the field shingles on both sides of the roof (Fig. 10–32). For this reason they are installed last. An extra layer of felt should be used for maximum protection. This felt should be a minimum of 6 in. wide and run the length of the building.

Always start from the wayward wind side of the building and work toward the side with the prevailing wind. There will be less chance of the wind getting under and blowing off the shingles. Always use 7/8 in. or longer nails for securing ridge shingles, because you will be nailing through three layers.

Hip Shingles from Tab Shingles

Cut the hip shingles the same way as ridge shingles. See ridge shingles. The same instructions used for installing ridge shingles apply.

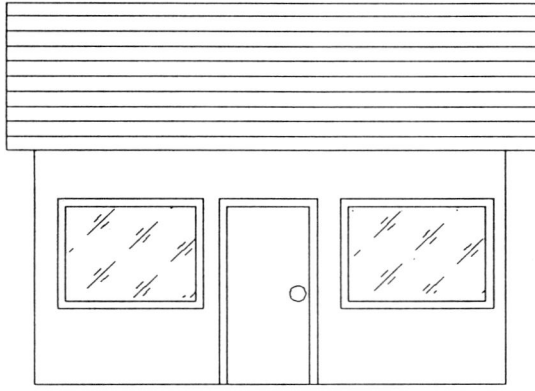

Figure 10–30 Nice horizontal alignment

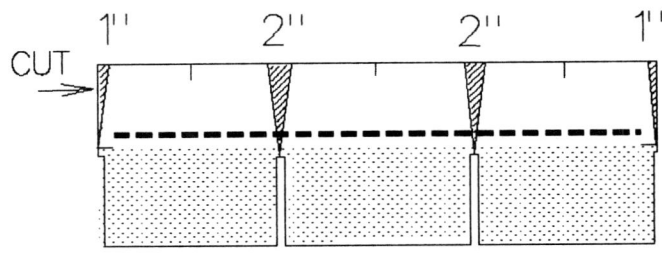

TO MAKE RIDGE CAP SHINGLES
CUT AS SHOWN.   1" ON SIDES
2" WIDE ON 'V'

MAKES THREE

Figure 10–31   Making ridge cap shingles from three-tab shingles

Hip Shingles at Lower Edge

The first hip shingle should be raised with a starter shingle. The second hip shingle is placed over this starter and nailed. Both are to be trimmed to match the eaves of the two intersecting roof lines (Fig. 10–33).

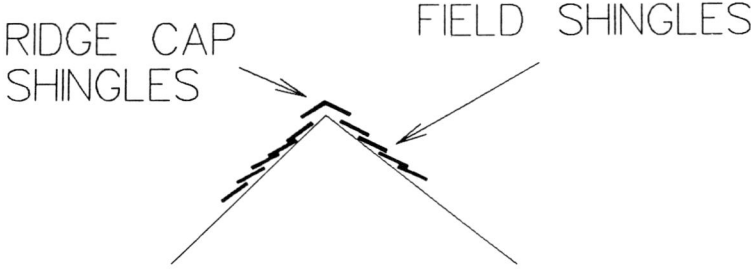

RIDGE CAP OVERLAPS FIELD

Figure 10–32   Ridge cap shingles

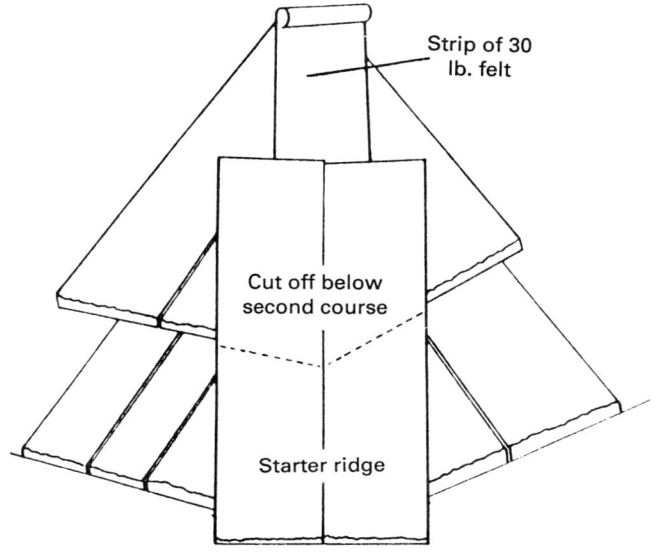

Strip of 30 lb. felt

Cut off below second course

Starter ridge

Figure 10–33 Starter used at hip (Reprinted with permission from *Roofer's Handbook,* William E. Johnson, 1976, Craftsman Book Co.)

Hip Shingles at Upper Ridge

The hip shingles are to be installed before the main ridge shingles are installed. The installation is shown in Fig. 10–34.

SHINGLING METHODS

There are several different shingling methods in current use. Each has its own advantages or disadvantages.

Straight Across Method

Each course of shingles is installed before the next. This method produces a good looking roof. Keeping the shingles in alignment from rake to rake is easy and adjustable from course to course. This method minimizes color noticeable variations of the shingles. The disadvantage of straight across shingling is that the installer is constantly moving and this takes time.

Stair-Step Method

The stair-step method of shingling is the most popular and produces a good quality roof at minimum expense of time and effort. Up to eight shingles are applied to the first course. Then, each additional course receives one less shingle, offset by the shingling pattern used.

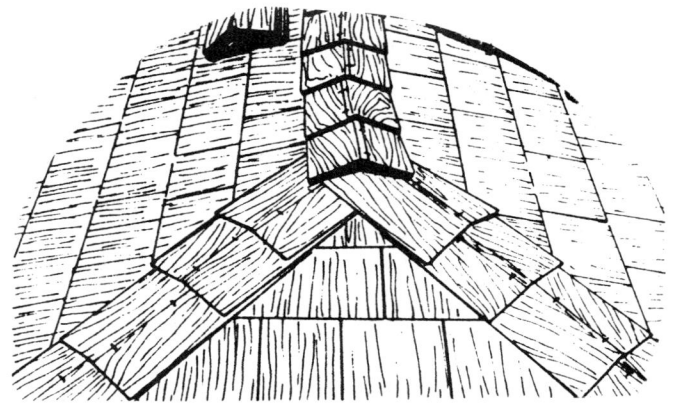

Figure 10–34 Ridge shingle overlaps hip shingles (Reprinted with permission from *Roofer's Handbook,* William E. Johnson, 1976, Craftsman Book Co.)

The shingles will overhang the rake and must be cut. The leftover pieces can be used on the opposite rake.

The maximum number of courses is eight before you must start at the eave line again (see Fig. 10–35). The disadvantages are in keeping the courses in alignment and the possibility of noticeable color variations across the roof.

Straight Up Method

Straight up shingling is used by many roofers for its ease of alignment along the vertical roof plane. The straight up method allows two roofers to work the roof simultaneously, one to the right of center, one to the left of center. The advantage of this method is speed of

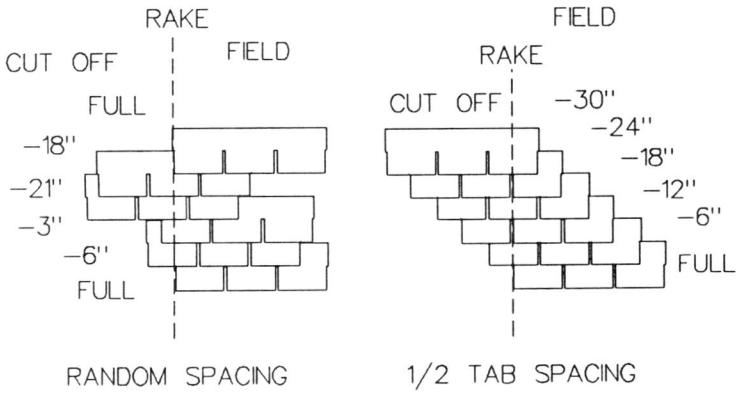

Figure 10–35 Starting it out correctly

installation. The disadvantages are in keeping course to course horizontal alignment and the possibility of noticeable color variation bands from rake to rake.

Thus far, the biggest problem I've seen is that many who use the straight up method do it incorrectly. They place the shingles directly one above the other. There must be an offset of at least one tab (Fig. 10–36). To do otherwise results in roof leaks.

I personally do not like the straight up method for the following reasons:

First, you must be continuously lifting and bending the edges of the already applied shingles. This is so you can nail the next shingle under the prior shingle's tabs.

Second, there is a potential shingle color problem. Shingles are manufactured in batches. When you purchase them for a job, you purchase all from the same batch number. From experience, I find that even with the same batch numbers, there can be differences in color. Installing shingles conventionally, in horizontal rows, usually will mask this color difference. Installing shingles straight up emphasizes the color difference. You will see blocks of different color shades on the roof.

The third reason I do not like the straight up method is because of the side to side spacing. When you have to lift and slide a shingle under another, then hold both and nail, there is a good possibility of shingle movement. Getting a good butt joint is difficult. Of 50 straight up shingle jobs I checked in 6 different localities, only 2 were properly installed. The rest showed vertical displacement of shingles. Some had only one or two vertical rows out of alignment, others all rows. Just by looking from the ground, you could see where each bundle began and ended.

The Skylight

Figure 10–37 shows a good skylight installation. See the chapter on flashing and vents for typical installation instructions.

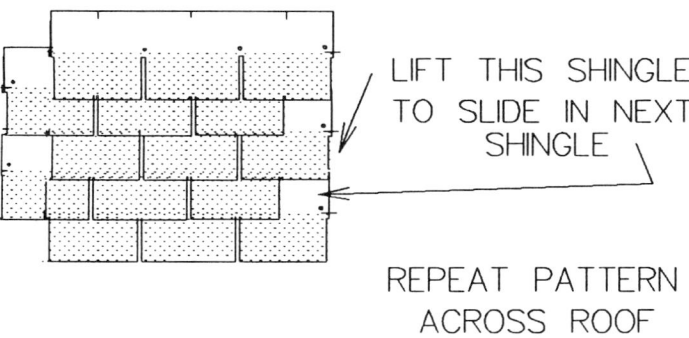

STRAIGHT UP SHINGLING

LIFT THIS SHINGLE TO SLIDE IN NEXT SHINGLE

REPEAT PATTERN ACROSS ROOF

Figure 10–36 Straight up method of installing shingles

Figure 10–37 Skylights

The Vents

See the chapter on flashing and vents.

The Full-Lace Valley

A full lace is where the entire valley is covered with shingles interwoven together. Full lace should not be used when the following conditions are present. One roof pitch is 4:12 and the opposing roof pitch is greater than 7:12, or if the distance from the eave to the ridge is different for both roofs. These conditions make it unlikely that the installer can obtain a leak-free lace since he is dealing with a different number of rows of shingles. Besides, it will not give a good professional appearance.

Full-lace valleys are somewhat easy to install since there is a minimum of cutting of shingles. The shingles in the valley should all be full-length shingles. If you find partial shingles installed, suspect a leak. The drawback to full-lace valleys is that finding a leak and repairing it is difficult.

The Half-Lace Valley

A half-lace valley is one in which one roof surface has its shingles brought to and through the valley onto the other roof surface. The second roof surface has its shingles brought to within 3 in. of the middle of the valley and then trimmed off at that point.

A half-lace valley can be used on roofs of different slopes or eave to ridge lengths with few problems. What you are looking for is the placement of the shingles in the valley. If both roof slopes are the same and both roof ridges are even, then it does not matter on which side the shingles are cut. If one roof ridge is higher than the other or one roof has a greater slope, then the cut shingles should be those of the higher sloped roof. The reason is this: Water will be running off both roof surfaces into the valley. The roof with the longer eve to ridge length will collect the most water. Water has weight and thus this side will be spilling a greater weight and volume of water into the valley. The steeper roof will be a problem since the water will be flowing down it at a greater force than the water flow of the lower roof. This force of the water entering the valley is somewhat counteracted by the force of the water coming off the lower roof. But, since its force is greater, it will have a tendency to travel up the lower roof for some distance. If the lower roof has the cut shingles, then this force of water will lift them and water will enter onto the sheathing under them.

Both full-lace and half-lace valleys can be made using strip or tab type shingles. Do *not* use locking shingles such as T lock or hex lock shingles for this application, because leaks will occur.

Open Face Smooth Valley

See chapter on flashing for details. An open face, smooth valley may not be to current code in your area. Check with your local building inspection department.

W-Flashing Valley

See chapter on flashing for details. The splash diverter used must be 1 in. high on all except metal roofs, in order to meet code.

RESHINGLING A ROOF

When installing asphalt shingles over an existing roof, you must take some care in how the shingles are laid. A common mistake is to lay the new shingles over the old, as in Fig. 10–38 (left). What happens is that over time the shingles lose their stiffness and attempt to conform to the underling roof shape (Fig. 10–38, right). In areas subject to children throwing things or areas subject to hail storms, the new roof may be severely damaged. Repairmen walking on the new roof can damage it. The reason for the damage is that the voids present under the shingles. Asphalt shingles need a good, solid supporting surface on which to rest. Without this support, hail, rocks thrown, and people walking on the roof will damage it.

See Fig. 10–39. The proper method of reshingling a roof is to use the butt-up method of shingling. The first course of shingles must be trimmed to fit properly, the field shingles do not have to be cut. This reshingling method gives the new shingles a good base to rest against. You should fill all voids in the old roofing before installing the new roof. All old loose or curled shingles should be nailed down flat. Water leaks should be repaired, rusted flashing cleaned and painted or replaced. One additional reroofing can be applied this way,

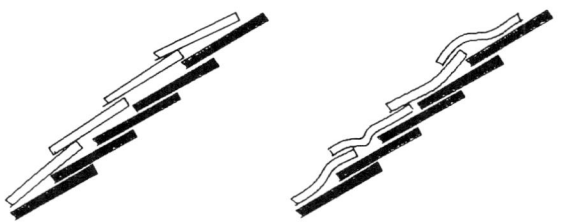

ASPHALT SHINGLES CONFORM TO
THE UNDERLYING SURFACE

Figure 10–38   Improper method of installing second roof

to a maximum of a three-layered roof. If you exceed three roofs, you will be in violation of the building code.

Tie-in to an Existing Roof

Figure 10–40 illustrates someone's idea of a good roof to roof tie-in—just lay the new shingles on top of the old! This is not correct. Shingles at a tie-in should be laced together the same as if you are installing a normal field of shingles. To do this at a tie-in, you must

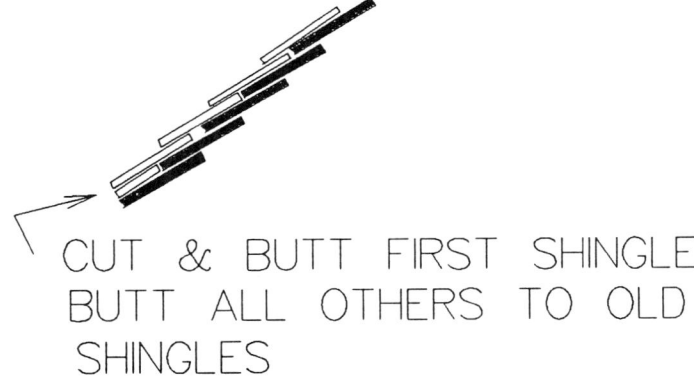

Figure 10–39   Proper method of installing second roof

Figure 10–40   Improper tie-in

pull or cut the nails under the tabs of the existing shingles. A pry bar is used for this purpose. Once the nails are removed, the new shingle can be side butted into place and the shingles renailed. You can do the same when replacing missing shingles (Fig. 10–41).

Clean Up After Yourself

During the roofing process, you will have an amount of trash left over. These can be pieces of cut shingle, the shingle packaging, the protective plastic strips from the seal downs, etc. These items frequently end up all over the yard and the neighborhood. Pick them up and dispose of them properly. Your roof job may be perfect, but if you leave trash at the site, it is a sign that you are a sloppy worker. The customer will suspect that you did a sloppy roofing job and you may get nitpicking complaints.

Be sure that during the installation, items are not being blown all over the neighborhood. It is bad customer relations and will prevent you from getting other roofing jobs in that neighborhood.

Be sure to account for all the nails and sharp pieces of metal you drop. If one person gets a cut or a nail through his or her foot, you will have a lawsuit on your hands. The law allows for an injured person to sue for medical damages and up to three times the actual losses. If a person gets injured and is out of work for 2 weeks, he or she may lose 2 weeks' salary. You, if proven to have caused the injury, will be liable for up to 6 weeks of the person's salary. If proven irresponsible, you may have to pay other punitive damages that can run into many thousands of dollars. If the person is permanently injured, the damage awards can run into the millions. CLEAN UP AFTER YOURSELF.

Figure 10–41 Missing shingles

MISCELLANEOUS POINTS

Trimming and Cutting Methods

To cut asphalt shingles, turn the shingle over and cut from the asphalt side. Do not attempt to cut from the granule side. A utility knife or a roofer's knife will work; keep the blades sharp and use a straight edge as a guide. The first cut should be a score mark. Then bend the shingle slightly and make the cut deeper. Now, bend the shingle sharply and it should break off clean.

Batch Cutting

You can batch cut roll roofing or rolled roofing felt with a circular saw equipped with a carbide tipped blade. You will wear out the blade every three to five rolls, but it does save time. Eye protection is essential. Secure the rolls while cutting. To keep the roll from moving while sawing, place a 2 by 4 under each side of the roll.

Painting the Sheathing

See Fig. 10–42. On slopes greater than 4/12 and in rather dry climates, i.e., in desert areas, it is acceptable to omit a drip edge. The shingles become the drip edge. But, during heavy rains, there may be some traveling water, traveling in that it flows around the drip edge and

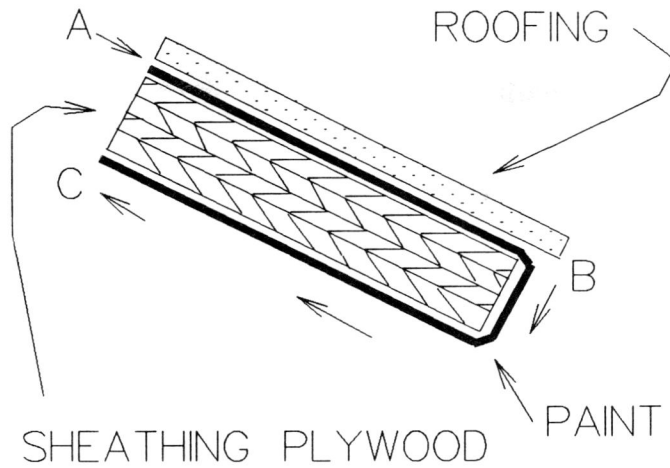

Figure 10–42   Painting will help prevent water absorption into sheating

back to the sheathing. It is for this reason that I recommend that you paint the sheathing or otherwise seal it when not using a drip edge.

Paint is an effective barrier to moisture. Paint from point A, about 1.5 to 2 ft from the edge, to point C, the wall. Give the cut side, point B, two or more coats. All the fascia and sheathing edges should be painted before the roofing and siding are installed. Since this can be accomplished with an airless paint sprayer or roller, it takes less than an hour. The siding and roofing will cover any over spray.

Safety on the Job

Here is a list of safety violations that I feel are commonly ignored:

- No safety hard hats
- No scaffolding
- Improper use of a ladder
- Throwing tools to each other
- No use of a rope lifted, tool carry bucket
- Improper use of frayed and worn out rope
- No pedestrian-covered walkway (An 8 ft high covered walkway is required by code whenever working above a public walk.)
- No pedestrian warning signs
- No safety rails on the so-called scaffold
- No solid decking on the so-called scaffold
- Lack of adequate securement for tools and paint cans

Flat and Built-up Roofs

INTRODUCTION

Flat roofs, I think, must be one of the worst types of roofs in the business. Not because they are difficult to install, but because they do not hold up as well as other types of roofs. Being flat, under a 3:12 slope, they generally do not drain water very well. Any little dips or irregularities in the surface flatness will result in a puddle. Being flat, they also have a tendency to collect large volumes of leaves, dead branches, and bird excretions. The sun bears down on them all day and melts the asphalt and its binding tars. Peaked roofs have few of these problems that tend to wear out a flat roof quickly.

Flat roofs are more subject to rot, blistering, splitting, and general deterioration than any other. To you, the roofer, they are a money-making dream, because they need replacing or recovering more often. The work is easy, as you are working on a level surface rather than a slope. The materials used to recoat a flat roof are inexpensive compared to other roofing materials. So why are there so few roofers doing flat roofs?

The reason is that you have the equipment cost for a heated tar kettle, gas cylinders, roofing torches/rollers, and lifting equipment. But all this should not exceed $10,000. Yes, this is more than a roofer's hatchet and some nails, but then again not much for a good paying business. The reason many of you don't do flat roofs is you haven't learned how. This chapter will help you to learn the proper methods used by the professionals.

WHERE TO USE A BUILT-UP OR FLAT ROOF

You will find roofs on residential buildings, industrial buildings, and commercial buildings. In the residential sector, many homes are built with too much square footage. The practicality of attempting to frame out a peaked roof just isn't there. Also, there are height restrictions, usually 16 to 18 ft from ground level. Rather than build a single-story house with a peaked roof, the developer decides to build two-story homes with flat roofs. Then, there are certain designs such as the lean-to style home that require a near flat roof.

FLAT ROOF DESIGNS

Here are a few designs of some residential flat roofs. As we proceed through this chapter, you will see that these are simple compared to the commercial built-up roofs.

Figure 11–1 illustrates a white gravel covered built-up roof with entry sun screen. I do not recommend this design for use in northern states. Rain and snow will cover the entryway frequently. In Southern California, rain is an occasional inconvenience, but this type entry does add a little style to an otherwise dull roof.

Figure 11–2 shows eave detail of a flat roof with an overhang. The blocking is used to keep the joists from twisting. The flat, horizontal, blocking is used as a nailer for the interior ceiling drywall.

Materials Used

In working on flat roofs, you will be using techniques and materials that are somewhat different from other roofing. Instead of individual roofing units, shingles, you will be work-

Figure 11–1 Gravel roofing

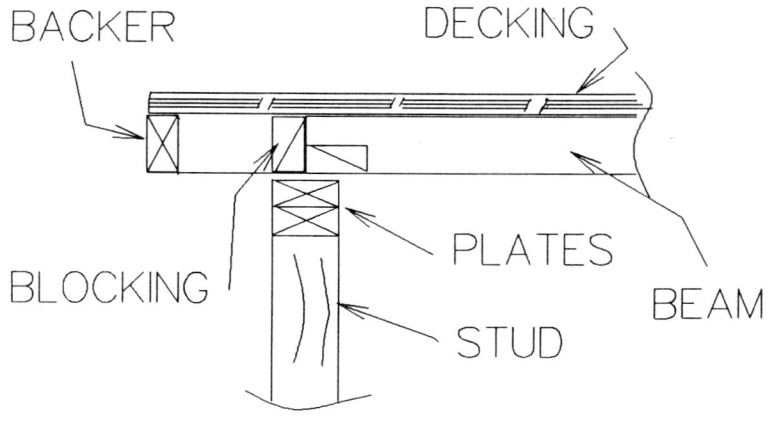

FLAT ROOF DETAILS

Figure 11–2   Flat deck construction

ing with rolls of webbing, matting, and roofing material. You will be working with hot asphalt or tar. You will be working on surfaces that may not be wood, but rather steel or concrete, and the fastener methods are different. The roofing application isn't by nailing but rather by heat rolling. Here are some materials you will use.

To bind the roof into a seamless unistructure, you lay down a coating of binder and then a layer of matting or webbing material. This matting material can be fiber glass, polypropolene, polyethylene, or other thermoplastics. The binder is asphalt or coal tar.

Fiber glass is just that, glass that has been spun into very thin fibers. These fibers are then woven into threads and then further woven into mesh cloth, or placed at random into mats. Fiber glass is rot- and insect-proof, fire-resistant, moisture-resistant, and dimensionally stable.

Polypropolene and polyethylene are thermoplastics manufactured from petroleum. They are highly resistant to rot, insects, moisture, and chemicals. They are, though, low-temperature plastics and begin to melt at 140 degrees Centigrade. They are dimensionally stable at the temperature ranges most roofs must endure. Both materials are inexpensive and can be spun and woven into cloth, mesh, and mats.

Tar and asphalt products tend to contract and stiffen with cold and melt and flow with heat. This makes these products susceptible to cracking and blistering. Woven mesh and mats are used to provide a certain degree of dimensional stability that often keeps the cracking and blistering to a minimum. A layer of binder, tar, or asphalt, is applied to the roof, the mats layered on it, and then pressed, rolled down. A second coating of binder impregnates the mats and forms the binder for additional layers of matting or roofing.

You can purchase *bitumus mats*. These are mats that already have been impregnated with tar, a bitumus. Bitumus means made from coal or a petroleum base, hydrocarbon. The advantage of the bitumus mats is the elimination of mopping on hot liquid tar. Bitumus mats require a special roller that heats the mat to the proper application temperature as it is rolled onto the roof surface.

Our top coat, weather exposure coat, is asphalt roll roofing or a layer of aggregate. Roll roofing is much the same as an asphalt shingle except that it is a continuous strip 36 in. to 39-1/2 in. wide by approximately 33 ft long. Weights are from 45 lb to 110 lb per roll. Coverage is one half to one full roof square per roll. Size, weight, and coverage will vary with manufacturer and product. Roll roofing may contain a layer of fiber glass or polyester embedded in it. It usually contains a top layer of rock or mineral granules.

The top layer does several things:

- It helps protect the bottom layers from the weather.
- It provides fire resistance from airborne sparks that may land on the roof.
- It acts as a reflective layer to keep heat from entering the building.
- It gives texture and color.
- It provides wear resistance to abrasion from people and animals walking on it or things being thrown or dropped on it.

Asphalts

Asphalt is a byproduct of petroleum distillation. It is a thick brown to black hydrocarbon tar. To make asphalt suitable for roofing, it is modified with plastics and air. Air blown through the asphalt mix loosens it up and allows it to flow better.

See Fig. 11–3. Asphalts are rated by the temperature at which they will no longer remain in place. Type IV or steep asphalt is just that, asphalt that can be used on a steep surface. Its melting point is higher than regular asphalt. See Appendix B, table 20.

If a low melting point asphalt were used on a steep roof, the sun's heat would make it flow and it would slide off the roof. We do not use steep asphalt for flat or low slope roofs. On these, we want a low melting point asphalt. Using a low melting point asphalt makes the roofs self-healing. Any small cracks that develop will melt back together when heated by the sun (Fig. 11–4).

Aggregate

See Fig. 11–5. On low slope built-up roofs, we use aggregate as the final coating. Aggregate is small stones or crushed tile or rock. See Appendix B, table 22.

The aggregate is applied while the top coating of the asphalt is still hot. The aggregate holds the hot asphalt in place. When the asphalt cools, it holds the aggregate in place. Normal application is 400 lbs gravel or 300 lbs slag embedded in a flood coat of 40 to 60 lbs of asphalt. Aggregate size is to be from 1/4 to 5/8 in. when applied to a built-up roof, per ASTM D1863.

Aggregate also reflects or absorbs U/V radiation from the sun. U/V radiation is very

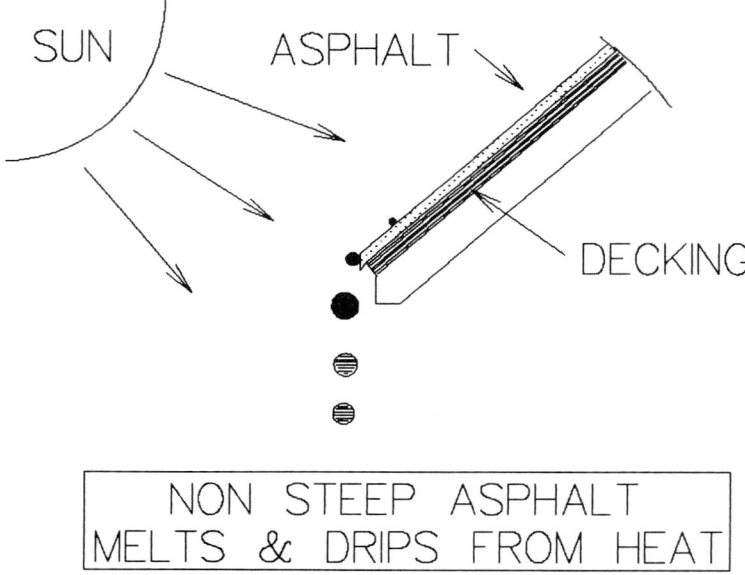

Figure 11–3   Wrong type of asphalt will melt and run off roof

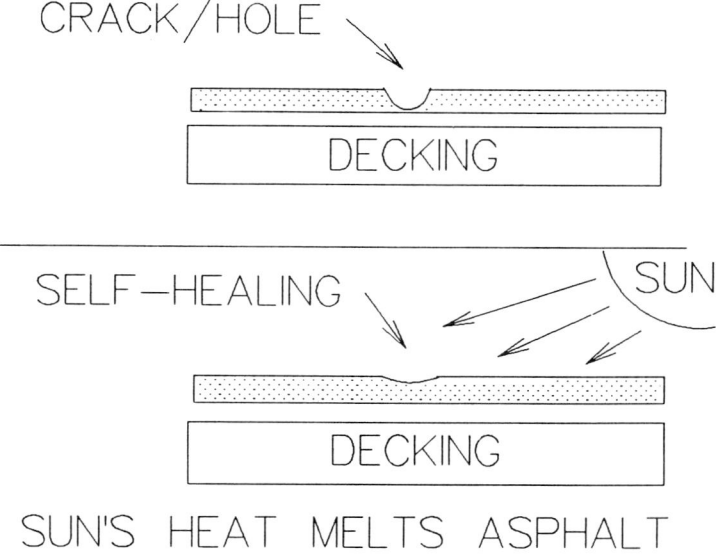

Figure 11–4   Some asphalts are self-healing

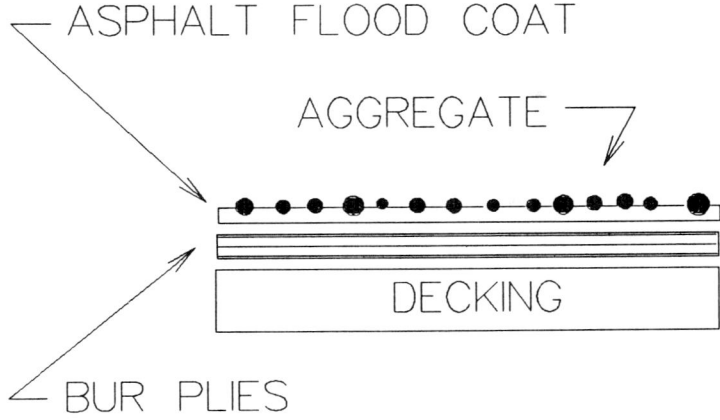

Figure 11–5 Typical built-up roof construction

destructive to the asphalt and roofing felts. U/V radiation often changes the chemical makeup of the materials to the point that they no longer afford protection to the building.

Aggregate does act as a reflector of heat in summer and a collector of heat in winter, which helps to stabilize the roofing materials. It helps maintain constant temperature within the building and lowers the heating and cooling cost of the building.

See Fig. 11–6. Aggregate must be washed clean of loose dirt and dust before being applied to a roof. Asphalt will not stick to dirt and dust for any long period. Dust covered aggregate will come loose and be washed or blown from the roof by wind and rain. The now exposed asphalt will deteriorate at an increased rate and the roof will leak.

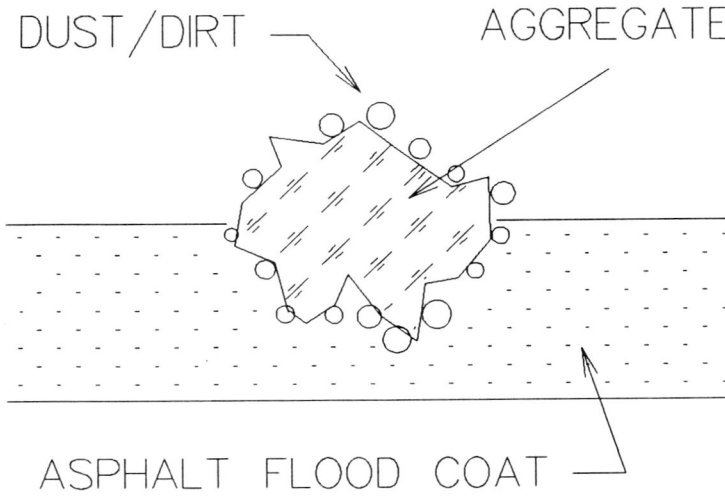

Figure 11–6 Dirt prevents proper adhesion of aggregate

See Fig. 11–7. Aggregate must be opaque, it is not to be translucent or transparent. Opaque objects block passage of light, translucent objects pass some light. Transparent objects allow all light to pass through them. If the aggregate used is translucent or worse, transparent, it will allow U/V radiation to pass through it to the asphalt. The asphalt will be destroyed, the aggregate to asphalt bond will fail, and the roof will fail.

Elastomer Coatings

Many built-up roofs are top coated with elastomers. These are sheet or liquid synthetic rubbers that self-heal. Small cracks that develop in the material disappear when heated by the sun. The sun's heat will have melted the material to the flow point. The material becomes liquid, flows into the cracks, and solidifies in the night air.

See Fig. 11–8. Liquid roofing is becoming popular due to its ease of application, good protection characteristics, and low cost. Liquid elastomers are brushed, rolled, or sprayed on to the roof decking in layers. Each elastomer layer is selected for a particular property; color, self-healing, elasticity, etc. Here are some liquid elastomer coatings currently in use.

- Chloro sulphinated polyethylene (Hapalon ™)
- Chloroprene (Neoprene ™)
- Polysulfide
- Polyurethane
- Silicon
- Vinyl

The fastest application method for applying liquid roofing is with an airless spray outfit. By adjusting the viscosity, rate of flow of the elastomer, you can apply liquid roofing to roofs ranging from flat to steeply sloped. Viscosity is adjusted by the addition of solvents and thinners.

As the elastomers cure, they form a continuous one-piece membrane over the entire roof surface. This membrane is thin and can be damaged by abrasion, such as people walking on it. Walkways should be provided if they are necessary. Prefabricated walkways can be purchased or one can be fabricated on site from mineral coated roll roofing. I prefer the purchased ones, they look, work, and last longer.

The thinness of the membrane coating is what keeps it from cracking and splitting. Thin materials expand and contract much easier than thick or multilayered materials. The thin membrane will closely follow the movements of the decking on which it is installed.

There is one major problem to the use of liquid roof material. It will crack if the deck on which it is installed cracks. All joints in the deck must be sealed with an elastomer sealer before the liquid roofing is applied. This keeps the liquid from seeping through the deck before it has cured. The sealer provides a base under the cured liquid that expands and contracts without opening up a void. See chapter on caulks and sealers.

Liquid elastomers can be colored to suit most customers. CAUTION: The use of Hapalon over Neoprene may discolor the Hapalon. The Neoprene layer(s) should be al-

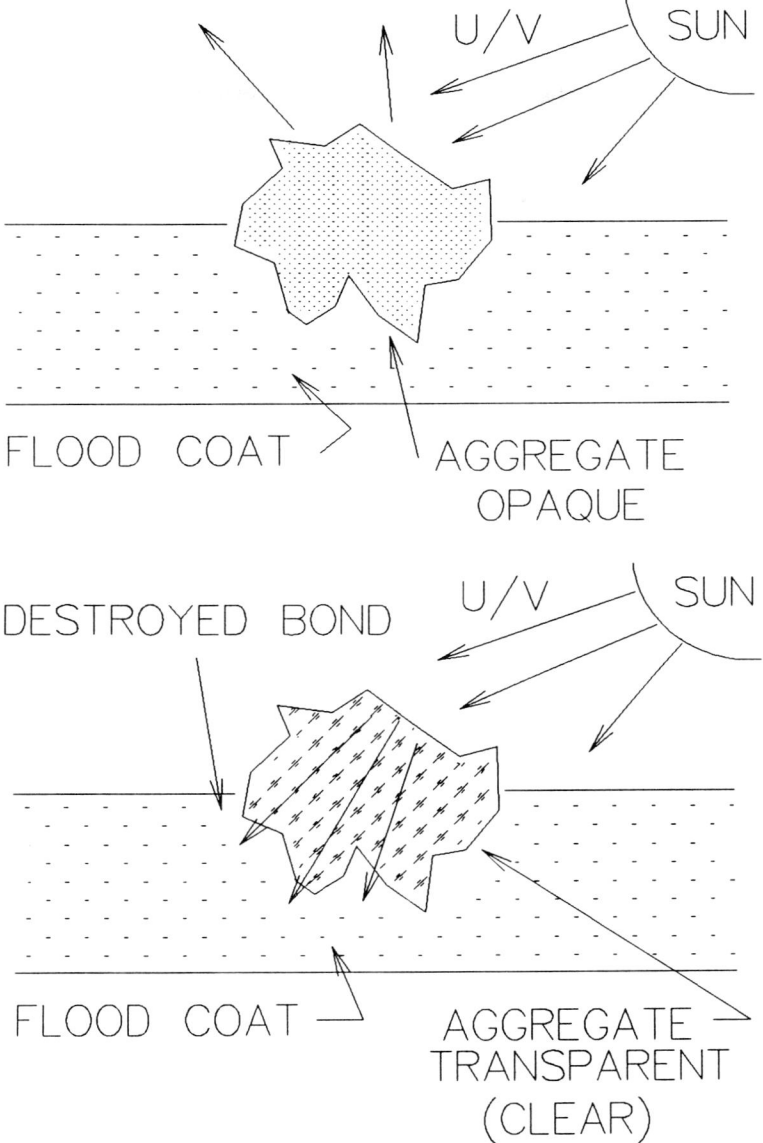

Figure 11–7   Opaque aggregate reflects heat and U/V. Transparent aggregate should not be used.

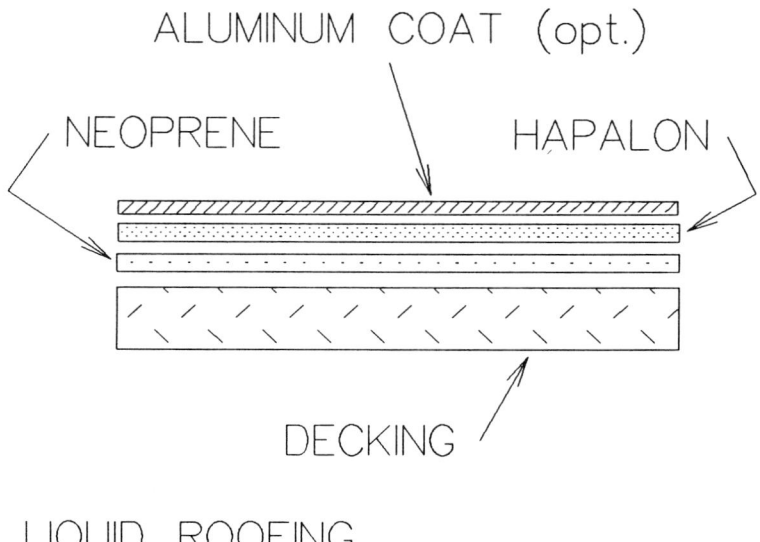

Figure 11–8   Liquid roofing

lowed to cure as long as possible before applying the final Hapalon layer. Two coats of Hapalon may be required for color consistency. Always use products with the same batch number for color consistency across the roof surface.

Liquid elastomers may be stiffened or extended with fillers. Common fillers are sand, cork, and ground walnut shells. Fillers added to liquid roofing will hold the liquid on the higher sloped roofs. Some liquid elastomers require the use of a surface primer before the elastomer is applied; consult the manufacturer's specifications.

Elastomer Membrane Roofing

Membrane roofing is considered to be one-ply roofing. The thin elastomeric roofing membrane is attached to the roof deck with glues, asphalts, or other adhesives. Consult product manufacturer for their recommended installation instructions.

The materials used are:

- Bituthene   0.062 in. thick
- Butyl (isbutylene-isoprene)   0.031 to 0.062 in. thick
- EPDM (ethylene propylene diene monomer)   0.045 in. thick
- Hapalon   0.045 in. thick
- Neoprene (chloroprene)   0.062 in. thick
- Silicon rubber   0.025 in. thick

Figure 11–9 illustrates membrane roofing installed on an industrial building. Note the discoloration on the side of the building. This is from the roof drain. Separate downspouts will prevent this from happening.

Membrane or Liquid Elastomers on Concrete

Membrane and liquid membrane roofing seals the entire roof with a layer of rubber or plastic. This covering is impervious to water and will hold water on or under it. New concrete or gypsum has an abundance of water that must be dissipated into the air. Cover the concrete or gypsum with a membrane coating and the water will not escape. What will happen is that on a hot sunny day the trapped water will turn to steam and will expand and blister the roof coating.

Membrane coatings will work on aged, dry concrete and gypsum. Consult the product manufacturers for their recommendations. Some will say it should not be done, some will say it should, and some will recommend a method of application that will work.

BUILT-UP ROOFS

Roof Felt

Built-up roofing installation is done with the use of asphalt impregnated roof felts and with hot liquid asphalt. A base sheet is nailed or spot mopped to the roof deck and then covered with hot asphalt. A felt is then added and it too is covered with hot asphalt. This process repeats until three to five layers of felt have been applied. The last felt receives a flood coat of hot asphalt, and aggregate is then embedded into the flood coat. Here are some rules for installing roof felt.

Figure 11–9   Membrane roofing

See Fig. 11–10. Roofing felt, when used for intermediate layers on a built-up roof, should be completely coated with asphalt. This coating is called solid mopped and the asphalt is applied at a rate of 20 lb per roof square. If used, coal tar pitch is to be applied at a rate of 30 lb per roof square. Be completely sure that you cover the felts with asphalt, as felt upon felt will not form a solid bond. The uncoated felt areas will slip on each other and cause delamination of the roof. Delamination is one cause of blisters, splits, and eventually leaks.

See Figure 11–11. The edges of the felts must not fall directly in line over prior applied felts, as this makes the felt joints two layers thick. Joints directly above each other will be four, six, eight, or ten layers thick depending on the number of roofing plies installed. This extra thickness will show up as a bump in the roof. It can prevent the proper drainage of water from the roof. Additionally, the expansion and contraction stress will be concentrated at the joints. This will result in eventual cracking and splitting of the roof plies, and water leaks will result.

Built-up roofs, with spans over 20 ft, should not have their first felt layer solid mopped. The expansion and contraction of the roof deck will pull the roofing apart, the felt plies will split, and leaks will result. The proper application of asphalt to the deck is by spot mopping. This attaches the first layer of felt to the deck while still allowing for different rates of roof and deck expansion. The roofing floats on these asphalt puddles, the spots. Spot mopping is done at a rate of 10 lb of asphalt per roof square. Each 1-1/2 ft square area of the roof should have a dabbed-on spot of heated asphalt.

See Fig. 11–12. On very large built-up roofs, spot mopping of the first layer of felt may not prevent splitting of the roof. The accepted roofing practice on large roofs is to use expansion joints. In essence, you are installing several smaller roofing sections, sections

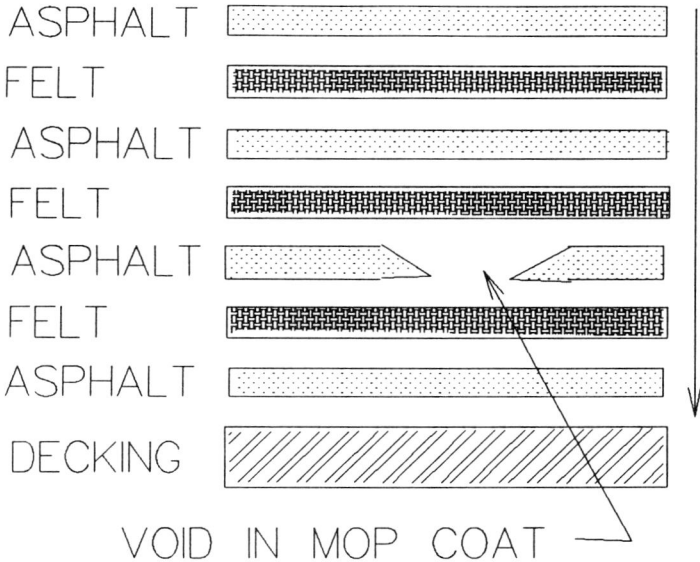

Figure 11–10   Void in coating will lead to blister

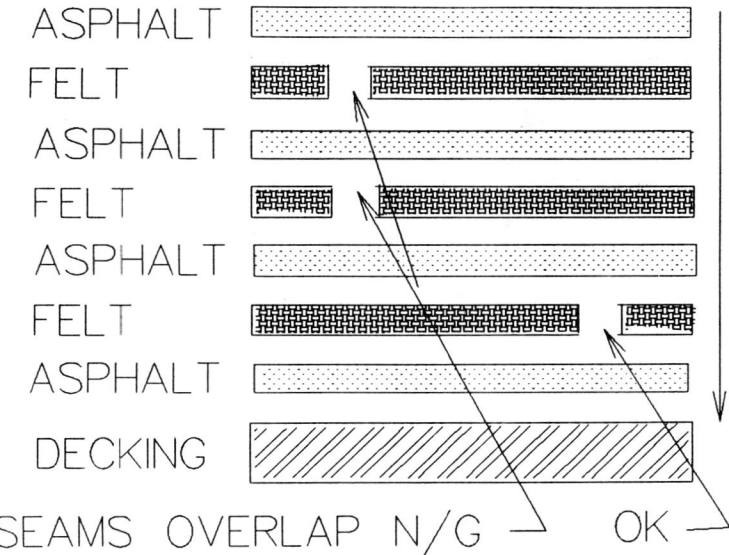

Figure 11–11 Overlapping seams will lead to splitting

that can expand and contract independently of the other sections. Expansion joints must be used every 200 to 300 ft along the length of the roof. Here are the recommendations of several manufacturers. Use an expansion joint at:

- Every 200 to 300 linear feet of roof.
- Wherever the decking material changes.
- Wherever the decking material changes direction.
- Wherever there is an addition added.

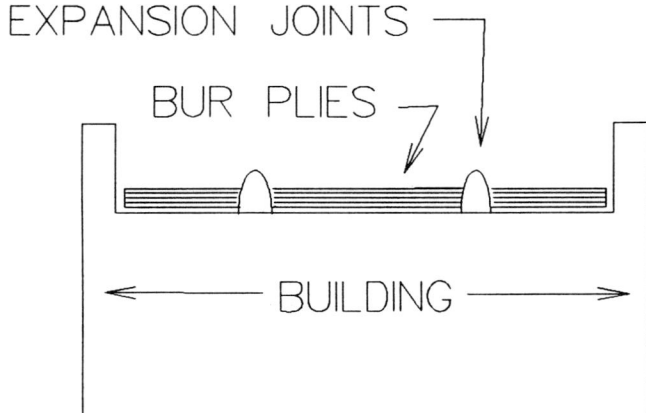

Figure 11–12 Expansion joints should be used in several areas

- Whenever the deck's shape changes to a L, U, or T.
- Wherever the deck is not over an interior (on an overhang).

The last item is because of temperature changes at the overhang that are not equal to the temperature changes on the roof's field. Cold air blows over the roof field, it blows over and under the overhang.

Expansion joints are to be a minimum of 4 in. higher than the roof decking. This applies to the side walls of skylights as well. This change from flat to vertical is called a curb. Cant strips should be used at all curbs for best roofing results.

See Fig. 11–13. Do not allow heavy equipment or materials to rest on built-up roofing for more than a few hours, less if it is a very hot day. The weight will compress the felt/asphalt layers and create low points in the roof, and these low points will collect water. Standing water will eventually seep into the roofing plies and destroy the felts and a leak will result. Equipment that is to be permanently installed, such as air handling equipment, should be installed on their mounting pads. These mounting pads should be attached solidly to the decking before the built-up roof is applied; Fig. 11–14 shows one method of attachment. Proper flashing is required to prevent leaks in and around the mounting pad. Consult the equipment manufacturer for alternative mounting methods. Install umbrella flashing around the equipment support legs.

Equipment Needed

We will need:

- A few knotted roofing mops
- A few 5 gal. buckets

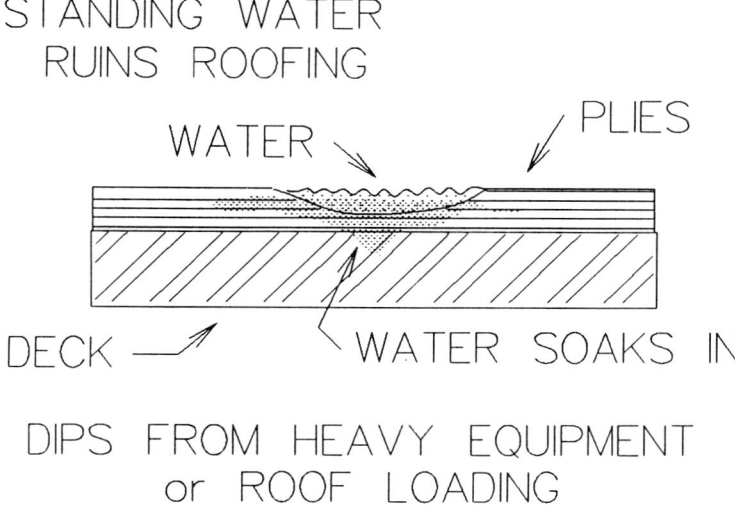

Figure 11–13 Ponding caused by heavy weight crushing the BUR

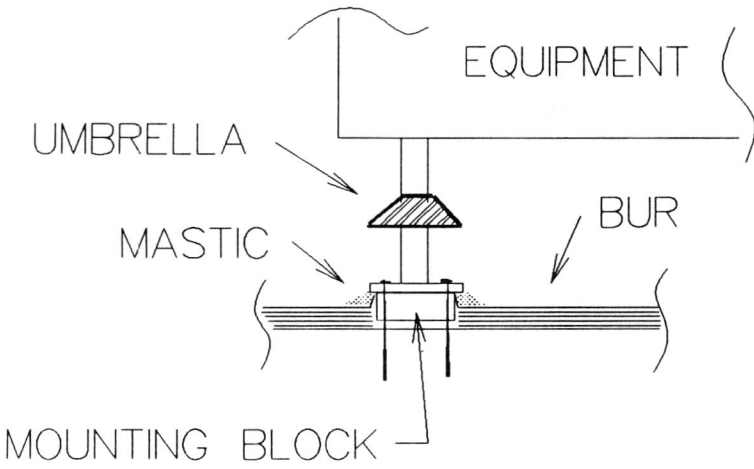

Figure 11–14 Provide mounting blocks for equipment

- A hot pot or kettle as a source of heat for melting asphalt
- An assortment of hand saws
- A circular saw with a wood cutting blade
- A hammer and a mallet
- An assortment of screwdrivers
- A 3 ft and a 6 ft level
- A tape measure
- An assortment of ladders
- Tin snips for flashing work
- A respirator
- Strong gloves
- Safety eyeglasses
- A hard hat

It is advisable to wear long-sleeve shirts. You will be cooler since the shirt will hold in moisture and, as the breeze blows, this moisture will evaporate, cooling off your body. It has been proven that prolonged exposure to the ultraviolet rays of the sun will cause sun burn and possible skin cancer. To top it off, you will be working with hot tar and other chemicals. A long-sleeve shirt will help keep these items from splattering on your body. A little common sense will go a long way toward preventing serious burns from sun, chemicals, and hot tar.

CORRUGATED DECKING

You will find that most of the larger commercial flat roofs are decked in corrugated steel. Steel decking is normally welded to the steel roof trusses and provides a firm, fire-resistant base for the built-up roofing. Here are some examples of corrugated decking steel in use. See the chapter on insulation for allowable flute to flute spans.

- 16 gauge, 1.5 in. standard rib, prime coated, 4.1 lb per sq ft
- 18 gauge, 1.5 in. standard rib, prime coated, 3.3 lb per sq ft
- 20 gauge, 1.5 in. standard rib, prime coated, 2.5 lb per sq ft
- 22 gauge, 1.5 in. standard rib, prime coated, 2.1 lb per sq ft

The listed decking steel is supplied in lengths up to 42 ft. The center to center flute spacing is 6 inches (Fig. 11–15). The installed flatness must be within 1/16 in. across every third flute (see Fig. 11–16).

Steel decking is to be a minimum of 25 gauge CRS (Cold Rolled Steel) and have minimum depth of flutes at 1-1/2 in. Suggested gauge to use is 18 gauge CRS in sheet widths of from 24 to 36 in. wide.

Corrugated Decking Flanges

Corrugated decking must be overlapped from sheet to sheet. When walked on, this connection must not separate (Fig. 11–17). To be correct, the flutes should overlap each other by a minimum of one flute; the overlap is for structural strength and for fire protection. In the event of an interior fire, the overlap aids in preventing the fire from getting to the insulation or the built-up roofing plies.

INSULATING CONCRETE DECK

Poured concrete can be used as a roof deck. This concrete can be left as is or covered with additional roofing. If it is to be covered with foam insulation or other nailable decking, I

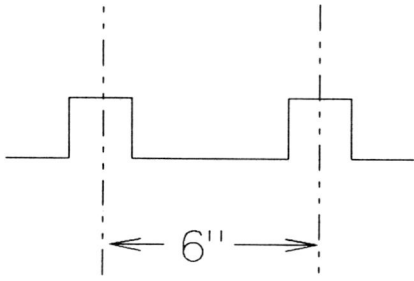

FLUTE SPAN   **Figure 11–15**   Example of flute span

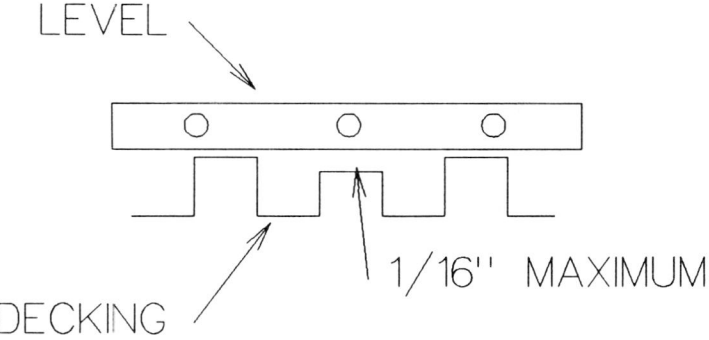

Figure 11–16   Make sure that decking is level

suggest that nailer strips be incorporated (Fig. 11–18). The nailer strips should be pressure treated lumber. Spacing will depend on the type of roofing or insulation being installed and shall conform to building codes.

Built-up roofing may be applied to a concrete deck. The concrete should be well cured and dry. But, even dry concrete will emit some moisture, which can become trapped under the built-up roofing plies. To help prevent moisture buildup, use a vent sheet for the first roofing ply layer (Fig. 11–19).

The concrete should be top vented a minimum of once for every 1,000 sq ft of roof surface. Top venting means that the barrier between the steel deck and the cement is vented to the atmosphere.

Prime the Concrete

Concrete decks are not generally porous to the hot asphalt being applied to them. The asphalt is too thick to soak properly into the concrete. The roof plies may delaminate at the

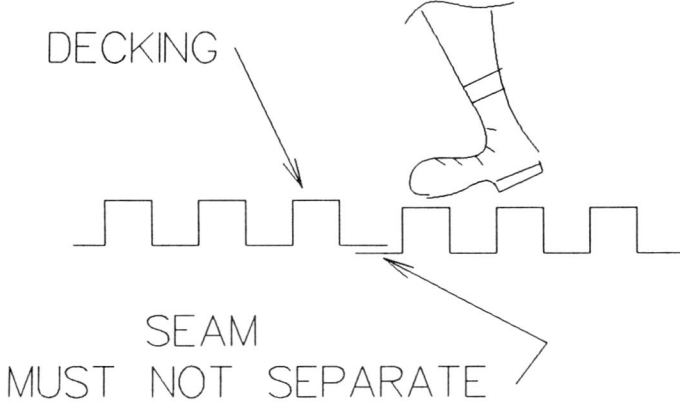

Figure 11–17   Seams must not separate

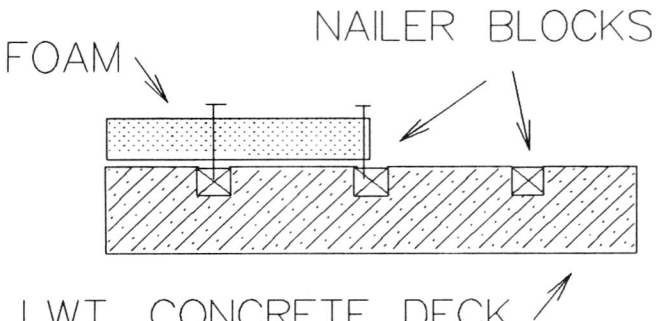

Figure 11–18 Use nailers in concrete decking if insulation is to be installed

concrete to first felt layer bond. To solve this problem, we must prime the concrete (Fig. 11–20). There are several commercially available primers for this purpose. The most common is cut back asphalt. This is asphalt that has been thinned by the addition of liquid hydrocarbons, mineral spirits, or by water and emulsifiers. Emulsifiers are chemicals that allow water to disperse within the asphalt. Cut back asphalt is thin and can sink into the small cracks and pores of the concrete. It forms an even top skin to which regular asphalt will adhere. Drying time is required before application of the standard roofing. The solvents, water, or mineral spirits, must completely evaporate if a good bond is to be made.

ROOF SHAPES FOR DRAINAGE

Most flat roofs are not flat. You can have a dead flat roof, 0:12 slope, but you would not have water drainage. Flat roofs are generally considered to be any roof with a slope of 3:12 or less. That is not to say that all flat roof work will be on a low slope roof.

Most flat roofs are domed roofs; they start high in the center and arc downward to the

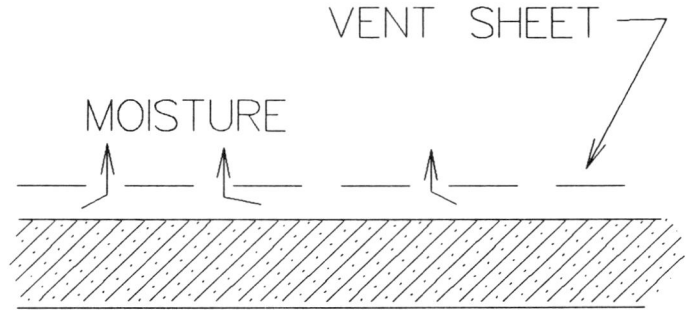

Figure 11–19 Concrete decks must have a vent sheet if covered with BUR

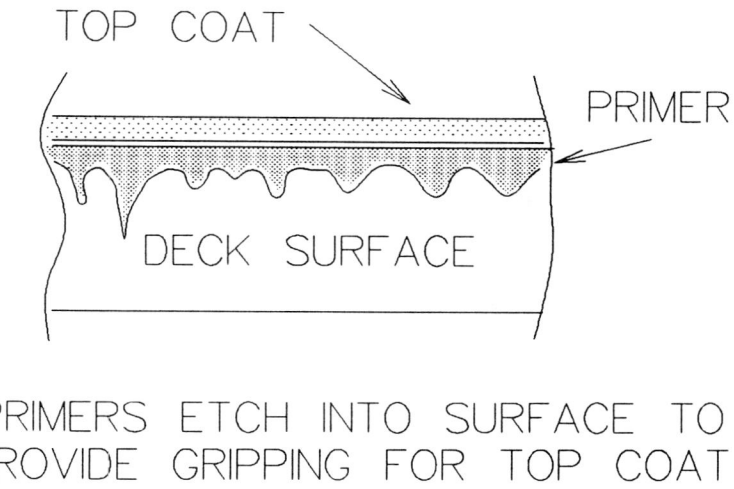

Figure 11–20   Asphalt primer is a must for proper adhesion of top coats

edges or overhangs. Some designs are just the opposite, they start high on the edges and arc down to the center. A center drain is provided on these roofs. This drain usually runs down a pipe column in the center of the building.

You may come upon a cricket type flat roof. A cricket roof may have several pyramid-shaped elevations across it, with each pyramid flowing to a roof drain at the corners of its base.

These types of sloped roofs are, in actuality, built with dead flat roofing decks. The drainage slopes are formed by using tapered blocks of foam insulation under the built-up roofing. See the chapter on insulation.

Some flat roofs are built with the decks sloped. These roofs are used on lean-to construction and add-ons. By add-on I am referring to the roofs that are added to an existing structure, such as a patio, storage area, or carport roof. The main roof in these add-on situations is usually a standard peaked roof. These require a low slope since you want to maintain head clearance at the end of the overhang. Most sloped roofs slope outward to the edges, but you will find complex designs where the roof slopes toward the building's center.

FORCES THAT DESTROY A FLAT ROOF

Forces build up in a roofing deck and eventually destroy the roofing. The main forces are expansion and contraction of dissimilar materials (Fig. 11–21). These materials are the deck material, the insulation material, the felt sheets, and the asphalt. Minor forces are created by the flashings and the fasteners.

If everything could be held at a constant year-round temperature, there would be no

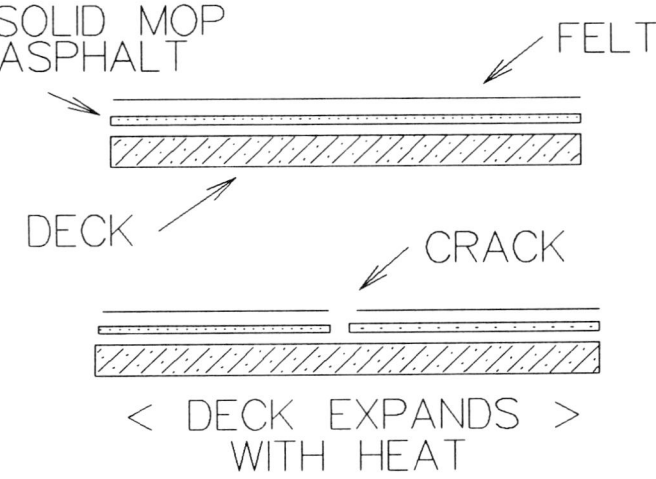

Figure 11–21   Deck expansion can crack BUR

problem. But roofs are exposed to variations of temperature from hour to hour, day to day, week to week, and month to month. The slow changing temperatures, week to week and month to month, generally are not a problem. All the materials tend to settle in and conform to each other if given enough time. It is the day to day and hour to hour temperature variations that cause the most problems. The materials just don't have enough time to adjust to each other.

A summer rain storm on a 100 degree day is an example. The roof has heated to near the melting point. The inner roof plies have heated up and expanded. Then comes a cool rain. This cools the top ply back to a solid while the inner plies remain warm and act as a heat source. The inner plies may still be expanding when the colder top sheet contracts (Fig. 11–22). The plies separate from each other and delamination occurs.

On a cold winter day, the interior of a building is heated to 78 degrees. The heat collected near the under side of the roof deck is about 90, and the decking expands. Outside the temperature is −20. There is now a temperature change of 110 degrees within the roofing material. Asphalt layers become brittle, felts contract, getting smaller, the decking expands and gets larger. Something has to give, usually the now very brittle asphalt, cracks form, and delamination occurs.

To counteract these problems, we must understand the principles involved. This section of the manual does just that. It gives you proven, but not perfect solutions. The perfect roof, the one that never fails, just has not been invented.

PREPARE THE SURFACE

Surface preparation is essential. Dirt, water, moisture, oils, leaves, bird droppings, nails, or whatever, will prevent a proper seal. If there is anything loose on the roof's surface, it must

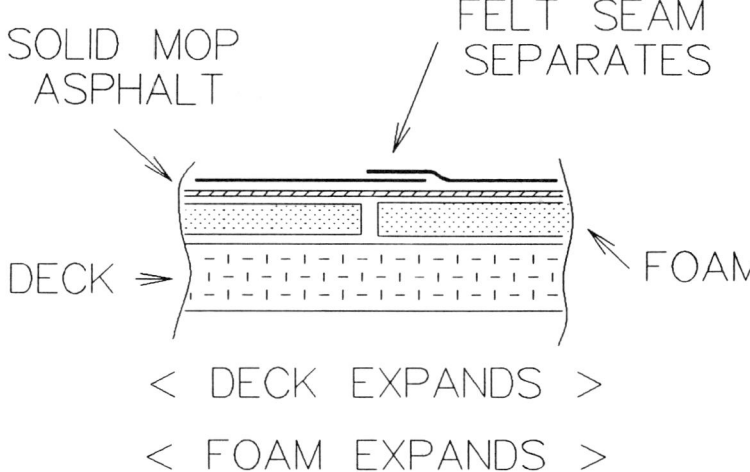

Figure 11–22 Do not lay seam over seam; it will pull apart

be removed before you lay down that first binder coat. If loose material is allowed to remain on the roof, the roof will blister in the future.

BLISTERS

I recently worked on a shopping center roof problem. The shopping center is 10 years old and has over 900,000 sq ft of built-up roof. A full 50% of the roof has failed so far. It has cost the owners some $550,000 in roof repairs and they are suing the general contractor and the roofing subcontractor for damages. The roofing materials carried a 20-year warranty. The materials manufacturer was sued, but he won as it was proven to be the contractor's installation, not the materials, at fault.

The area, New York State, is subject to overnight moisture in the form of dew. The subcontractor laid out the foam insulation and fastened it to the corrugated metal sheathing. He then applied the binder coat to the roof insulation early in the morning. Then, he applied the first layer of roofing. The next day he applied the second binder coat and the second layer. Day three and day four were repeats of the first two days. The fifth day he applied binder and stone. The term for this type of application method is phased application and it is *not* to be done. The contractor made several other mistakes during his roofing application.

- First, he did not allow the early morning dew to dry completely.
- Second, he attempted to cover the roof one layer at a time.
- Third, some foam insulation fasteners missed the corrugated sheathing.

- Fourth, dirt, dust, leaves, and bird droppings were deposited on the roof during and after each roofing step.
- Fifth, the binder material was not heated to the proper temperature.
- Sixth, his workers carried moisture and dirt to the roof on their shoes.
- Seventh, there is a suspicion that he intermixed asphalt felts with coal tar pitch impregnated felts.

Here is what happened. Six years after this roofing job was completed, the roof began leaking. An independent engineering laboratory was called in to inspect and make recommendations. They used an infrared camera to find the water leaks. They then extruded samples of the built-up roofing at several locations. Their findings were that the roof delaminated from the insulation and between each layer.

The killer of this roof was water, in the form of moisture. When water is heated to 212 degrees Fahrenheit it becomes steam. A drop of water 1/8th in. in diameter expands to fill a space of 187.5 cu in. when it becomes steam—steam occupies a space 1,400 times its liquid volume. Since the water was trapped under the layers of binder and felt, it had nowhere to go. The steam produced great pressure and this pressure tore the layers apart. A good percentage of the moisture detected was at seams (Fig. 11–23), under small particles of dirt and leaves, and in bird droppings.

Of course, neither New York, nor anyplace else, experiences temperatures of 212 degrees Fahrenheit (the temperature needed to heat water to steam). However, black, the color of roofing materials, absorbs heat. Several hours of 95 degree sun beating down on a flat roof surface can and does produce 212 degree heat within the roof layers.

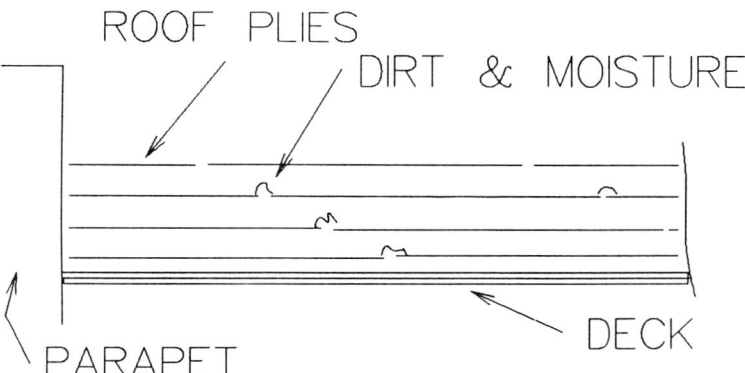

SEAM AREAS MUST BE FREE OF DIRT AND
MOISTURE, ELSE BLISTERS WILL OCCURE

Figure 11–23   Trapped dirt and moisture will result in roof failure

To further complicate matters, night air is cool and moist. The water vapor that expanded during the day and started a blister now condenses back into water. As it cools and condenses, it forms a vacuum within the blister. If there is a small crack in the membrane to the outer air, the moist outer air is sucked in. This adds to the amount of trapped moisture. Let's say that we now have a volume of water of 2/8 or 1/4 cu in. The next hot day will expand this water to a volume of 375.0 cu in. The roof blister has now doubled in size. Night falls and an additional 1/4 in. volume of moisture is absorbed. There is now 1/2 cu in. of water and it goes on night by night, day by day. The blister gets bigger and bigger.

In theory, the largest obtainable blister is 8 ft in diameter. This is due to the weight of the roof material exceeding the pressure of the steam that is forming the blister. But, in the case of the seam blister, this was not true. Several blisters formed along various portions of the seam and when they grew large enough, they joined. This roof had blisters several feet across and a hundred feet long.

One blister was detected where the roof insulation fastener missed the corrugated metal. Since the fastener was not secure, it didn't adhere tightly to the insulation. A minute void formed and trapped some morning dew.

I said that the entire roof was laid layer by layer. What should have been done is the following. All four layers should have been applied simultaneously and in the same day. This is of course impossible due to the large surface area of the roof being covered. But, if the layers were staggered, applying only what could be done during a given period, the moisture problem could have been avoided. By stagger I mean lay down, say 100 sq ft of the first layer, 80 sq ft of the second, 60 of the third, and 40 of the fourth (Fig. 11–24). Continue doing this day by day. This way there is always clean hot material being bonded to clean hot material.

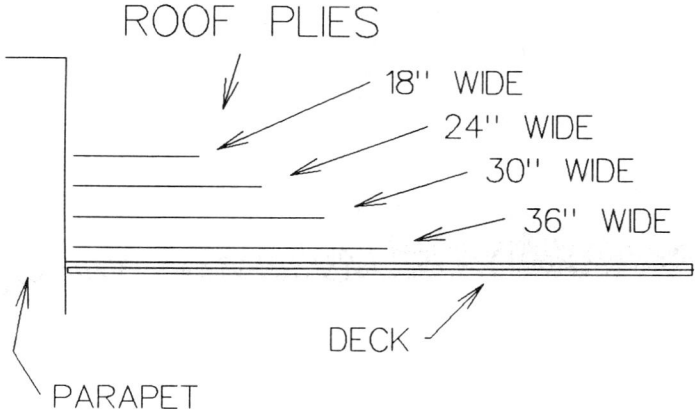

ROOF PLIES

18" WIDE
24" WIDE
30" WIDE
36" WIDE

DECK

PARAPET

BEGIN WITH CUT ROLLS SO THAT
SEAMS ARE STAGGERED

Figure 11–24 Proper installation of a BUR

As for the heated material problem, Type II asphalt binder was used. This requires an application temperature of 350 to 425 degrees Fahrenheit. In the kettle, the asphalt was 385 degrees Fahrenheit. When the asphalt was placed into carry buckets, carried to the installation area, mopped on, and then covered with felt, it cooled to less than 330 degrees. A good quality bond was not formed. It would have been better to use Type I asphalt that has a mopping application temperature of 300 to 375 degrees Fahrenheit. When heated to 385 degrees Fahrenheit, it would still be acceptably hot at the time of application.

Don't get the idea that putting 330+ degree Fahrenheit asphalt on the roof will dry up the moisture on the roof. It does not. It may dry some surface moisture, but it will not dry up trapped moisture. It will only keep it trapped. Plus, the heat of the asphalt will instantly turn some moisture to steam and start the first blisters. In addition, the cool moist roof will decrease the temperature of the hot asphalt at the binding layer to below the proper adhesion temperature of the asphalt. This will start a blister.

It was found that the membrane material and the binder used may not have been compatible with each other. Coal tar impregnated membranes may not be used with asphalt binder. You must be sure to use only what the manufacturer recommends if you wish to maintain the manufacturer's warranty.

The solution to the shopping center's roofing problems was to tear off the old roofing and reroof properly. Two sections have been completed as of this writing. An additional $750,000 must be spent to correct the entire roof. This cost was passed on to the shop owners in the form of rent increases. They, in turn, passed it on to their customers in the form of higher retail prices. The general contractor and his subcontractor were forced into bankruptcy and each lost his business. The contractors have lost years of building up their business and their family's livelihood. Their employees went elsewhere after being unemployed for several months.

Unfortunately, the new roofing contractor didn't do the reroofing job correctly. He did everything right except demand that the shopping center owners pay for a complete job. They didn't provide enough money for a top reflective coat. The sun has already, in less than 2 years, started to deteriorate the jet black single layer membrane that was applied.

PARAPET WALLS

Many flat roofs will have what is called a parapet wall around their outside perimeter. This nonstructural wall is used to hide the roof and those items on or coming through it (Fig. 11–25). The items can be vent pipes, air conditioning units, skylights, etc. Flat roofs are hidden because they are normally not very attractive. As a roofer, you must learn how to make this roof to parapet wall connection leak-free. It is a major source of water leakage into buildings. Sometimes, once you have made it leak-free, someone else, say the sign man, hangs a business sign on the outside of the parapet wall, causing water leaks through the connections he made. Guess who gets the blame for a roof leak?

Figure 11–26 illustrates a flat roof with a parapet wall. The parapet wall has been tiled with red mission tile.

See Fig. 11–27, typical light construction of a parapet wall—light in that it is all

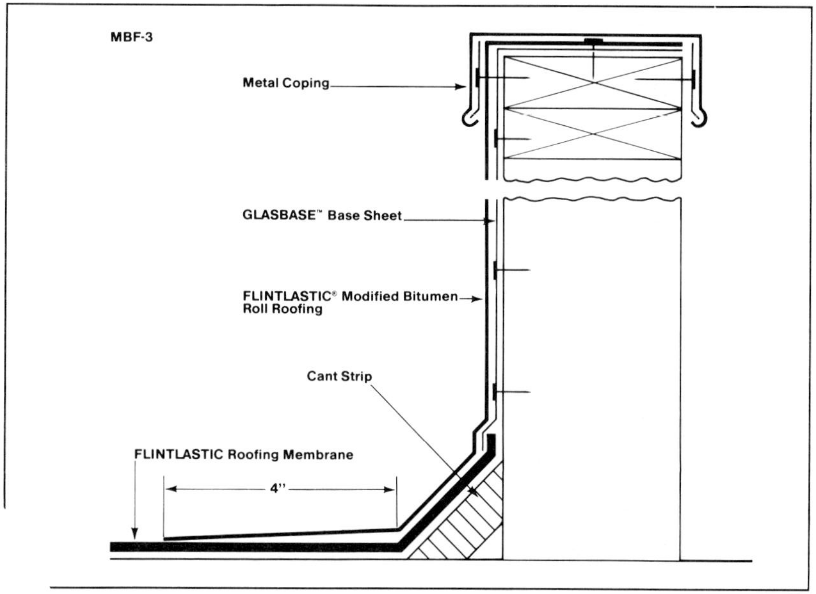

BASE FLASHING AND WALL COVERING ON WOOD PARAPET WALL

Over the completed membrane at wood parapet walls, apply MBF-3 Base Flashing and Wall Covering consisting of one ply of Base Sheet and one ply of FLINTLASTIC Modified Bitumen Roll Roofing. Completed membrane shall be extended 2" above top edge of cant and adhered to cant strip only. Install one ply of Base Sheet extending from top outside edge of parapet wall, across top and down the vertical surface to top edge of cant, nailing 12" o.c. through tin-discs in all directions. Torch apply one ply of FLINTLASTIC Modified Bitumen Roll Roofing (FLINTLASTIC GMS shall be set in hot asphalt) extending from top outside edge of parapet wall, down the vertical surface and out onto the roofing membrane a minimum of 4". Nail 9" o.c. through tin-discs along top of parapet wall. Side and end laps shall be consistent with field membrane application.

MBF-3

Metal Coping

GLASBASE™ Base Sheet

FLINTLASTIC® Modified Bitumen Roll Roofing

Cant Strip

4"

FLINTLASTIC Roofing Membrane

Figure 11–25 (Reprinted with permission of GS Roofing Products Co.)

305

Figure 11–26   Parapets may be angled

wood framed and used primarily for residential or small commercial buildings. Lumber used for the top sill plates, the parapet wall top sills, the blocking, and the cant strip should be pressure treated lumber. This area is subjected to much rain and snow buildup. The framing is protected by the outer wall covering, the drip cap, and the roofing material. But, these items wear out in time and they all allow some moisture penetration. Eventually, the framing lumber will get wet. Pressure treated lumber is highly recommended for these reasons. Commercially available cant strips are sometimes made of wood pulp fiber impregnated with wax, which are acceptable.

PARAPET WALL DETAIL

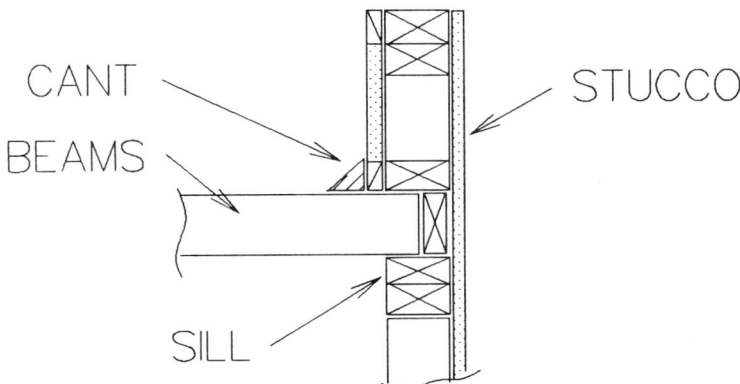

Figure 11–27   Construction of a parapet wall

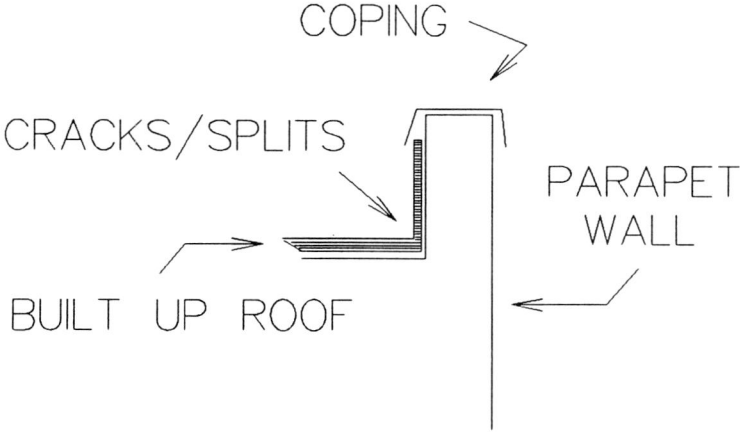

WRONG CONSTRUCTION

COPING

CRACKS/SPLITS

PARAPET WALL

BUILT UP ROOF

Figure 11–28 Lack of cant will crack BUR and cause leaks

Figure 11–28 shows the wrong way to bring roofing material up a parapet wall. Almost all roofing material can be bent to conform to a curved shape; cement and clay tiles are the exception. The materials that can be bent without much harm are metal shingles, flashings, and asphalt felts. When it comes to the asphalt materials, you must be careful.

Bends should not exceed 45 degrees. Figure 11–29 shows a 90 degree bend. This will crack and split and form a water leak. Since the material expands and contracts, it will tend to pull away from the corner, thus forming a blister or void that is highly subject to puncture damage (Fig. 11–30).

CANT STRIPS

The proper method of making a 90 degree bend is by using two 45 degree bends and a solid backer plate (Fig. 11–30). Here we show a cant strip, as in "can't collect water there," installed at the junction of the wall and roof deck. The roofing material then runs from the roof, over the cant strip, up the wall and under the drip cap or Reglet flashing.

When installing a flat roof you may find that there are many 90 degree corners. These can be at equipment mounts, skylights, and roof to parapet wall connections. A 90 degree corner will provide a spot for snow, ice, or water to collect. The cant strips serve the purpose of making a smooth connection, as sharp corners tend to crack tar and flat roof materials. Figure 11–31 shows a cant strip on an equipment mounting base.

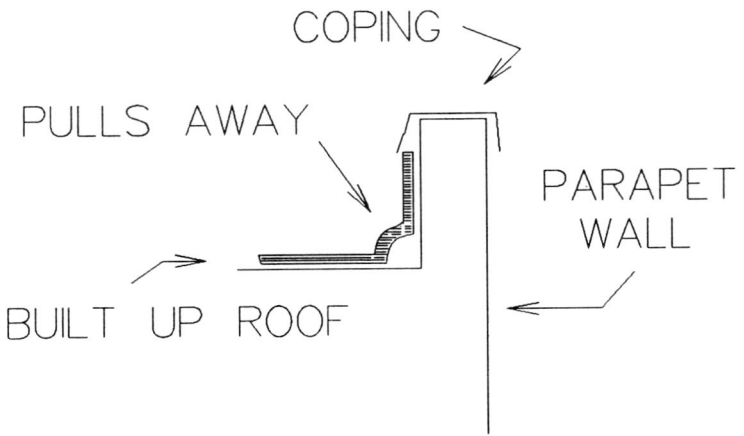

WRONG CONSTRUCTION

COPING

PULLS AWAY

PARAPET WALL

BUILT UP ROOF

Figure 11–29   Lack of cant may cause BUR to blister at corners

Ruberoid© MB Flashing Detail Type 3MB PLUS Torched

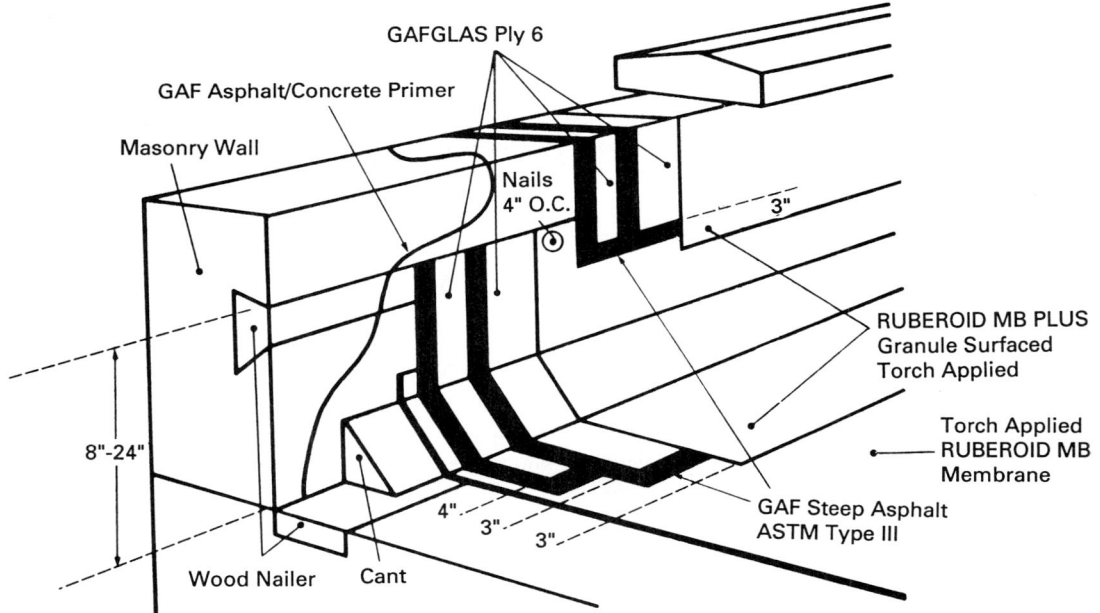

GAFGLAS Ply 6

GAF Asphalt/Concrete Primer

Masonry Wall

Nails 4" O.C.

3"

RUBEROID MB PLUS
Granule Surfaced
Torch Applied

Torch Applied
RUBEROID MB
Membrane

GAF Steep Asphalt
ASTM Type III

8"-24"

4"   3"   3"

Wood Nailer   Cant

Figure 11–30   Proper construction of parapet with BUR (Reprinted with permission of GAF Building Materials Corp.)

CURB - EQUIPMENT PLATFORM

Completed membrane shall extend 2″ above top edge of cant strip, adhered to cant strip only. For nailable surfaces, install one ply of Base Sheet extending from top edge of cant to top edge of equipment platform, nailed 12″ o.c. through tin-discs in all directions.

Cover entire platform with one ply of Base Sheet extending 4″ down all sides, set in asphalt over concrete platforms and nailed 12″ o.c. through tin-discs in all directions to wood platforms. Install one ply of FLINTLASTIC Modified Bitumen Flashing Roll extending to top edge of platform and 4″ minimum out onto roofing membrane. Nail 9″ o.c. through tin-discs along top edge. Install metal inverted pan.

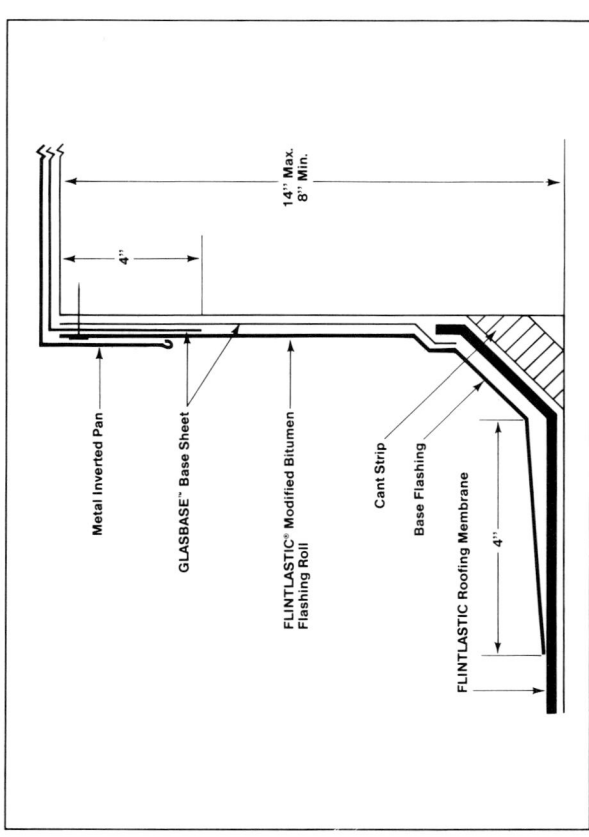

Figure 11–31 Factory equipment platform (Reprinted with permission of GS Roofing Co.)

VENTING

Like most roofs, we have protrusions through our roof decking and our roofing material. These are usually air or sewer vent pipes. Figure 11–32 shows you one method used to seal these pipes against leakage. In this instance, lead flashing was installed.

You may want to "umbrella" or put hoods over these pipes. The hoods keep snow, ice, rain, birds, and small animals from entering the building.

The vent pipes can be made from plastic such as ABS or PVC, or from clay or metal. You will need to flash the base, seal the roofing, and install a boot to provide a leak-free interface. For more information on venting, see the chapter on new construction and the chapter on flashing and vents.

EQUIPMENT STANDS

Equipments stands are used to keep equipment off the roof surface. The supports are through the roof and secured to the roof's beam system. Some equipment such as air handlers can weigh a thousand pounds or more and require this support. Elevating the equipment helps keep the roof surface dry and prevents small animals from nesting under the equipment. Elevating the equipment makes it easier to install, maintain, or remove. Most heavy equipment is air lifted to the roof via helicopters or large cranes. Figures 11–33 and 11–34 show two methods of sealing a BUR to the equipment supports.

GRAVEL STOP OR EDGE FLASHING

As with standard peaked roofs, we have a certain amount of metal flashing on our flat roof. Most of the flashing is formed at the factory and sold in 10 or 20 ft lengths.

Edge flashing is used to dress off the BUR. It provides a drip edge, a gravel stop, and trim all in one. Figure 11–35 shows one such installation.

VENT SHEETS

Vent sheets are specially manufactured to do one or two things, depending on the type. The standard vent sheet does not allow moisture escape from under the roofing plies. It does provide air space for the moisture to reside in. This air space gives the moisture a place in which to expand or contract without harm to the roofing.

The second type of vent sheet is of a waffle design. This sheet provides air spaces the same as the standard vent sheet does. The difference is that the waffles are channels. These channels allow for escape of trapped moisture into the atmosphere at the roofing edges or via installed vents (Fig. 11–36). All escape paths must be protected from water entry via rain or roof flooding. Water entry under the sheet will soak through the roof decking and into the building.

PIPE FLASHING

Where projections extend through the roof surface, install a collar of FLINTLASTIC Modified Bitumen Roll Roofing over the base ply to extend a minimum of 4" beyond flanges. Install metal flashing with a minimum 4" wide continuous flange on top of the FLINTLASTIC flashing collar. (For nailable surfaces, flange must also be nailed 3" o.c., 3/4" from perimeter). FLINTLASTIC Field membrane shall be fully adhered to the flashing collar and metal flange. Seal flashing around projection with a bead of melted modified bitumen for torch applied systems, and a cant of flashing compound for hot asphalt applied systems.

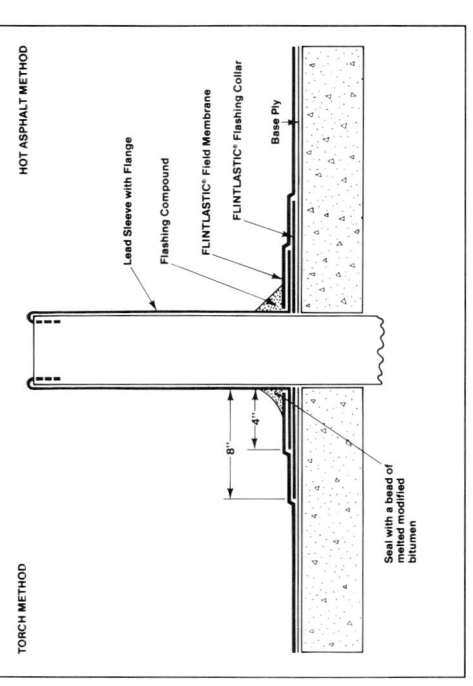

PIPE FLASHING

Figure 11–32 Factory vent pipe (Reprinted with permission of GS Roofing Co.)

311

WOOD DECK EQUIPMENT SUPPORT

Over the completed membrane at wood support, install one ply of Base Sheet to cover entire support extending to top edge of cant strip nailed sufficiently to hold in place. Follow with one ply of FLINTLASTIC Modified Bitumen Flashing Roll extending to top edge of support and a minimum of 4″ out onto roofing membrane. Nail 9″ o.c. through tin-discs along top edge. Install one ply of FLINTLASTIC Flashing Roll over the top of support extending 4″ down all sides. Apply metal cap.

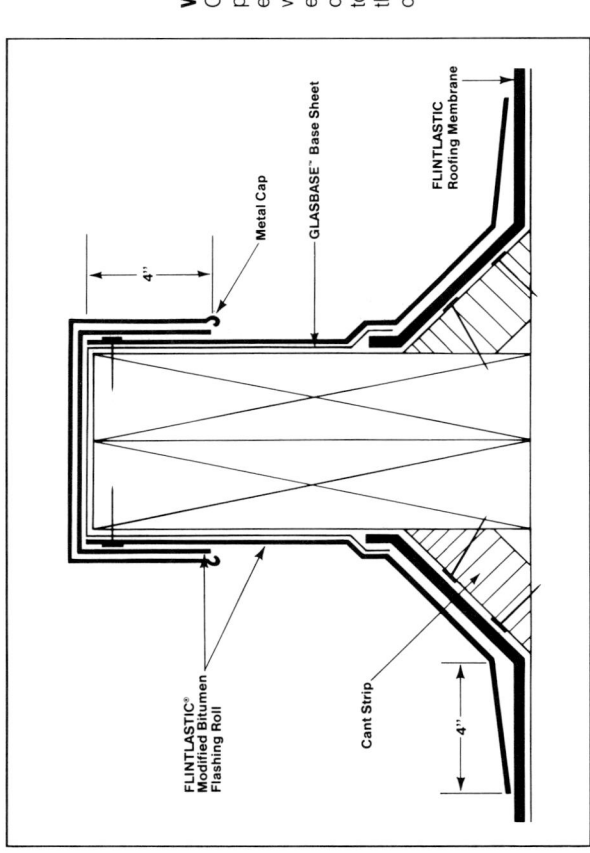

Figure 11–33   Factory wood deck stand (Reprinted with permission of GS Roofing Co.)

EQUIPMENT STAND - PIPE SUPPORT

Where projections extend through the roof surface, install a collar of FLINTLASTIC Modified Bitumen Roll Roofing over the base ply to extend a minimum of 4" beyond flanges. Install metal flashing with a minimum 4" wide continuous flange on top of the FLINTLASTIC flashing collar. (For nailable surfaces, flange must also be nailed 3" o.c., 3/4" from perimeter.) FLINTLASTIC Field membrane shall be fully adhered to the flashing collar and metal flange. Seal flashing around projections with a bead of melted modified bitumen for hot asphalt applied systems, and a cant of flashing compound for torch applied systems. Install hood with clamp ring extending 1" below top edge of flashing sleeve and tighten. Apply a bead of caulking around top edge of hood.

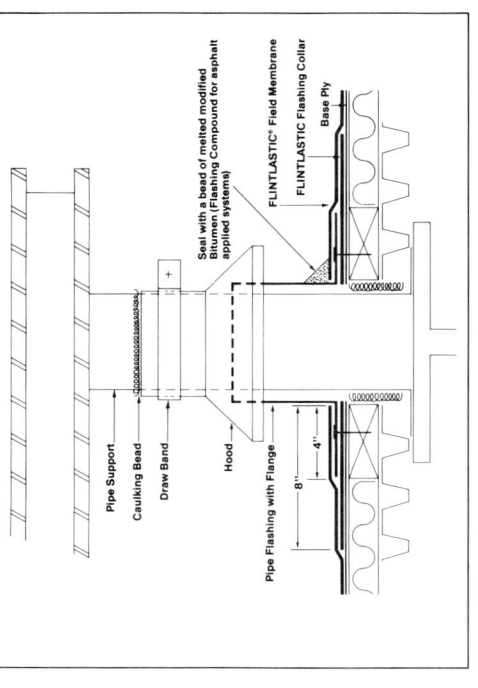

EQUIPMENT STAND - PIPE SUPPORT

Figure 11–34 Factory pipe stand (Reprinted with permission of GS Roofing Co.)

EDGE FLASHING

Apply field membrane at perimeter edges, extending 2″ beyond edge and fold down. Install a minimum 26 gauge gravel stop or metal edging, lapping 4″ on ends, with end laps set in a complete bed of flashing compound. Nail flange 6″ o.c. staggered (end laps shall receive two nails). Follow with one ply of FLINTLASTIC Modified Bitumen Flashing Roll extending a minimum of 4″ beyond Flange and out onto roofing membrane. Metal edging shall have a ¾″ rise for optional gravel surfaced membranes and ⅜″ rise for unsurfaced membranes.

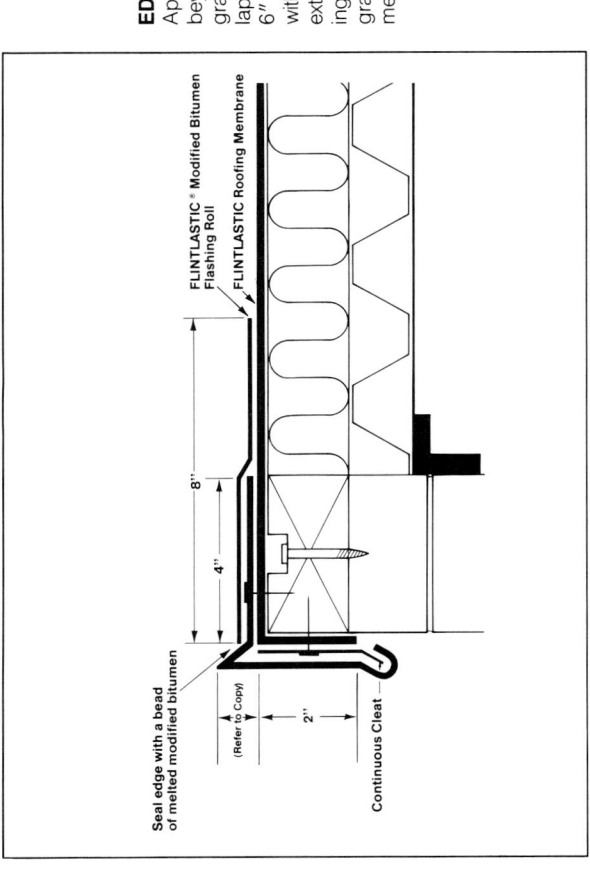

Figure 11–35 Factory edge flashing (Reprinted with permission of GS Roofing Co.)

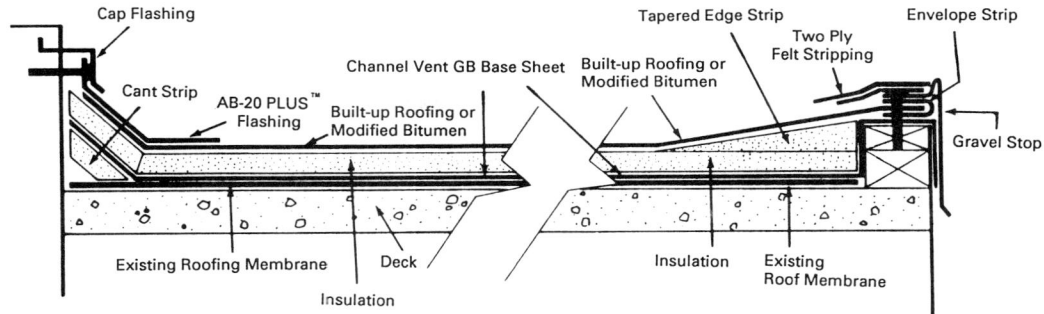

Figure 11–36 BUR cross section showing vent sheet placement (Reprinted with permission of Celotex Corp., Roofing Products Division)

Both the standard and the waffle type vent sheets provide protection against delamination and splitting. The sheets do not form a solid mass with the decking and can expand and contract independent of the decking. The built-up roofing plies float on the vent sheets.

PROVIDING WATER DRAINAGE

There are times when you have a flat roof and want to dome it for better water drainage. Foam insulation sheets are used for this purpose. The center of the building's roof is plied two, three, or more sheets high. As you work toward the edges, you use fewer sheets in the pile. So that a stepped look doesn't result, you add tapered insulation sheets or wood wedges.

Roof Drainage Systems

Figure 11–37 illustrates a proper roof drain in a built-up roof. The built-up roof slopes from all directions to the drain. Figure 11–38 shows a common problem you might encounter. The roof drain was installed too high above the sheathing. The built-up roofing plies were not thick enough to provide a slope to the drain. The result is that the drain forms a circular dam that keeps standing water on the roof. This sometimes occurs on what was a good draining roof when a second or third roof is installed. Use a level to be sure you are draining to the drain. Follow manufacturer's instructions when installing drains (Fig. 11–39).

Another roof drain used on flat roofs is the *scupper*. A scupper is a drain hole cut through the parapet wall. A sheet metal flange is used to guide the water off the roof and into the air, away from the building's walls. Figure 11–40 is a typical BUR to scupper installation.

Tapered Edge Strips

There will be times when you must build a drainage slope from some item to the BUR. Tapered edge strips are used for this purpose. Figure 11–41 shows a flexible expansion joint with tapered edge strips installed. These strips can be foam plastic or wood.

CORRECT SLOPE TO DRAIN

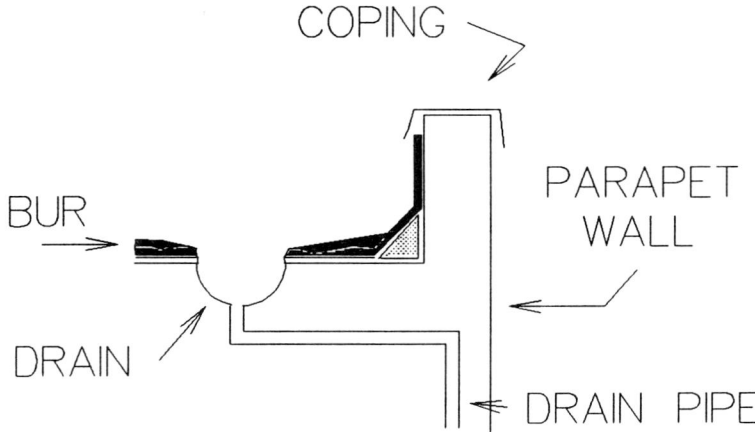

Figure 11–37   Roof must slope to drain

WRONG CONSTRUCTION

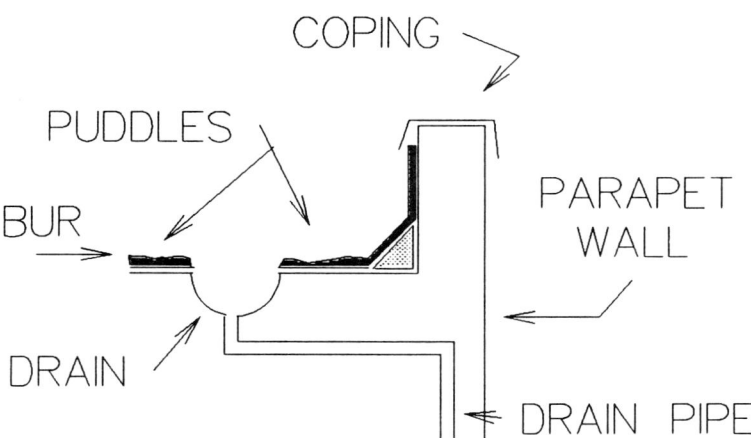

Figure 11–38   Buildup of asphalt around drain prevents proper drainage

DRAIN DETAIL

Install base ply, trimmed even with drain flange. Follow with an approximately 40" × 40" collar of FLINTLASTIC Modified Bitumen Roll Roofing extending into, and fully adhered to drain flange and interior surface. Install FLINTLASTIC Field membrane fully adhered to base ply, flashing collar and extending into drain. Install clamp ring and tighten while membrane is still hot. GRPC recommends the use of a minimum 30" square, 4 lb. lead or 16 oz. soft copper Flashing installed between the flashing collar and field membrane. FLINTLASTIC flashing collar shall extend 4" beyond lead or copper flashing.

NOTE: For hot asphalt applied membranes, base ply shall be extended into drain and set in Flashing compound 9" wide around ring and flange.

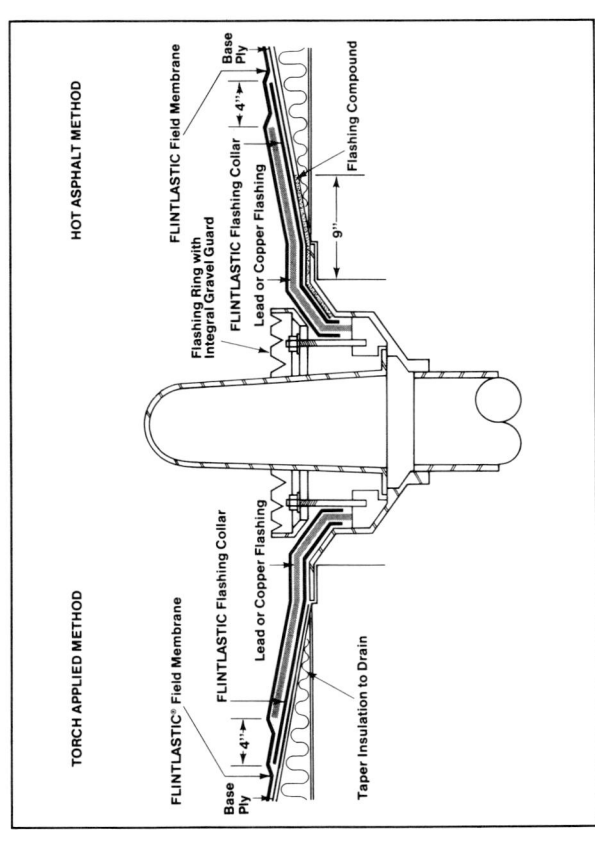

Figure 11–39   Factory drain detail (Reprinted with permission of GS Roofing Co.)

SCUPPER

Over the roofing membrane base ply at parapet walls, install one ply of Base Sheet set in asphalt to prepared and primed surfaces. Install base sheet extending from top outside edge of parapet wall, across top and down vertical surface to top edge of cant strip. Base sheet shall be installed to cover wall area directly behind scupper and extend a minimum of 6" beyond scupper flanges.

Install metal scupper and nail 3" o.c. through tin-discs along top and side edges to top of cant. (For nailable surfaces, flange must also be nailed 3" o.c. along bottom edge, ¾" from perimeter). Follow with FLINTLASTIC Field Membrane extending 2" above top edge of cant strip. Install one ply of FLINTLASTIC Modified Bitumen Roll Roofing extending from top outside edge of parapet, down the vertical surface and a minimum of 4" out onto roofing membrane. Nail 9" o.c. through tin discs along top of parapet wall. Trim flashing around scupper opening and seal while membrane is still hot. Sheet metal flanges shall be continuously soldered at joints.

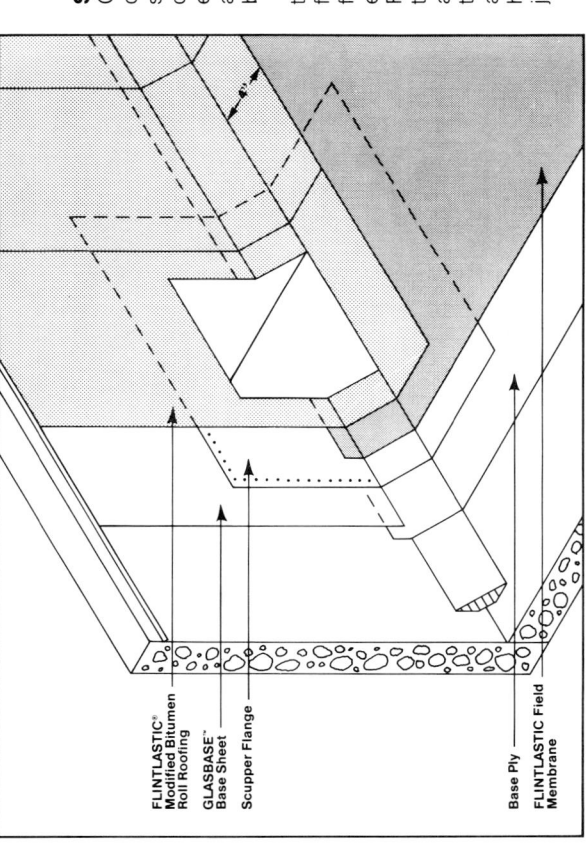

FLINTLASTIC®
Modified Bitumen
Roll Roofing

GLASBASE™
Base Sheet

Scupper Flange

Base Ply

FLINTLASTIC Field
Membrane

Figure 11–40   Factory scupper (Reprinted with permission of GS Roofing Co.)

318

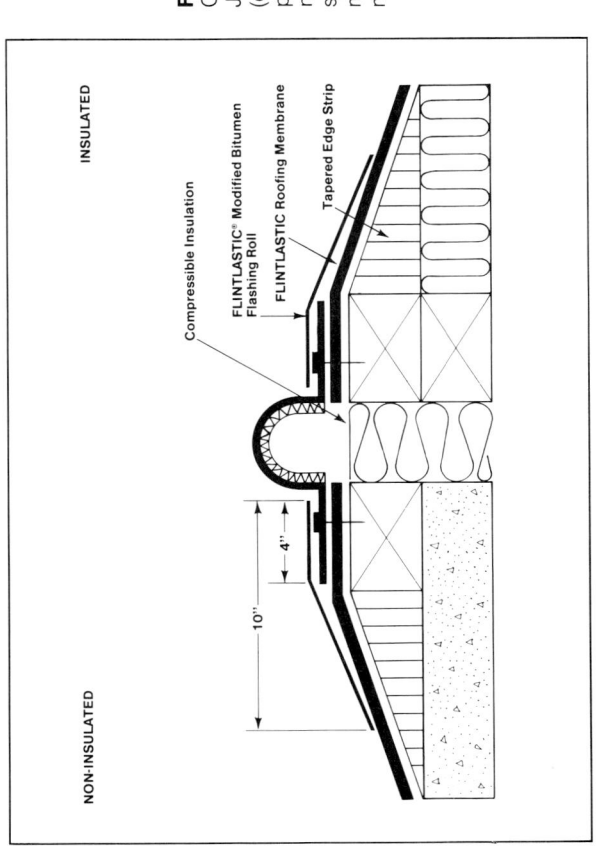

FLEXIBLE EXPANSION JOINT

Over the completed membrane, install a Flexible Expansion Joint with a minimum 4" flange. Nail flange 6" o.c. staggered (end laps shall receive two nails) with end laps set in a complete bed of flashing compound. Seal end laps and terminations of expansion joint according to Manufacturer's specification. Install one ply of FLINTLASTIC Modified Bitumen Flashing Roll lapping 6" on ends and extending a minimum of 6" beyond flange and out onto roofing membrane.

Figure 11–41   (Reprinted with permission of GS Roofing Co.)

BASE FLASHING ON CONCRETE OR MASONRY WALL WITH METAL COUNTERFLASHING

Over the completed membrane at parapet walls and other vertical surfaces, apply MBF-1 Base Flashing consisting of one ply of FLINTLASTIC Modified Bitumen Flashing Roll to prepared and primed surfaces. Completed membrane shall be extended 2″ above top edge of cant strip and adhered to cant strip only. Torch apply one ply of FLINTLASTIC Modified Bitumen Flashing Roll (FLINTLASTIC GMS/fi™ shall be set in hot asphalt) extending a minimum of 8″ up vertical surface and 4″ out onto roofing membrane. Nail top edge 9″ o.c. through tin-discs with concrete fasteners. Seal top edge with a bead of melted modified bitumen for torch applied systems, and a three course flashing of YELLOW JACKET and flashing compound for hot asphalt applied systems.

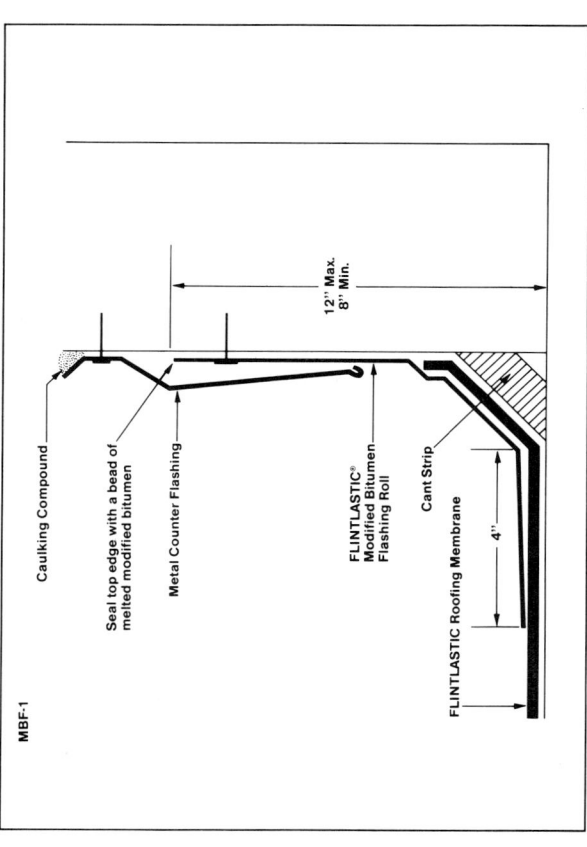

MBF-1

Caulking Compound

Seal top edge with a bead of melted modified bitumen

Metal Counter Flashing

12″ Max.
8″ Min.

FLINTLASTIC®
Modified Bitumen
Flashing Roll

Cant Strip

4″

FLINTLASTIC Roofing Membrane

Figure 11–42 Factory base flashing (Reprinted with permission of GS Roofing Co.)

320

Counterflashing

These will be times when you will need to use counterflashing during the installation of a BUR. Figure 11–42 shows one of these instances.

Grease Guard™s

Over restaurant facilities you will find a stove vent. This vent empties a considerable amount of cooking grease onto the roof. Cooking grease will destroy a BUR. It will eventually eat through the roofing plies and cause leaks and possible structural damage to the building. Cooking grease is unsightly, a fire hazard, and a worker hazard. To prevent this problem, grease traps are installed. These are filter systems designed to catch the grease and keep it contained and off the BUR. One source is U.S. Intec., P.O. Box 2845, Port Arthur, Texas, 77643. The factory is Grease Guard, Inc., 931 South Roselle Road, Schaumburg, Illinois, 60193.

MISCELLANEOUS POINTS

Here are some small but useful items:

Felt Weight

Roofing felts are sold by weight in pounds per roof square as installed. This means that there will be some extra square footage to the roll. The rolls cover 108 sq ft with the extra 8 sq ft being used for the head and side laps. Fiberglass felt weighs about 50% less than asphalt felt.

Acrylic and Aluminum Coatings

Some BUR manufacturers require a 30 day curing time before acrylic or aluminum reflective coatings are applied.

Some BUR manufacturers forbid the use of solvent-type aluminum reflective coating. Water-based aluminum coating is permitted.

Asphalt Emulsion Coatings

Some BUR manufacturers require a 30 day curing time before applying asphalt emulsion products.

Torch and Kettle Equipment

All OSHA safety rules apply. See the chapter on safety.

- LP gas cylinders are to have a vapor withdrawal valve and a pressure gauge that can read 0–60 PSI.
- Pressure regulator is to be able to adjust from 0–60 PSI.
- Hoses must have a working pressure rating of 350 PSI, a burst pressure rating of 1750 PSI.
- Torches must be equipped with a shut-off valve, a pressure release trigger, and a support stand.
- Torches required for hot roof application are a detail torch, a field torch, and a torch wagon.

This equipment can be purchased from several companies. Here is one that I know of: FLAME Engineering, Inc., P.O. Box 577, La Crosse, KS 67548, 1-800-255-2469.

Non-Torch-Applied BUR Systems

There are a few heat-applied BUR systems that do not require flame or torch heating. They do require specially heated rollers and membranes manufactured to be heat applied. One such company is: Mweld/Inc., 700 Highway 365, P.O. Drawer 1288, Nederland, Texas, 77627.

Hot, Cold, Electric, and Torch

You will come upon these terms frequently when doing built-up roofing work. Each term references the same thing, the method of applying the roof coating to the roof. Cold mopping is where the asphalt is applied to the roof plies in an unheated state. Special asphalt emulsions are designed for this purpose; see the chapter on coatings. Hot mopping is done with heated asphalt applied to the felt plies. Gas and electric torch application is where the felts are impregnated with asphalt and then heated in place on the roof. The heat melts the asphalt and binds the plies.

12

Sheet Metal/Plastic Roofing

INTRODUCTION

This chapter will introduce you to the sheet and corrugated metals and plastics commonly used for roofing. I do not discuss metal tile in this chapter. Metal tile is in the chapter on tile. The last part of this chapter contains information on preparation and painting of metal and metal roofing.

WHY USE A METAL ROOF?

Why would anyone want to use a metal roof? The answer is rather simple—metal is one of the most durable substances we have. It can withstand attacks by insects, wind, rain, ice, fire, rodents, and man with ease. Metal is reasonably inexpensive. Although not used for residential construction that much anymore, especially in the Western United States, metal roofs have been around for hundreds of years. Most of the older farm buildings in the New England countryside were roofed with metal and are still going strong.

When one thinks of a metal roof, one thinks of galvanized or painted steel, but this is not so. Actually, there are millions of metal roofs made of aluminum, copper, and other alloys as well as steel. An alloy is a mix of various metals to obtain a material for a particular desired purpose.

MATERIALS AVAILABLE

Aluminum

Aluminum is the newest material in this group. Aluminum oxidizes almost as quickly as it is manufactured. This oxidation protects the metal from the effects of weather, and allows aluminum to heal itself within hours.

Scratch a piece of aluminum and you will see bright metal. A few hours later you will be hard pressed to see the scratch mark, as the oxidation has already reformed. Aluminum is lightweight, easy to form into complex shapes, abundantly available, and inexpensive.

Aluminum roofing may be supplied with an anodized or bonded coating applied. Anodizing is where the coating is applied with electrolysis or electroplating methods. Bonded is a glue or painted on method of applying the coating. Of the two methods, anodizing is considered the best.

Copper

One metal roofing material that has been around for centuries is copper. Copper doesn't rust, it's easy to form into odd shapes, and it is available in various thicknesses and sheet sizes. Copper, if coated with a clear varnish, will retain its brilliant golden bronze look for generations. Allowed to weather without a clear coating, it turns pale green due to oxidation. Many churches, government buildings, expensive estates, and schools are roofed with copper.

Copper is not sold by metal gauge, it is sold by the number of ounces per square foot. Standard weights of copper for roofing use are 16, 18, and 20 ounces. It can be supplied with various plastic coatings for prevention of oxidation.

Both copper and aluminum have advantages of being ductile, easily worked into various shapes, fire-resistant, insect and rodent resistant, and not having to be top coated. Both have the disadvantage of being easily dented and costing more than steel.

Steel

Steel roofing is the least expensive of the three main roofing metals. It is structurally strong. But, steel does need to be coated by either zinc galvanizing or paint to prevent rust formation. Rust is oxidation very much the same as the oxidation of copper and aluminum, but with one major difference. The oxidation of copper or aluminum protects the base metal and stops quickly. Rust, on the other hand, eats into steel and consumes the base metal.

Steel roofing is supplied as raw uncoated steel, factory primed steel, or factory finished steel. The factory finish may contain a rust-proofing primer, a color coat, and an abrasion layer.

Tin and Tinplate

Tin is a soft metal with a white-blue tinge. Tin is corrosion-resistant and melts at somewhat a low temperature, about 850 degrees Fahrenheit. It is used with zinc as solder and as a coating for terne.

Tin is malleable and can easily be shaped to fit difficult shapes. Most tin is sold as foil and not used for roofing. Tinplate is steel with a bonded layer of tin on both sides and is used for roofing. Many older metal roofs are stamped tinplate. The stampings are of various geometric designs and nonstandard from manufacturer to manufacturer.

Galvanized Steel

Galvanized steel is steel with a coating of zinc. Terne and tinplate are often improperly called galvanized steel. There is a difference. Tinplate is coated with tin, terne with lead and zinc, galvanized with zinc. Some steel is coated with an alloy of tin-lead-zinc and is considered galvanized. Some nails and sheet metals are coated with zinc and chromate and are considered galvanized. You be your own judge on what is or isn't galvanizing.

Lead Sheet

Lead sheet is very expensive and is not normally used for roofing. It is sometimes used as a flashing material. Lead is soft, malleable, and easily formed to complex-shaped surfaces. It is used with tin to make solder and used by itself or as an alloy for coating steel.

In certain situations lead is the only roofing material that will do the job. If a building houses radiation-sensitive equipment or equipment that produces harmful radiation, then lead is the answer for blockage of this radiation. The radiation blocked is in the X-ray and gamma-ray ranges that are harmful to people and sensitive electronics.

The roof should be engineered by a licensed civil or mechanical engineer. This person should specialize in this type of radiation protection.

Composites

See Fig. 12–1. Many metal shingles and some metal sheet roofing are composites. A composite is a material made up of several different materials. Typically, the base or core material is steel or aluminum. To this core a protective coating is applied. Application of the coating may be by pressure lamination, electroplating, hot dipping, gluing, or spray painting. The protective coating may be the final color or another color coating may be applied. Some manufacturers use a clear binder coat sprinkled with mineral granules. The granules give U/V protection, abrasion resistance, and color to the panels.

Monel

Monel is a substitute for stainless steel and is very corrosion resistant. See Appendix B, table 13 for more information.

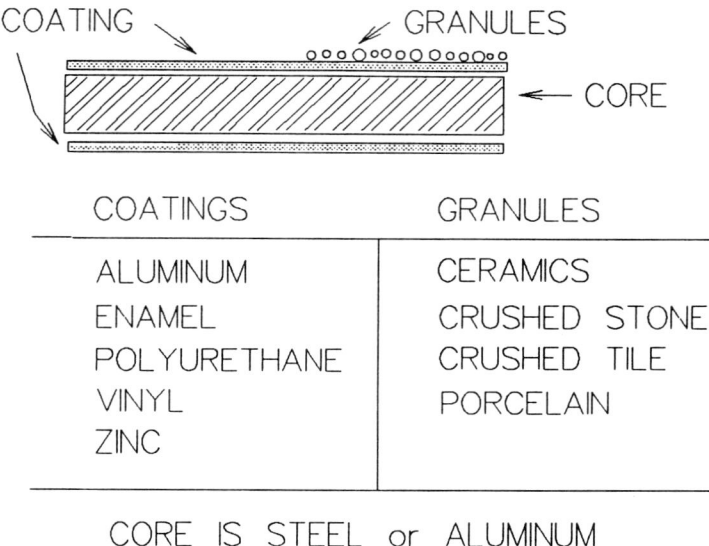

CORE IS STEEL or ALUMINUM

Figure 12–1   Composite metal roofing material

Terne or Terneplate

Terne is steel hot dipped in a tin-lead alloy. Terne is used for steel roofing that is to be painted. See Appendix B, table 13 for more information.

Metal and Plastic Composites

Some companies produce a composite material that has a plastic core sandwiched between two light gauge aluminum or steel sheets. The material is thin, lightweight, and can be scored and bent to fit around corners or other angles. It is approved for some construction, but not all. Check with your local building inspection department before purchasing.

Gold and Gold Leaf

A very expensive roof covering is gold, which costs about $350 per ounce at this writing. Gold is elegance at it finest and is used on roofs of very expensive private homes and many government buildings. Gold does not corrode or discolor. It does not rot or mildew and is resistant to most salts and chemicals. It is an ideal roofing material as it is maintenance-free and super long-lasting.

The gold used to cover a roof is sheet gold, real gold pounded or rolled into foil sheets only thousands of an inch thick. These sheets are then hammered into place over steel or another metal roofing surface. No one has ever attempted to steal the gold off a roof because it would mean peeling 1/10000 in. thick, hammered in place, gold from a roof.

As a note of interest, that $350 an ounce price goes a long way when you are using super-thin foil sheets. It covers dozens of square feet of roof surface.

PHYSICAL SIZES AND WEIGHTS

All metal roofing offers advantages to the roofing contractor, the first being coverage. Metal sheets are sold in 2, 3, and 4 ft widths, 8, 10, 12 ft or more in length. The standard width for aluminum corrugated sheet is 26 in. The standard for corrugated, galvanized steel sheet is 27-1/2 in. This gives each sheet a lot of roof coverage per sheet as compared to shingles or roof tile. Installation time is greatly reduced. Metal sheets will not soften and melt in extreme heat or become brittle and crack in extreme cold. To the roofer, the advantage is that metal roofing can be applied in almost any kind of weather. Whereas asphalt shingles and other roofing materials must be installed in warm dry weather.

LOWER INSURANCE COST

To the building's owner, a metal roof usually means lower insurance cost, due to metal roofs being almost fireproof. Of course, any material that gets hot enough will melt, and metal roofs are no exception. Metal roofs do tend to prevent combustion and retard flame spread, whereas asphalt and wood roofs do not.

METAL ROOFING ADVANTAGES

One definite advantage of metal roofs is that they can be sealed against the elements if designed and installed properly. This is not so with most shingle type roofs. Shingle roofs require a 4:12 or greater slope so that water will run off without backing up under the shingles and entering the building. Metal roofs can be installed with as little as 1/4 in. slope in 10 ft and still be leak-free.

Due to the size of shingles, most shingles work best when used on large, smooth surfaces. Metal is malleable and can be used to roof small odd shapes. Copper sheets can be used to roof over very small, high slope steeples, an almost impossible job to do with shingles.

Like everything, there are exceptions. In many areas of the country, the use of wood shake shingles is outlawed, because the fire danger is too great. This is especially true of Southern California, with its yearly rash of brush fires. Los Angeles County banned wood shake shingles in 1989. To counter this and still present the look of shakes, many manufacturers have now started making imitation shake shingles from coated steel, aluminum, clay, and concrete. These shingles have the advantage of being fire-resistant and the disadvantages of requiring large flat installation surfaces. What you will find is a mixture of shingles and sheet metals on many buildings. This gives the advantages of both.

To the architect, metal roofing can make life easy. Free-form shapes, that are at best difficult to form with other materials, are somewhat easy with metal. You will see pyramids, geometric domes, cones, sawtooths, and dozens of other shaped roofs, and sheet metal covers most of these roofs. If the designer can imagine it and the salesperson can sell it, it can be roofed with metal.

Where to Use It

There are several building types where metal roofs are put to their best advantage. I'll start with the rural areas, mainly farms and out-buildings.

- Animal shelters
- Equipment sheds
- Feed and grain storage buildings
- Garages
- Hay barns
- Patio covers
- Silos

These are classified as utility buildings and are not built for beauty, just efficiency. Sheet metal or corrugated metal roofing fits the bill.

A silo is a tall round building used to store chopped up corn stalks and grains. You will see similar buildings called grain elevators (silos) used to store oats, barley, corn, etc. for the general consumer market consumption. Since the tops of the silos are dome shaped, the sheet metal roof has several formed wedges interlocked together to form the dome.

In the northern states, sheet metal roofing has been around for generations, both on the farms and in the cities. The metal itself is of a past era. It was for the main part embossed with various designs to resemble shingles and geometric artwork.

In newer homes, you will find sheet metal roofing used for steeples and used over patios, carports, and storage sheds. You may find sheet metal roofs used over pools and garden areas. Metal sheet and shingles are quickly replacing wood roofs as the desired roofing for fine quality homes. They increase the resale value due to good looks, low maintenance, high fire resistance, and lower insurance premiums.

In the public arena, sheet metal roofs are used over walkways, as part of store fronts, and to cover low-cost school and government buildings. This includes their maintenance yards. Sheet metal roofing is used to cover equipment storage yards, bus stops, and shade areas in parks and other public areas.

The military uses metal buildings called quonset huts for low-cost living areas, storage areas, maintenance areas, and the like. Commercial use is for store fronts, store roofs, shopping center rest areas, building fascias, and loading-unloading areas. One must use a little care though. I've seen drive-through fast food stores that didn't account for large trucks turning the corner into the take-out window line. The roof corners were destroyed. Most trucks require at least an 11 to 14 ft clearance.

Factories may use metal for their roof covering. This includes their equipment storage sheds, temporary product storage areas, and other out-buildings.

Auto and boat dealers use metal roofing to protect their inventory from the effects of weather and sun. The dealers build free-standing metal awning type structures.

Many hotels and motels use metal roofing over their entryways to protect their customers from the weather. Apartment owners frequently build carports with metal roofing. Gas stations use metal roofing over their gas pumps with allowance for truck height clearance.

Although it is usually an aluminum siding job, many roofing jobs will require you to install aluminum soffit under the eave areas. It is easy to do and results in not only a better looking roofing job, but also extra profits for you.

As you can see, there is a big market for the roofer who knows metal roofing. There is a bigger market for the roofer who can do a good job installing it.

Additional Advantages and Disadvantages

I covered some advantages and disadvantages of metal roofing, here are a few more:

Advantages

- Metal roofing can be painted to match the mode or desire of the building's owner. Do not use self-cleaning or chalking paints. The color will run onto the building being covered.

- Metal roofing can be repaired with patching compounds and paint. Sheet removal is unnecessary in all but the worst instances.

- Metal roofing is a good conductor of heat and will quickly melt off snow and ice. Many homes in the New England area use metal roof edging to prevent ice and snow buildup. When the sun's rays start to warm up the metal, the snow slides off the slippery surface.

- Most metal roofing does not require sheathing or underlayment felts. This is a cost saving in material and labor.

- Metal roofing can cover greater spans that most other roofing.

- Metal roofing is long lasting and will last upwards of hundreds of years if properly maintained. This can be a super cost saver.

Disadvantages

- Metal roofing can be noisy, especially in a hail or rain storm. The newer products are either coated or laminated with other materials to help minimize this noise.

- Metal roofing can be slippery to walk on, especially when wet. One must take care during maintenance performance.

- Metal roofing does conduct heat and cold very well. This requires good ceiling insulation when used for containment of living areas.

- Metal roofing does conduct electricity and MUST be properly grounded or equipped with a grounded lightning rod or two (Fig. 12–2).
- Metal roofing will sometimes react poorly with redwood or cedar. Redwood and cedar in contact with metal will release tannic acid. Tannic acid will attack the metal and stain the wood black. Always use a butyl rubber strip or caulking bead when interfacing redwood or cedar with metal.
- Metal, steel, or iron will rust and must be coated with paint or another material. Be sure the coating is compatible with metal roofing as metal has a high rate of expansion and contraction. Coatings that do NOT match this expansion/contraction rate closely will soon crack and peel.
- Improperly installed metal roofing will leak like a sieve. Follow the manufacturer's directions to the letter. Use the proper recommended fasteners and weather stripping.

Tools Required for Metal Roofing Work

You will need certain hand tools and certain specialized metal working tools to be successful in the metal roofing field. For installation of factory finished, corrugated metal and plastic panels, you only need the following:

- A tape measure
- A wood hand saw

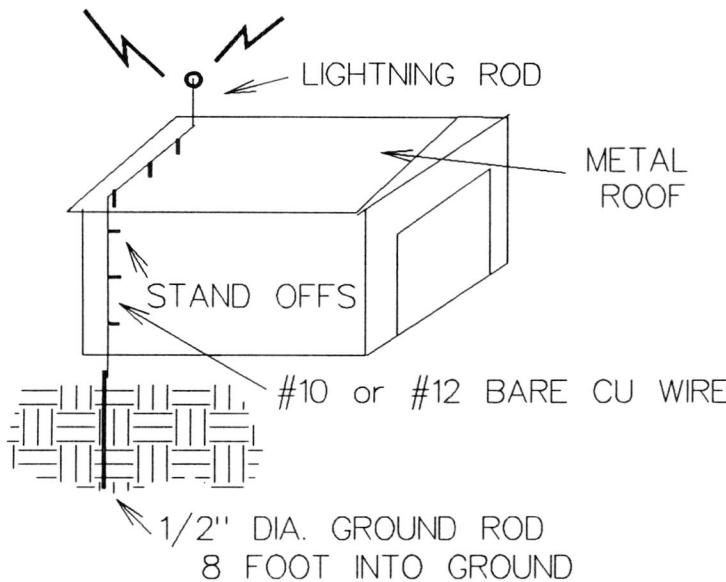

Figure 12–2   Metal roof should be grounded against lightning

- An assortment of screwdrivers
- A caulking gun with clear RTV sealer
- A wood rasp and a metal file
- A claw hammer
- A power drill with carbide-tipped bits
- A 3 ft and a 6 ft straight edge

When you get into the sheet metal roofing, you will need the following as well:

- An assortment of straight, right hand, and left hand cutting shears
- A metal bender or brake, either manual or power driven
- A metal shear, either manual or power driven
- An assortment of bending jigs
- A hole punch set
- A circular saw with an assortment of carbide-tipped wood and metal blades.

That should just about do it for the tools. Some manufacturers may have specialized sheet to sheet interlocks that require special tools, but your sheet metal supplier can inform you on this.

I mentioned using carbide-tipped drills and saw blades. There are two good reasons for this. First is that carbide-tipped cutting tools last several times longer than noncarbide-tipped tools. Second, if you are installing plastic sheets, the corrugated plastic panels are reinforced with fiber glass. Fiberglass reinforcement is an abrasive and will wear drill and saw blade tips very quickly.

The power drill and power saw equipment should be variable speed equipped. Plastics must be sawn and drilled at slow speeds. The reason is that plastic has a low melting temperature and the high-speed drills and saws will exceed this temperature. The plastic will melt back together as fast as you cut it.

Preformed Metal Roofing

Sheet metal roofing will generally be supplied with built-in joining lips or seams. These lips are brake or roll formed at the factory and should not be cut into if it can be avoided. In situations where cutting is required, there will be a raw edge, i.e., uncoated metal. This uncoated metal should be painted before being installed. Care must be used to seal the seam area of the cut properly. Use RTV as a sealer.

Metal Roofing over Metal Roofing

If the old metal roof is in such a state of disrepair that it needs a new roof, then I recommend a complete tearaway. The ribs and seams of the old roof will hamper your new installation if the old roofing is allowed to remain in place. Besides, the old metal is probably well rusted and may contaminate the underside of the new roofing material.

Exception: some old metal roofing was in the shape of large sheet tiles. There are no standing seams, just overlapping seams. These roofs can be roofed over with new sheet metal roofing with a minimum of problems. All rust and corrosion should first be removed from the old roof. The old roof should be sealed with a spray coat of paint or polyurethane before installing the new roofing.

Asphalt Shingles over Metal Roofing

I will advise you to remove the old metal roofing. The reason is this: In past months I have seen several asphalt roofing jobs prematurely failing. When questioning the building's owners, I found that the asphalt shingles were installed approximately 10 to 12 years ago. All the buildings were in the same area and all had roofs that needed replacing, because the asphalt shingles were dried out and curled.

What happened is that the underlying metal roofing is acting as a heat sink and as U/V reflector. This is creating a very hot boundary layer between the asphalt shingles and the old metal roof. This heat coupled with the reflected U/V radiation is drying out the asphalt.

As a solution, I reviewed several options. First is to paint the metal black or another nonreflecting color, but black and other dark colors absorb heat and add to the problem. The second idea was to install sheet foam insulation on the metal and cover it with plywood sheathing. This will work, but is more expensive than a tear-off. Plus, it has been proven that insulating directly to sheet metal and then roofing over the insulation cuts the life of a roof considerably.

This applies to BUR systems, but in this situation it may not apply. The third idea was to sheath over the metal with plywood and no insulation. But this wouldn't work because the bottom of the sheathing will not receive air circulation and will eventually deteriorate. The last solution is to tear the old metal roofing off and start fresh.

Roofing over Odd or Geometric Shapes

We can fill another book with all the possibilities involved. I will offer these suggestions if you intend to do sheet metal roofing. Find reliable sheet roofing suppliers and ask them for their installation instruction manuals. The supplier or manufacturer should have a complete library of designs and problems with solutions.

Second, attend a course or two at your local trade school or community college. There are courses in sheet metal drafting and in sheet metal fabrication. These courses are low cost and should give you the basics of sheet metal design.

Third, work for a sheet metal fabricator or sheet metal roofer for a minimum of 6 months. Keep detailed notes of the problems and solutions. Six months should give you enough time to study methods and workmanship. It is ample time to learn about the supplier and customer base that you will be working with.

Here are some general tips and construction methods used in installing sheet metal roofing:

Hips and Ridges

Hips and ridges of corrugated or sheet metal roofs must be closed in just as if you were applying shingles to the building. There are formed ridge caps for the purpose. These caps come in two different shapes. The first is a smooth edged cap and the second is a corrugated edged cap. Ridge vents can be installed if the framing allows. See chapter on flashing and venting.

Valleys

Valleys are standard W-flashing valleys. The exception is that the splash divider only need be 3/4 in. high. If you are doing several different types of roofing, then I advise purchasing only flashing with 1 in. high splash dividers. These are required on all other roofs. Stocking or purchasing only one will cut inventory and paperwork costs. See chapter on flashing and vents.

Joining Metal Sections

Figure 12–3 illustrates one design of metal sheet roofing, where the doubled areas are the joints. The nail alignment shows where the purlins are located.

Figure 12–4 shows another sheet metal roofing pattern. It also shows that the owner is prepared for sun or snow, the two plastic panels provide light to the interior. The odd-looking standoffs are roof heating cable holders and are located over the main entry. The

Figure 12–3 One type of seam in a metal roof

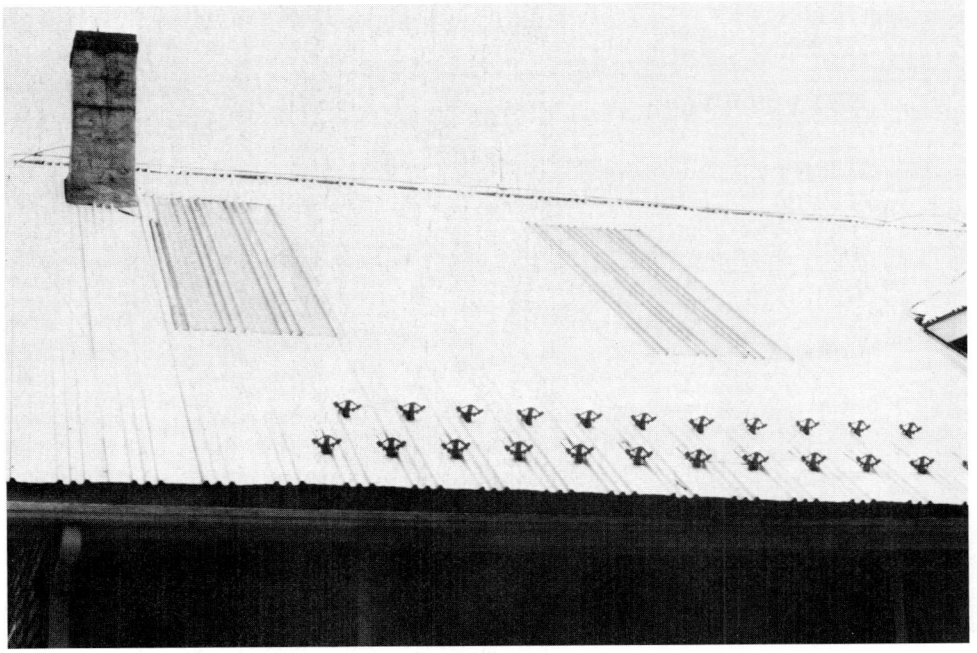

Figure 12–4   Sheet metal roof

roofing and the gutter system should be grounded to earth ground whenever roof heater cables are used.

Figure 12–5, top left illustrates an older, not generally used, method of securing sheet metal roofing. Each sheet is flanged and butted together. A cap is then placed over the butt joint seam and then soldered its full length. This method makes a good watertight expansion joint. The disadvantages are soldering times and the cost of the solder.

Sheet lead flashing can be used for seaming panels. The panels are bent with an L shape lip on the connecting sides and butted together. A strip of lead is then placed on top of this joint and pounded around the joint in an inverted U shape.

A standing seam is used for joining noncorrugated metal roofing (see Fig. 12–5, lower right). The seam can be filled with sealer if an airtight seam is required. The panels can be slipped together lengthwise or they may be connected by tilting and inserting from the sides. If the roof section from eave to ridge is longer than the panel length, then you must use two or more panels. Be sure that the uppermost panel overlaps the lower panel.

Fascia and Gutters

One advantage of sheet metal roofing is that you can form gutters and fascia from the roof sheets. For fascia, just bring the sheets over the edge by the width of the required fascia and bend into place. For gutters, mount a 2 × 4 across the length of the eave. Bring the sheet

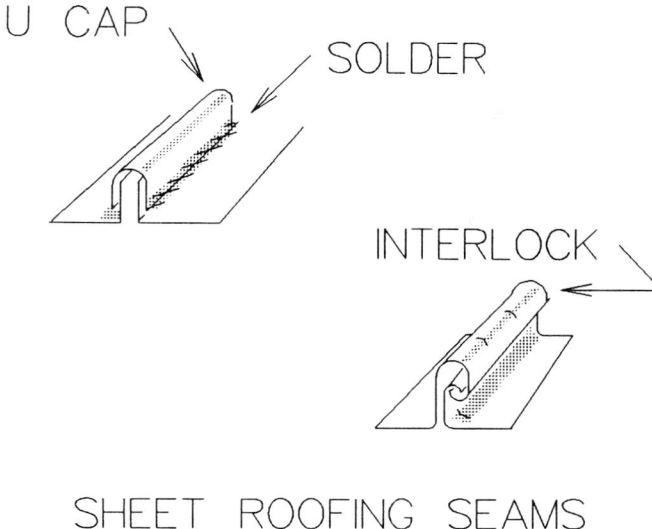

SHEET ROOFING SEAMS

Figure 12–5   Two types of seams

metal into the valley formed and up and over the 2 × 4. Be sure to add a little slope as you proceed from one rake to the other. Add a downspout and you're in business. To be on the safe side you can line the gutter valley with roofing felt and a U shaped piece of flashing. Do this before extending the roof sheets into the gutter valley.

Corrugated Roofing Installation

See Fig. 12–6. Corrugated roofing sheets require support under the corrugations. Wood, plastic, or foam rubber gasket strips are used to provide this support. These gaskets should be used at both ends of the sheet and at all purlins or other places where the sheet is nailed. These gaskets, termed *closure strips,* provide another function. Sheet roofing often vibrates in a wind and it tends to rattle during rain and hail storms. The gaskets help to prevent transmission of vibration and sound to the building's other structural members, namely the purlins and rafters.

Used at the eaves, the gaskets help seal out moisture, wind, insects, birds, and the like.

Wood closure strips should have a coating of clear RTV on them. The RTV is applied to the mating surface of the closure strip and the roofing to dampen noise transmission.

Blocking and Purlins

Corrugated roofing is not normally installed over a sheathed roof. Roof sheathing serves two prime purposes. First, it gives a solid base for the roof to be installed on. Second, it helps prevent the rafters from twisting. In an open rafter design, no sheathing situation,

FOAM RUBBER FOR CORRUGATED ROOFING

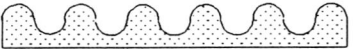

SEALS END AND GIVES FINISHED LOOK

METAL FOR CLAY TILE ROOFING

KEEPS BIRDS AND INSECTS OUT

Figure 12–6   Closure strips and bird stop

there exists a potential for problems. Will the rafters twist? To what do we nail the roofing? Figure 12–7 solves both problems. First, use blocking between the rafters. The blocking should be on 4 ft centers. This will help prevent the rafters from twisting. Second, add nailer strips across the rafters and the length of the roof. These nailers are called purlins. Purlins should be on 2, 3, or 4 ft centers. The corrugated roofing sheets are nailed to these purlins.

Vent Space

Whenever a shed-type roof is butted to a building and the sides of the shed are not enclosed, a vent must be provided. This vent is to be 8 in. wide and run the full width of the shed to

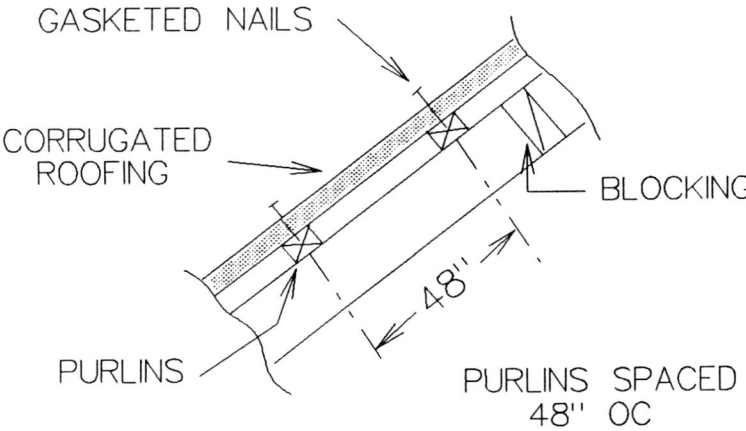

Figure 12–7   Use of purlins for corrugated roofing

building intersection. Example: A patio cover fabricated from corrugated plastic panels and attached above the exit door of the building. The vent space does two things. First, it removes excessive heat buildup from under the roof. Second, it helps prevent excess wind uplift under the roof that may blow off the roofing.

If you are worried about it, you can build a rain shield over this opening. Install the rain shield 8 or more inches above the vent opening.

Wind Uplift

An open shed-type roof such as a patio roof will be torn off if not properly vented and fastened. Many contractors and homeowners install corrugated roofing with as few fasteners as possible. They think that because the material doesn't weigh much, it doesn't require many fasteners. This line of thinking is 100% wrong. It is because the material is so light that it needs as many fasteners as you can put in it. Wind uplift can exert forces that will rip a 45 lb per square foot BUR roof off a building. We are dealing with 2 to 4 lb per square foot roofing, when roofing with corrugated metal or plastic.

My advice is to install cross braces or purlins every 36 in. along the beams or rafters. To these purlins, screw closure strips with the screws on 12 in. or less centers. The corrugated roofing is then fastened with gasketed nails or screws through every rib that contacts the closure strips. To be doubly safe, the closure strips should be glued to the purlins and the corrugated roofing sealed to the closure strips with clear RTV. This is in addition to the nails and screws used.

Height of Shed-Type Roofs

The building code states that covered walkways must have a clear head room height of 8 ft from the floor. Be sure your finished roof project meets the code requirement.

Rake Treatment Using Metal

Metal L flashing can be used along the rakes. The sheet metal panels should be bent at the rake edge. The 225 degree bend forms a 45 degree V with the face of the metal roofing. This edge now acts as a channel to direct water runoff down the roof to the gutter or eave. Chipped or exposed metal should be touchup painted to prevent future rusting or poor visual effects. An L metal flashing is then installed over the bent-up channel to dress it off properly.

Welding of Corrugated Metal Roofing

Welding of corrugated metal roofing is not recommended. The metal used is normally too thin to be welded properly and without burn-through. I do not recommend that welding be attempted on metal gauges thinner than 20 gauge. Aluminum and stainless steel of any gauge should be welded by a professional welder trained to weld these materials. Soldering or brazing of seams can be done on all gauges of sheet metal due to the lower temperatures involved. In doing residential work, I recommend that seams be sealed with commercially available sealers rather than by welding, soldering, or brazing. See the chapter on sealers and caulks.

Venting a Metal Roof

The best attic vents for a metal roof are ridge vents and gable vents, as they will be the easiest to install. Flat roof vents will require flashing and flashing requires shingles or another roof covering to protect it. You do not have this option on a metal roof.

If you must flash around pipes or other vents, then use flashing that can be soldered or brazed to the roofing. Elastomer sealers can be used but may not last as long as they should. Solder or brazing will last as long as the remainder of the roof lasts. You cannot braze, but you should be able to solder aluminum to steel or to aluminum. The alternative is to use lead flashing sheet and pound it into place with a big mallet.

Mobile Home Roofs

Most mobile homes and recreational vehicles use corrugated sheet metal for their roofing (Fig. 12–8). These roofs are subjected to physical damage from people walking on them and packing baggage on them. There may be physical damage from road dirt, blown dust, and bird and tree droppings. There may be chemical damage from salt water spray and cleaners.

Most of these roofs are constructed with a very light frame of 2 × 3 lumber or with hat sections of aluminum, steel, or plastic. A hat section is a U shaped piece of sheet used to

Figure 12–8   Metal roof on an RV

strengthen and support a thinner material. Due to this light construction design, the corrugated roofing will bend and distort if walked on. Standard auto body or sheet metal repair methods must be used for repair. Auto body fillers can be used to repair small dents, holes, and creases. Use automotive primers and paints for base protection. See chapter on sealers and caulks for reflective top coatings.

Roofing Over a Mobile Home

You can roof over a mobile home that is permanently fixed in place. Roof trusses can be designed to fit over the existing metal roof, and then the owners can install the roof of their choice to these trusses.

Snow Removal

See Figure 12–9. Very little snow accumulates on this roof. The high slope coupled with the metal roofing prevents the snow from sticking. Check the center window area, a drain is present. What I don't like is that the drain empties onto the lower roof and allows too much water and dirt to flow onto the lower roof. This eventually erodes the shingles and discolors them as well.

Concealment Areas

Figure 12–10 shows a nicely designed commercial sheet metal roof. Behind the sign are slats. Behind the slats is a small built-up roof where the air conditioning unit and roof vents are concealed.

Figure 12–9   Snow has little chance of sticking to this roof

Lightning Rod Installation

Lightning, which generates a lot of electricity, loves to hit metal roofs. When lightning hits, it does a lot of damage. Metal roofs should be grounded for best prevention of the building and its occupants (see Fig. 12–2). This is done with the use of lightning rods and grounding stakes.

The lightning rod is installed on the highest peak of the roof. A AWG (American Wire Gage) #12 or larger wire is then brought from the lightning rod to the ground. Stand-offs are used to hold the wire away from the building. At the ground, a 1/2 in. diameter rod, 8 ft long, is driven 6 to 7 ft into the earth. The wire is attached to this ground rod with a wire rod connector. Most building supply houses can provide the lightning rod, wire, standoffs, ground rod, and connector. Estimated cost is $45 to $85, which is money well spent if it prevents the building from burning down or prevents an occupant from getting electrocuted. Check local building codes for proper installation of the lightning rod and its grounding. Local codes vary and may be different from what I described.

Radio and TV Reception

Metal roofs conduct electricity, and this includes microwaves received from radio and television stations. The roofing acts like a big antenna and this can interfere with other sensitive equipment in the building. To avoid this, the roof should be properly grounded. But, there is a catch. A grounded metal roof will block radio and television reception if it is desired. A roof mounted antenna will be necessary for best reception.

Figure 12–10   Commercial metal roof

PLASTIC ROOFING

Fiberglass Panel Alternative

Thus far we have covered metal sheet goods as roofing material in this chapter. There is a problem with metal that I have not yet mentioned—it does not pass light. Suppose you are covering a storage barn with metal sheet and want to have a sun/daylight lighted building interior rather than having to use electric lights. The solution is to install fiberglass panels.

Fiberglass Panels

Fiberglass roofing panels are generally the same size and shape of the metal panels, 2×8 ft. Larger or smaller panels can be purchased.

The panels are made of polyester resin and fiber glass. They are weather resistant, rodent- and insect-resistant, and inexpensive. With the correct additives, by the manufacturer, they are U/V- and fire-resistant. They can be obtained in opaque, translucent, or transparent configurations. Opaque is where very little to no light passes through. Translucent passes light but blocks detailed viewing, the light that passes is diffused or scattered. Transparent lets light through and is almost glass-like.

The panels can be purchased in flat or corrugated sheets. They can be purchased as neutral color, clear, one color, or multicolor. Fiberglass panels are a good insulator of heat, cold, and electricity.

The panels can be intermixed with metal panels or used by themselves to cover an entire roof, if code allows. Figure 12–4 showed a building with fiberglass light panels. The installation techniques are very close to those used to install metal roofing.

Selecting Color

By selecting the correct color, you can control plant growth. Color will give coolness or warmth to an area, or give the impressions of gayety, businesslike, or richness to an area.

Green panels will provide coolness and calm, and will aid in green foliage growth. Red panels will provide warmth and aid flowering plants. Multicolored or yellow or orange add a festive feel. Gold or bronze will provide a businesslike or rich feel. Blues provide a friendly feel, and browns, an earthy feel.

The only thing I must warn you about is color fastness, the ability to maintain the bright richness of color over time. It generally does not happen and most panels will fade with exposure to sunlight. So, I suggest the brighter, darker colors be installed, as they will eventually become pastels. The lighter colors will fade to off-color whites.

Light Transmission

Plastic roofing panels will transmit light through them. The color and transparency of the panel control the amount of light. Clear panels will transmit up to 94% while darker colors may only transmit 10%. The average colored panel transmits approximately 30%. Check the manufacturer's specification for the color and light transmission you desire.

Heat Transmission

Plastic roofing panels will transmit heat through them. The color and transparency of the panel control the amount of heat. Clear panels will transmit up to 84%, while darker colors may only transmit 14%. The average colored panel transmits approximately 25%. Check the manufacturer's specification for the color and heat transmission you desire.

Where to Use Fiberglass Panels

The primary usage is over outdoor areas where filtered light is desired. Examples of outdoor areas where fiberglass panels are used are as follows:

- Carports
- Entryways
- Gardens
- Greenhouses
- Patios
- Sun decks
- Swimming pools
- Walkways

They are also used for skylights, room dividers, and as fencing.

Lexan™ Sheets

Lexan is a polycarbonate plastic with exceptional strength. It is used for display case and store front windows, boat windows, and wherever a clear, break-resistant panel is required. Lexan is frequently used for skylights and roof panels. The proper thickness of Lexan will stop a bullet or a sledge hammer blow with ease.

The material is sold in sheets from a fraction of an inch thick to several inches thick. It is not an inexpensive plastic and ranks high in cost compared to other plastic sheet goods.

Lexan can be obtained as crystal clear or with various colors in the tan or bronze range. U/V protection is generally built in. Lexan is not resistant to scratches and certain household or commercial chemicals and cleaners. Cleanup should be with soap and water.

Lexan will burn and give off toxic fumes while burning. Consult your local building code department before purchasing.

Physical Sizes and Weight

Fiberglass corrugated roof panels are available in widths of 26, 27-1/2, 33, 33 3/4, 35, 36, 40 and 42 in. Standard available lengths are 3, 4, 5, 6, 7, 8, 10, 12, and 14 ft. Special-order sizes are 6 in., 9 ft, and 13 feet. Weight is from 4 to 6 oz per square foot. The most commonly sold width is 26 in. The most commonly sold lengths are 8 and 12 ft. Common size for ribs is 4 in. center to center and 5/8 in. high.

Caution on Weight Loading

Fiberglass panels will not support a person's weight. The panels will break and personal harm may result. Do NOT install fiberglass panels where a person must walk on them. Example: As a patio cover under a fire escape or window used as a fire exit. If used this way, adequate blocking must be provided.

Caution on Fire Possibilities

Fiberglass panels are plastic and plastic burns. While burning, plastic will give off toxic smoke and will melt and drip. These drops of melted plastic will be very hot and will stick to clothing, furniture, and skin. The drops may be on fire and cause spreading of the initial fire to other areas. It is for these reasons that the building codes restrict the use of plastic as roof and wall coverings. Plastics are covered in U.B.C. Chapter 52. In essence, it states that a single pane of plastic glazing may not be more than 16 sq ft when used above a first story. The vertical height of the pane may not be more than 4 ft and the plastic may not cover more than 25% of the wall area of any story. The 16 sq ft dimension can be increased to 24 sq ft if there is an approved, working sprinkler system installed in the building. U.B.C. 5206 covers corrugated plastic panels and treats them as if they were skylights. Glass skylights are covered in U.B.C. Chapter 34.

Cleaning Plastic Panels

Plastic panels will scratch if cleaned with abrasive cleaners. Do not use steel wool, scrubby pads, or dirt-filled rags for cleaning. Use soft, lint-free cotton cloth or paper towels. Do not use harsh cleaners or chemicals for cleaning. Do not use ammonia or hydrocarbon or chlorinated solvent-based chemicals for cleaning. Use soap and water followed with a clear water rinse. Naphtha and kerosene can be used on most acrylic, methacrylate, and Plexiglas™ plastics to remove grease and adhesive glues.

PREPARING PLASTIC FOR PAINTING

If you decide that you want to paint over a plastic panel, then the panel must be cleaned and etched. Paint does not adhere well to plastics. Wash the panels with soap and water and follow with a clear water rinse. Etching can be done with hydrocarbon base solvents or with ketone base products such as methyl ethyl ketone, MEK. MEK can be used on acrylics and plexiglass to clean and etch them before painting. Also, MEK can be used to etch PVC, polyethylene, and polypropylene plastics. Use acetone or styrene monomer as an etchant for cleaning polyester base panels before painting, i.e., fiberglass panels.

PREPARING METAL FOR PAINTING

Bare metal sheet should be cleaned before being painted. New metal contains manufacturing oils and may be contaminated with oils, greases, and fingerprints. Paint will not adhere to a dirty or oily surface and blistering, flaking, and peeling will result if the metal is not cleaned. Start with a soap and water wash followed with a clear water rinse. Follow this with a rub down of denatured alcohol or paint thinner or auto body prep to remove any residual oils.

Rusted steel should be wire brushed and then cleaned with phosphoric acid, sodium bisulfate, or sodium hydrosulfate. Commercial plumbing supply stores should have these products.

Painted metal sheet should have all loose paint removed. This can be accomplished by wire brushing, sanding, or with a liquid paint remover. Methyl chloride base paint removers work best. Acetone and liquid sandpaper can be used to clean and etch painted surfaces that are not heavily damaged. Lacquer can be cleaned with commercial lacquer cleaner or with auto body prep.

NOTE: Some factory painted steel roofing is coated with lacquer or with polyurethane. The use of acetone and liquid sandpaper may remove or blister these coatings and cause a mess. Test the metal's surface before applying chemicals to the entire sheet.

PAINT TO USE

I recommend the use of low V.O.C. (Volatile Organic Compound) water-base acrylic latex paints for both plastic and metal sheets. Two thin coats will work better than one thick coating. As an alternative to the acrylic latex, you can use enamel or lacquer base automotive paints on the sheet metals.

A primer coat should always be used before top coating. Airless spray or rolling are the best application methods. Additives are available to make the coatings slip-resistant on the roof areas used as walkways.

Special industrial-strength paints are available by mail order from many companies. Be sure of what you are ordering, some products sold are no better than the consumer products you can purchase at your local paint store. Expect to pay $18–$28 a gallon for consumer paints, $40–$90 a gallon for automotive and commercial paints.

There are a lot of very good metal trim paints available. These are formulated to give a tough, abrasion-proof skin. They work very well on small metal surfaces such as hand rails. I advise against using these trim paints on sheet metal roofing. The coatings are very thick and may not expand or contract properly with the roofing.

REFLECTIVE COATINGS

Metal roofs can be painted over with reflective coatings if needed. The acrylic and aluminum coatings in the chapter on coatings will work fine.

INSULATING COATINGS

Proprietary coatings are available for insulating metal roofs. The coatings are sprayed on in 1/4 in. thickness to a depth of 1-1/2 in. Most are modified plastic foams that contain high quantities of glass fibers and aluminum. Nitrogen compounds may be incorporated. Nitrogen compounds swell when burning and provide an insulating barrier between the burning insulation layers and the layers under them. This tends to stop or slow down the spread of the fire.

The spray-on insulations are applied to the underside of the roof panels. Modified airless spray equipment is used for this operation. R-value is from 0.5 to about 5.5, depending on the product and its applied thickness. Check with your local building inspector to find out its use. The inspector should be able to inform you on code and who in your area might be authorized to apply the insulation.

A PAINTING TIP

Paint all wood or metal framing before you install the roofing panels. It makes the job much easier as you do not have to worry about getting paint on panels that are not there. Paint the entire framing, even where the panels will eventually be installed. If using S-Grn or S-dry lumber or treated lumber for the framing, only paint the lumber surface where the panels will be installed. Allow the remainder of the lumber to weather for 10 days before painting it.

If the under surface of the paneling is to be painted, paint it before installing it. This will help prevent rust formation at the mounting surfaces junctions.

If you are not going to paint the weather exposure surface of the panels, then use aluminum nails for panel securement. If you use steel or galvanized steel nails, you will have rust problems in the future.

Shake and Wood Roofing

INTRODUCTION

Please read the chapters on flashings and problems before reading this chapter. They contain information you will require. I will cover the specifics of wood and shake shingles in this chapter.

See Fig. 13–1. Shake shingle roofs are popular in some areas of the country, mainly the northwestern states where red cedar and redwood grow. This roof is starting to see the end of its useful life span. The roof is about 18 years old and the shingles are starting to split, curl, lift, and warp. Due to fire codes, it will not be replaced with wood shake, but with metal, asphalt, or some sort of tile.

WHY INSTALL SHAKE SHINGLES?

Even with all its faults, wood shingles or shake has a reputation for being high in quality and being long-lasting. This reputation goes back to the days when most other roofing materials were of poor quality or nonexistent. It is true that cedar and redwood deter insects, that they are very long-lasting woods, and that they weather to a pleasing color and shine. It is true *only* if you use the heartwood. The outer white wood is no better than pine or fir. Many people buy redwood for its name and not its quality, which is a costly mistake.

Figure 13–1 Shake shingle roof

Redwood and Cedar

Heartwood is difficult to come by and therefore very expensive. To the status seeker, the more costly a product is, the better it is. We have just about depleted the supply of redwood in this country. Why use an endangered species of wood when there are so many other, and many better, products on the market? Red cedar is an acceptable alternative, it is commercially grown, and resembles redwood in every way.

Another choice is pressure- and fire-retardant, treated fir, pine, or other commercially grown lumber. Pressure-treated wood will last upwards of 40 years. Fir and pine are replenishable, are commercially grown, and are inexpensive by comparison to redwood.

Insects and Fungi

Cedar and redwood are resistant to insects and fungi because of the natural oils and tannic acid contained within the heartwood. The softer white wood, sapwood, is no more resistant than fir or pine. The white wood is the wood from the bark to the heartwood and it is whitish in color.

Weathering Color Change

The content of oils, acids, and the minerals cause the color change. All trees drink water from the ground and all groundwater contains dissolved minerals. These minerals are carried into the tree and deposit themselves in the cells of the lumber. Cut lumber, exposed to the elements of wind, rain, snow, sunlight, and U/V radiation, changes these minerals into oxides. These oxides absorb and reflect light differently than the minerals from which they were produced. Color is the product of the light rays we see reflected from an object. If all light is reflected, we see white, no light reflected, we see black. Some reflected light and we see color. For lumber, we see light tans turn to dark browns or to silvery grays. The lumber then is said to be aged or weathered. It will take on a soft shine or sheen.

Tannic Acid

The tannic acid in redwood is a roofing problem that you as a roofer should know about. Tannic acid will attack steel. Galvanized flashing and nails are steel with a thin zinc coating. As the coating wears off, the steel is exposed and the acid attacks. This will result in black staining of the shingles and any walls under them.

The use of stainless steel, brass, copper, or aluminum flashing and nails will prevent this from occurring. If galvanized flashing must be used, then you must coat or gasket the flashing with butyl, neoprene, or silicon rubber everywhere it contacts the redwood.

Fire Caution

In fire-prone areas of the country, the use of nonfire resistant wood and shake shingles is prohibited, outlawed. Los Angeles is one such area. The reason is simple. One grassy field or house on the block catches fire. The sparks are then carried by the wind to the roof tops of surrounding buildings. The sparks lodge themselves in the cracks of the shingles and set the shingles afire. One such fire destroyed over 600 homes in less than 4 hours. Most of the homes left standing were those that had tile or asphalt shingles. Even some tile roof homes burned. The sparks blew under the tiles and set the roofing felt and sheathing on fire. Check your local codes before using wood roofing.

There are some new clear color, fire-retardant sprays on the market. At this writing, these products were still in the testing stages and not widely approved. Perhaps, in the future, we can just hose the roofs down with these chemicals and turn them into Class I roofs. That should please many insurance companies. It also will please many homeowners who are paying high rates for fire insurance because of their wood roofing.

Shake Versus Wood Shingles

The difference between a wood shingle and a wood shake shingle is that wood shakes are split from the log stock. Being split, they are rough surfaced. They are generally tapered split, tapered in that they are thicker at the weather exposure or butt edge than at the nailing or tail edge. The weather exposure edge profile is generally very rough. This is not true of wood shingles. The other difference between wood shingles and shake is that the wood shingles are all the same width. Shake shingles can be anywhere from a few inches to a foot or more wide (Fig. 13–2).

- Rebutted shingles are Grade #1 and Grade #2 shingles that have exactly parallel edges. These shingles will fit tightly together side to side and are used for tight jointed siding. The face side may or may not be sanded.

- Handsplit shake shingles are formed differently than wood shingles. Shake is split by hand and therefore is thicker and rougher than wood shingle. Grades are #1 handsplit and resawn, #1 tapersplit, and #1 straight split.

- The handsplit grade is formed from a two-shingle blank that is then sawn diagonally from end to end. This produces two shingles, each having one split side and one sawn side. Thickness varies from tail to butt and from shingle to shingle. Starter shake is handsplit shake with a length of 15 in.

- The tapersplit shake is all hand split. Both sides are rough and the thickness varies from shingle to shingle. Tapersplit shingles are tapered in thickness from tail to butt.

- Straight cut shake is the same as tapersplit except each shingle is cut from the same direction on the log. This produces a shake of near constant thickness of 3/8 in. from tail to butt. Siding shake is generally straight cut shake that is then sawn with parallel edges.

Spaced Sheathing Use

What do these differences mean to the roofing installer? Well, for starters, it means different roofing deck construction. Wood shingles must have air circulation under and between each shingle. If air circulation is not provided, moisture will be contained and fungi will grow. Fungi produces dry rot that will destroy the roofing.

The reason is that wood shingles are saw cut and lie flat. They can be nailed tightly together. To get air circulation under them, one must use a spaced sheathing.

Wood shake is split and is rough, thus shake cannot be nailed tight to the sheathing or to each other. Wood shake may be installed on a solid roof deck. In practice wood shake is frequently used with a spaced deck. Both the spaced and solid sheathed decks are approved and acceptable for wood shake.

Physical Sizes and Coverage

The thickness measured at the butt end of red cedar wood shingles is 0.40 to 0.50 in. depending on the shingle grade. Lengths of wood shingles are 16, 18, and 24 in. Coverage is four bundles per square for Blue, Red, and Black label shingles, two bundles a square for undercoursing shingles, one carton a square for rebutted shingles.

The thickness of red cedar handsplit shake varies from shake to shake. Shakes have a butt thickness of 1/2 to 1-1/4 in. in 3 ranges. Range #1 is 3/8 in. to 1/2 in., range #2 is 1/2 in. to 3/4 in., and range #3 is 3/4 in. to 1-1/4 in. Lengths are 18 and 24 in. for shingles, 15 in. for starter shingles. Coverage is 5 bundles per square except for the 18 in. by 3/8 in., straight side cut, siding shingles.

Coverage at Eaves, Valleys, Hips, and Ridges

A square of shake will normally cover 100 sq ft of roof surface at a weather exposure of 10 in. When the weather exposure is less than 10 in., the actual square feet of coverage will be less. Example: Weather exposure of 5-1/2 in. will result in a square of shingles only covering 55 sq ft of roof. For eaves, valleys, hip, and ridges, the following coverages apply.

- One square of wood shingle covers 240 linear feet of eave.
- One square of wood shake covers 120 linear feet of eave.
- One bundle of shake will cover 16 to 17 linear feet of ridge.
- Add one extra square of wood shingle for each 100 linear feet of valley or hip to be covered. You are cutting these at an angle and the cutoff pieces generally cannot be used elsewhere.
- Add two extra squares of wood shake for each 100 linear feet of valley or hip.

Shingle Overhang Recommendations

The rake overhang should be 3/4 to 1 in. The eave overhang should be 1-1/4 to 1-1/2 in. Properly place the shingles using the gauge on your roofer's hatchet. As an alternative, you can nail a 1 × 4 to the rakes and use it as a guide. Remove the 1 × 4 after all shingles are in place. A 2 × 4 can be used at the eaves for the same purpose.

Shingle Sidelap Recommendations

Sidelap of the shingles is to be no less than 1-1/2 in. Be careful not to have a third course shingle edge in line with a first course shingle edge. What may happen is that the second course shingle may split and in doing so, it becomes the source of a leak.

Shingle Side to Side Spacing

The side to side spacing should be 1/4 in. plus or minus 1/16 in. If you are having a hard time gauging this, you can make a gauge from a piece of 1/4 in. thick plywood. The plywood should be 8 in. long and 6 in. wide. Glue a piece of 1 × 2 on it for a handle. After awhile, you will find that you no longer need the gauge.

TYPICAL WEATHER EXPOSURES

Weather exposure recommendations are as follows. There are more but these are the exposures you probably will be using.

- 5 in. for roofs using 16 in. wood shingle.
- 5-1/2 in. for roofs using 18 in. wood shingle.

- 7-1/2 in. for roofs using 24 in. wood shingle.
- 5-1/2 in. for triple layer roofing using 18 in. shake.
- 7-1/2 in. for double layer roofing using 24 in. shake.
- 7-1/2 in. for roof pitches between 4:12 and 8:12 using 24 in. by 3/8 in. shake.
- 10 in. for double layer roofing using 24 in. shake. This is the maximum exposure allowed.
- 10 in. for roof pitches over 8:12 using 24 in. by 3/8 in. shake.
- For sidewalls, you can add 2-1/2 to 3-1/2 in. to the above.

In areas of high wind or considerable rain, I suggest you stick with the lower weather exposure numbers. Use 5-1/2 in. for 16 and 18 in. shingles and 7-1/2 in. for 24 in. shingles.

Slope Restrictions

Wood shingles and wood shake can be used on slopes from 3:12 to vertical. They are not recommended or allowed for slopes under 3:12 without special building department authorization.

Handsplit shake is not recommended for slopes under 4:12. Vertical or near vertical slopes are considered decorative sidewall rather than a roof. The weather exposures can be greater and the thickness can be minimum, i.e., 11-1/2 in. exposure with single layer, 24 in. long, 3/8 in. thick shingle or shake. For 18 in. shake use 8-1/2 in.

Underlayment

Underlayment is to be shingled 30 lb roofing felt, 18 in. wide. Start at the eave with a 36 in. wide felt. Apply 15 in. starter shingles to this and then the 18 in. felt over the starter shingles starting at the point of normal weather exposure and laying toward the ridge. To this, you install the first course of full-size shingles. Slip the tail under the 18 in. felt and nail in place. Over this shingle, you lay a sheet of 18 in. felt starting at 1 to 1-1/2 in. above the weather exposure and laying toward the ridge. You continue this shingle, felt, shingle pattern until you reach the ridge. This is called shingled or shingling the felt. See Fig. 13–5.

Wherever the felt ends at an opening or laps onto another felt, double the felt back on itself by 8 to 10 in. This folding technique (Fig. 13–6), prevents any water that may have gotten under the shingles from entering the building or sheathing. It also stiffens the felt so that, if used over spaced sheathing, the felt will not relax and sag into the voids between the sheathing.

Along the hips and ridges, install an extra lengthwise strip of #30 felt so that it overlaps the last shingled felts. This provides extra weatherproofing at these points.

In the valleys or at vertical corners, a #30 felt under the metal flashing is recommended. This keeps the building's escaping interior moisture from attacking the underside of the flashing.

Roofing felt required for most roofs is 1.5 rolls per square. In nonsnow belt areas, it

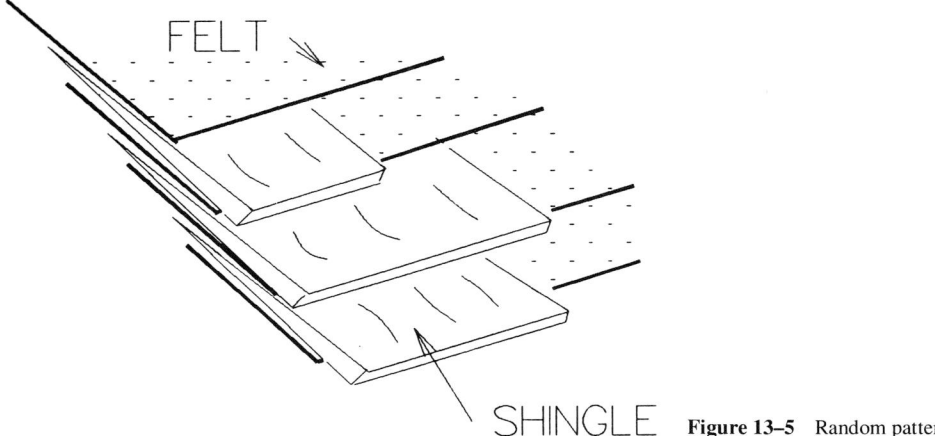

Figure 13–5 Random pattern

may be permissible to shingle with wood without using felt. Consult your local building department.

NOTE: All galvanized steel flashing, if used, must be painted before installation of the wood shingles. The recommended flashing materials for wood roofs are aluminum, copper, or brass.

REROOFING WITH WOOD SHINGLES

Most old roofs may be roofed over with wood or shake without removal of the old roofing, but some old roofs should not be. You can use wood shake over wood, asphalt, T-lock, and Dutch lapped shingles, but with these recommendations. Do *not* use wood shingles over wood roofs in continuously wet areas of the country, such as in the deep south. You can use wood shingles over Dutch lapped shingles, but removal of the old shingles will give you a smoother roof.

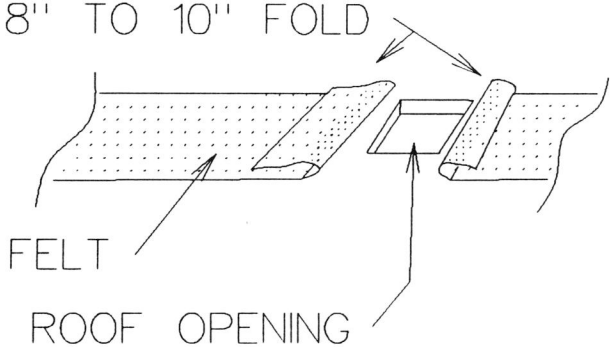

Figure 13–6 Foldback helps prevent water entry into building

Do not attempt to seal old wood shingles with plastic roofing cements, because the bonds will not hold.

Use step flashing for the connections to dormers and chimney sides. If continuous strip flashing is in place, leave it and step flash over it. Step flashing should go a minimum of 2 in. up the walls, 5 in. under the shingles. For best results, if you can do it, run the step flashing 5 in. up the walls and 6 in. under the shingles. Do not use continuous flashing, the water will carry dirt, leaves, and loose wood under the shingles and they will get trapped there, forming a dam. This dam will hold water under the shingles and the shingles will rot.

Chimney and dormer base flashings should be lapped over the wood shingles by a minimum of 5 in. and should remain exposed. Do not put a decorative layer of shingles over the flashing, it will trap dirt and the shingles will rot. Use a minimum amount of exposed nails when installing the base flashing. Cover exposed nails with plastic cement.

To reroof a wood roof, you should use horsefeathers to level the old roof surface. Butt these shims, thick side to butt side of the existing shingles. The reversed wedge shape of the horsefeathers should give you a smooth, even surface on which to work.

At the starter shingles of the hips and ridges, you will find multiple layers of shingle. These should be removed for a minimum of three courses back. If you do not remove them, you will end up with a huge bump at the starter edge when the new roofing is installed. This is especially true if the old shingles are wood.

The new roofing should be installed as a new roof complete with shingled underlayment. If the old flashing is beyond repair or use, it should be removed and replaced with new. Existing leaks should be repaired as best possible. Extremely warped shingles should be split lengthwise and renailed as flat as possible or replaced.

If you must remove the old shingles, use these tools. For wood shingle or shake, use a 2 to 3 ft long crowbar. Start at the top and work toward the eaves. For old asphalt shingle removal, use a flat bottom shovel. Slide it under the shingles and give a twisting upwards motion. This should pull the nails and flip the shingles out of your way.

Vents should have a minimum of 1 in. clear space around them. The bottom of the vent flashing is to lap onto the shingles by a minimum of 5 in. Leave the metal exposed. Place a wide, full-size shingle at the top side of the vent, centered on the vent.

If reroofing over a spaced deck, try to nail to the decking boards whenever possible. Do not nail the new shingles more than 3 in. above the weather exposure line in an attempt to hit the sheathing. Just nail as normal, and the old wood shingles should hold the new in place without problems.

Metal Ridge and Hip Caps on Old Roofs

Many older wood roofs have metal ridge flashing caps installed on them. When reroofing, with asphalt or wood shingles, you should attempt to remove this metal and replace it with new caps after the new shingles are installed. This is not always possible, because the removal process may tear up too many old shingles. If this begins to happen, leave the old metal in place. Clean it up and give it a new coat of paint.

Install the new shingles up to within 1/4 in. of the center divider of the old flashing. Be sure to dub the edges of your new shingles.

Rake and Eave Treatment of Old Roofs

The old shingles at the rakes and eaves should be trimmed back to the fascia boards. Install new drip edging over these cut edges. Nail the drip edge so that the nails go through the butt portion of the old shingles; this prevents wavy bumps. After the edging is in place, you can install your new shingles with the proper overhangs.

Valley Treatment of Old Roofs

Current codes require a W-flashing be used with wood or shake roofs. The old roof may or may not have this type of flashing installed. If it does, and it is in good shape, you can leave it in place. Clean it up, paint it, and shingle to within 3 in. of the splash diverter. You may have to cut the old shingles back an inch or two for appearance. Be sure that all shingles, old and new, are properly dubbed.

        If there is not a W-flashing installed, then one should be. Cut back the old shingles to the edge of the existing flashing or to the width of the new flashing being installed. Install the new flashing over the old and shingle to within to 3 in. of the splash diverter. Be sure to dub the corners of the new shingles. If more than one length of flashing is required, install the lower flashing first, then install the upper flashing, overlapping it 6 in. onto the lower flashing.

        Flashing should extend 10 in. from its center, outwards on both sides for roofs of 12:12 slope or lower. For roofs over 12:12 slope the minimum extension is 7 in. The recommended distance from the splash diverter to the shingles is 5 to 6 in.; 3 in. is used by many roofers for appearance. I personally like the wider gap since it provides a better water runoff channel with less chance of dirt or leaves getting caught. Use flashing that is a minimum of #26 to #28 gauge metal, the thicker (#26) is better.

NEW ROOF PATTERNS

You can pattern the roof by different shingle placements. For most, the straight line from rake to rake alignment of the butt ends is attractive. If you need a rustic appearance, then you must stagger the butt ends at random. Do not lower the shingles more than 1 in. from the main weather exposure line being used. A stagger pattern of plus or minus 1/2 in. from the rake to rake weather exposure line will produce a good looking rustic pattern roof. See Fig. 13–7.

        To produce a Dutch weave pattern, you lower every fourth or fifth shingle by 3/4 in. to 1 in. from the rake to rake weather exposure line. See Fig. 13–8. These patterns can be duplicated with asphalt shingles.

BEST QUALITY ROOF

The weather exposure and the shingle course lapping control the quality of the roof. You can shingle to maximum exposure to save money now. Over the years, the minimum exposure roofs will provide the best value. I recommend three-coursing of all wood shingles and

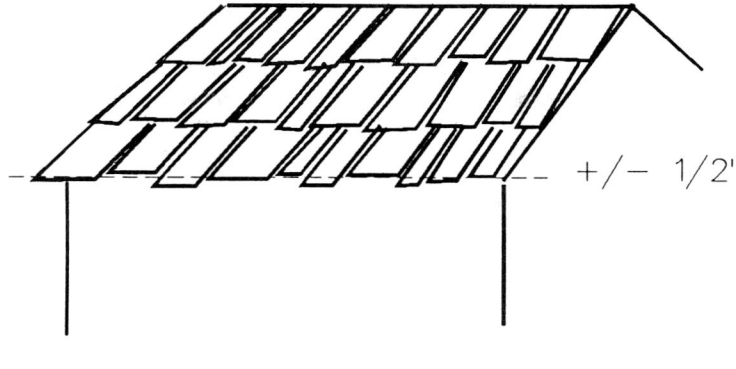

RANDOM SHINGLE PATTERN

Figure 13–7 Coursing for proper exposure

shake (see Fig. 13–9). This is where each course overlaps on to the next by two thirds. In extreme wind areas, I recommend an even smaller weather exposure.

All flashing should be aluminum, brass, or copper 26 gauge sheet. Valley flashing should be W-flashing, all rakes and eaves should have drip edging installed. Any steel flashing used must be primed and painted. All valleys should empty directly into gutters, a valley that empties on to the roof will create leaks. All shingles in the valleys should be dubbed. There should be a #30 felt under the metal W-flashing.

Step flashing is required along the sides of chimneys, skylights, and dormers. An extra strip of #30 felt is recommended at all turns, corners, and ends. Crickets are to be installed on the top side of chimneys. Base flashing is to be installed on the bottom sides of

DUTCH WEAVE

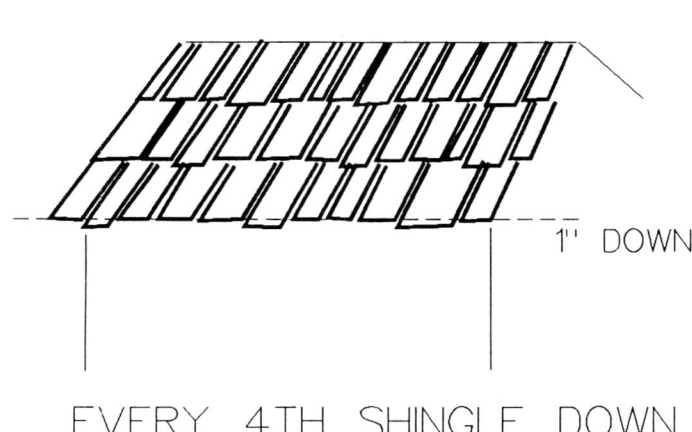

EVERY 4TH SHINGLE DOWN

Figure 13–8 Dutch weave pattern

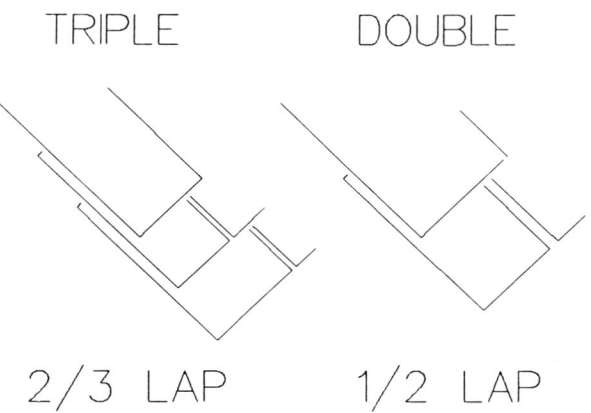

Figure 13–9

chimneys, skylights, and dormers. Base and vent flashing should empty on to the shingles. An extra strip of #30 felt is recommended under all base and vent flashings. No shingle joints are to be made directly over the top half of a vent flashing. For appearance, the flashings should be prime coated and painted to match the final weathered color of the singles. This is usually a shade of dark brown. Paint should be a top-quality latex.

Mastic, plastic cement, or other roofing cements are not to be used in place of the proper metal flashing. They may be used to cover exposed nails used to install metal flashing. They may not be used for wood to metal seals. Exposed nails should be gasketed nails for best leak resistance.

Hips and ridges should be shingled at the same weather exposure as the field shingles. Hip shingles are to be installed before the main ridge shingles. The main ridge starter shingles should overlap the tops of the last hip shingles by 1/2 in. to 3/4 in. An extra strip of #30 felt should be installed lengthwise along all hips and ridges. This felt should lap over the field felt. The hip shingle starter shingles should be cut at an angle to match the 90 degree turn of the roof line.

Shingle edge joints should have a minimum of 1-1/2 in. sidelap from course to course. Shingle edge laps are not permitted to line up vertically with each other within three courses.

RIDGE SHINGLES

You can purchase prefabricated ridge shingle for most wood roofs. If your supplier does not have prefabricated ridge shingles, you can make your own. You will need a table or radial arm saw capable of cutting a 35 degree angle. You will need a stapler capable of using wide crown staples, such as those in Fig. 13–10.

S2 Series 1" Wide Crown Staples

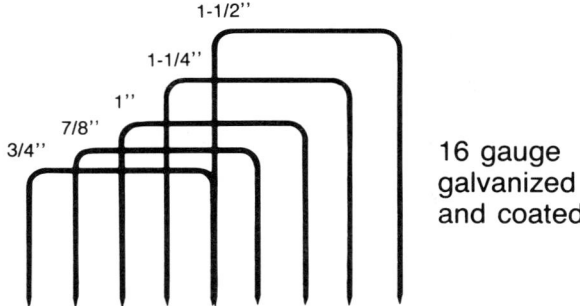

16 gauge
galvanized
and coated

CATALOG NUMBER	LEG LENGTH	QTY. PER BOX
BCS1097	3/4"	12.8M
BCS1098	7/8"	11.2M
BCS1099	1"	9.6M
BCS1101	1-1/4"	8M
BCS1102	1-1/2"	6.4M

Figure 13–10   Wide crown staples (Reprinted with permission of Stanley Bostitch)

Cut the edge of two mating shingles at a 35 degree angle and staple them together as shown in Fig. 13–11. Make two piles of ridge shingles from these cut shingles. The first pile is stapled right side up, the second pile is stapled right side down, alternately. Be careful to get the rough side of saw cut shingles facing up. When installing the shingles on the hip or ridge, stagger the cut lines as in Fig. 13–11. This helps to prevent water leaks as the roof ages.

REPAIR TO A BROKEN OR SPLIT WOOD SHINGLE OR SHAKE

The best repair is replacement, but on an older wood roof this will be very obvious. New shingles are light tan in color, the old weathered shingles will be black or dark brown. So we do the second best thing, we repair the shingle. If the shingle is split lengthwise, then what you have is two thin shingles; just renail so that there are two nails in each. If you have a split shingle and a water leak situation, then use a metal shingle under the split.

A metal shingle is nothing more than a piece of flat sheet metal cut to fit under the old shingle. Do *not* nail the metal shingle in place. Bent the lower corners of the metal shingle downward about 1/4 in. This will hold the shingle in place once it is slipped under the split wood shingle (Fig. 13–12). Use copper or aluminum, do not use steel.

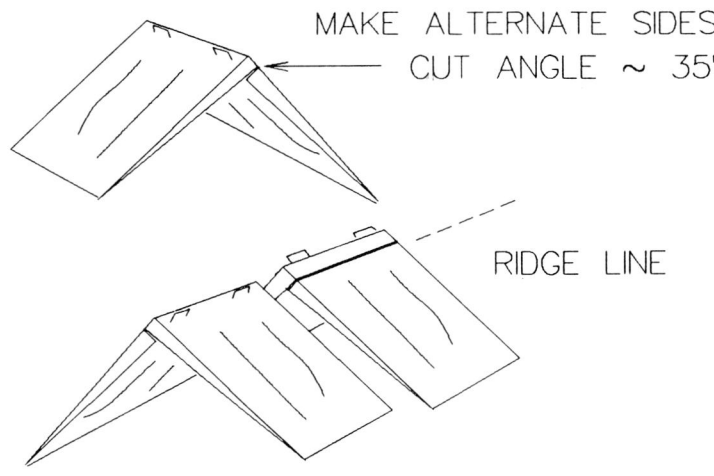

MAKE ALTERNATE SIDES
CUT ANGLE ~ 35'

RIDGE LINE

ALTERNATE SIDES DOWN RIDGE LINE

Figure 13–11   Ridge shingles

POINTS HOLD SHINGLE IN PLACE

SIZE TO FIT

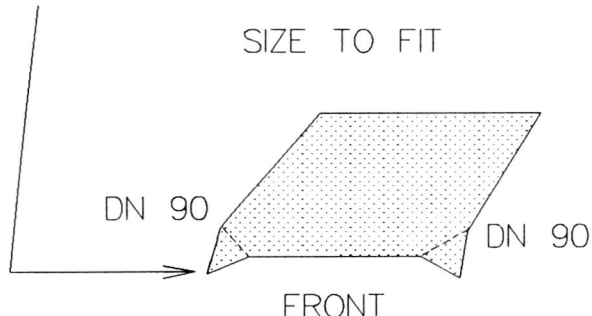

DN 90

DN 90

FRONT

METAL SHINGLE 28 GAUGE
NO NAILS REQUIRED

Figure 13–12   Metal shingle

As a tip, keep some old shingles from past jobs. Use these for repairing missing or split shingles on future jobs, as they have already weathered to the correct color.

Other Wood Roofing Repairs

Wood shingles can twist, warp, rot away, or be blown off. Missing and rotted shingles must be replaced. To match up the color to that of the existing shingles, you can try a water-base latex wood stain. There are many fine brands available. Paint, or better yet, soak the new shingles with the stain. When dry to the touch, they can be installed.

Rotted shingles are an indication that something is wrong. Inspect for the proper type of decking, proper flashings, proper nailing, and proper underlayment. If any problem is found, then it must be repaired before installing new shingles in place. If not done, the old problem of rotted shingles will return in a short period and you may have a callback.

Twisted or warped shingles may be salvaged by splitting them into two shingles and renailing as new. If the twist or warp is too great, then you must replace them. To nail a shingle under an existing shingle, you should use a technique called slant nailing. Place the new shingle in place but 1/2 in. lower than the existing shingles. Start your nail at a slant just below the butt end of the upper shingle and hammer. The new shingle should slip into place with the nail head covered (Fig. 13–13).

Lighten, Brighten, and Waterproof Old Shake

This can be done. The product is pentachlorophenol and is found in several clear wood finish products. It is a preservative and a bleaching agent. It will help make old wood roofs look and work like new. Application is via a low-pressure pump sprayer like the ones you find at the hardware store. Clean the roof surface and then flood all shingles with the prod-

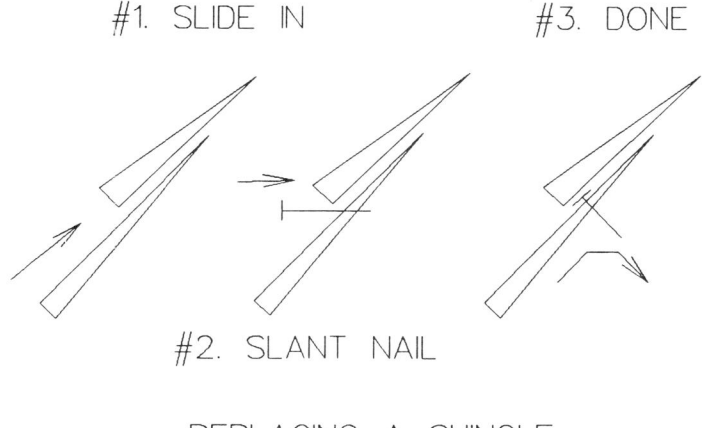

#1. SLIDE IN   #3. DONE

#2. SLANT NAIL

REPLACING A SHINGLE

Figure 13–13 Slant nailing

uct. Cover all plants, concrete, and other items at ground level as this chemical will stain and kill.

Vertical Wall Installations

Occasionally you must roof over a vertical or near-vertical wall surface. This can be a Mansard roof or a building's siding. If a Mansard roof is being shingled, just follow the directions for a new roof. If shingles are being installed as siding, then you have a few rules to follow.

You must use straight edge shingles that are butted tightly together to obtain a weather-tight surface. Careful vertical alignment is important; use a level and a plumb line, and use them frequently. All inside and outside corners should have #30 felt installed with 6 in. wide L flashing installed over the felt. Watertight sheathing paper or proper wall insulation should cover the building's framing. Shingle weather exposure should be a maximum of 8-1/2 in. for 3/8 in. thick shingle used double coursed.

Corner joints should be mitered, jointed, or woven (see Fig. 13–14, top down view). Nailing, flashing, and shingle patterns are the same as for a roof; see best quality roof section in this chapter.

MISCELLANEOUS TIPS

Wood roof installation is not difficult. The amount of tools required is small—a roofer's hatchet, claw hammer, pencil, tape measure, chalk line, tin snips, and a circular saw will do for most installations.

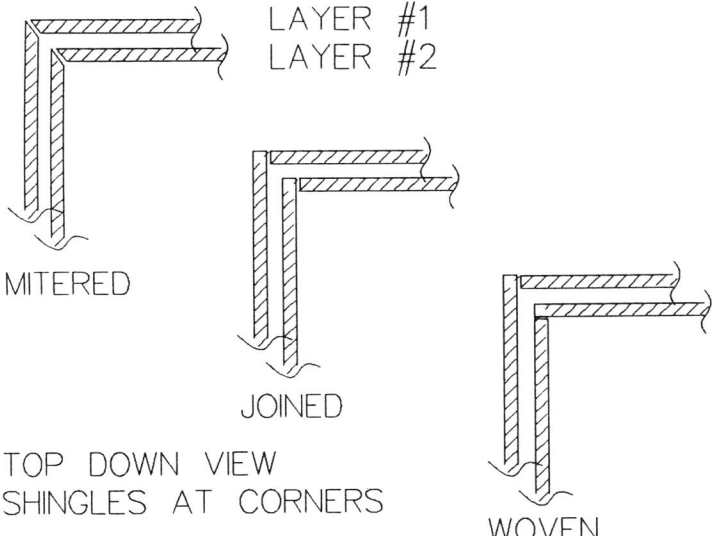

Figure 13–14 Sidewall corner joints

You may want to assure your safety on the roof. You can make a toe-board for yourself from a 10 ft long 2 × 4 and a few shingles. Nail the toe-board to the roof just above the eaves and use it as a resting place and a catchall.

For fast, easy installing of shingles, use a power nailer or stapler. Tie the electric cable or the air hose to your waist or to a convenient place on the roof. This keeps the weight of the cable or hose from pulling the equipment off the roof and makes handling easier.

When felting, roll out up to six rolls at a time. Each should overlap the next by the shingle weather exposure distance. This way you can shingle five or six courses at a time using stair-stepping techniques.

Keep your roofer's hatchet blade sharp, as you will need a sharp blade to cut or shape your shingles. The circular saw should be fitted with a 12 or 14 tooth carbide cross cutting blade. Most saw cuts will be on the diagonal or across the grain of the shingles. Make templates of any special shingles you need, these are the hip and valley shingles that are cut at an angle. Spray paint one side of the template shingles for identification as a template.

Be sure you have a good supply of the correct length fasteners with you. Remember that the hip and ridge shingles require longer nails than the field shingles. The metal flashings will require shorter nails and the felt will require short nails or staples.

Acid Rain

Factory fumes and acid rain problems will require a trip to your local shingle supplier and maybe a call to the shingle manufacturer. Some newer shingles may be formulated to combat this problem. You may have to switch to wood shingles. Wood is very acid-resistant; years ago, many acid containers were manufactured from wood. The problem here is fire and the fire codes, so check before installing wood.

Figure 13–15   Rent bins for trash removal

Rubbish Removal

Figure 13–15 illustrates workers removing old shake shingles from roof. Collection companies can generally supply you with large bins for trash. The bins are delivered to the site, left for as long as you need them, and then picked up. The cost varies from company to company, about $80 a week is normal. Call and get pricing and details. Some landfills are now charging by the ton, others will not accept many items.

14

Tile Roofing

INTRODUCTION

This chapter is about tile roofing and roofs. Tile roofs have been in existence for hundreds, maybe thousands of years. Tile is durable and has a beauty and richness of its own.

The northeast coast of the United States has an abundance of slate tiled roofs. The best single example I can think of is at West Point, New York. The West coast of the United States has an abundance of clay tile roofs in the Spanish tradition. These are called mission tiles since they are patterned after the early Spanish missions of the American Southwest. There are Mediterranean tiles patterned after those used on buildings around the Mediterranean Sea—these are two-piece barrel tiles. There is a mix of slate, clay, and concrete tiles used in other parts of the nation. These tiles are flat or near flat and may or may not have interlocked edges. Newer tile products include plastic tiles, fiberglass tiles, and metal tiles. The metal tiles are sometimes called bonderized metal shingles since they are a sheet of stamped metal with granules bonded to them. There are bright colored and glass-like tiles that are porcelain enamel tiles. The glass-like glaze coating is baked on at high temperatures.

TILE ROOFS

Tile roofs are considered by many as the ultimate in roofing. This may or may not be accurate. Tile roofs are durable, they will withstand severe weather and high winds very well; the metal tiles, being very lightweight, may not. Slate and metal tiles will withstand items being thrown on them. The cement or clay tiles will not withstand large hail stones or thrown rocks.

Fire Ratings

Although the tiles are fire-resistant, they are not fireproof. It is true that the material the tile is made from will not readily burn. Usually, it will not support combustion of any sort. So why do some tile roofs not carry a Class A fire rating? The answer lies in the fact that tile roofs are not sealed roofs. Sparks from a nearby grass or house fire can enter the cracks between the tiles and set the felt and roof sheathing on fire. This type of fire is difficult to find and extinguish since it is hidden from immediate view. It is for this reason that many insurance companies do not give their best rating, therefore, their lowest rates to many tile-covered buildings.

You can obtain a Class A rating if you select the proper weight of felt. According to many tile manufacturers, a 75 lb felt underlayment is required to obtain a Class A rated roof.

Color of Tile

The colors available for man-made tile range from white to black and throughout the rainbow. Slate tile is natural gray with some possible black-gray, purple, or other mineral hue. Typical colors of slate are:

- Black and gray-black
- Green and gray
- Green
- Green (mottled) and purple
- Gray
- Purple (marked or streaked)
- Red

Mission and Spanish tile are generally orange or red in color. Cement tile is commonly colored to match the Spanish tile red. Cement tile can be colored to any color desired, but you must special order and pay the premium for color.

Many roofers will intermix different colored tiles on the same roof. This is acceptable practice providing one doesn't attempt to mix blue and yellow, red and green, or any other unacceptable color combination, as follows:

- Blue and violet
- Green and yellow
- Green and red

- Orange and yellow
- Blue and yellow
- Green and blue
- Orange and red
- Red and violet

MATERIALS USED TO MANUFACTURE TILES

Tile can be manufactured from natural rock, cement, clay, and from metals or plastics. The metals and plastic tiles are becoming popular due to the ease of manufacturing and the lower cost involved. Many tiles have fillers added for extra bulk and strength. Some fillers in common use are ground-up seashells, ground walnut shells, mineral fibers, and organic fibers. Other fillers can be used to impart special qualities to the cement and clay tiles. Perlite, an expanded volcanic stone, when added, will lighten the tile and give it some insulation value. Liquid plastic resins, when added, will improve the internal bond strength and help prevent breakage.

Clay tiles and cement tiles are molded or cast tiles. A slurry is made of cements, sands, silicas, and clays. This slurry is then poured into a mold and allowed to cure. The mold may or may not be a compression mold. A compression mold is a two-piece mold that has male and female forms. The slurry is poured into the female side and then, under several tons of pressure, the male side is brought into contact. The pressure squeezes out the water and heats the materials until a solid bond is obtained. This produces a dense tile with a hard surface glaze on both sides. Some less expensive clay tile is painted or glazed after demolding. This paint is a surface color and may not be as long-lasting or as damage-proof as solid through coloring.

Clay mission tile, '~', is 18 in. long and 13 in. wide. Dry weight is 800 lb per square. Two-piece mission or two-piece barrel tiles are tapered from 7-1/2 in. at the tail to 9 in. at the butt end. The length is 18 in. Dry weight is approximately 1,000 lb per roof square. The glaze applied to both tiles can be matt, semi-gloss, or high-gloss finished. Color is normally throughout and is controlled by the clay colors used. Blended colors are available.

A lightweight mission tile by MaxiTile™ is made from Portland cement™, silica, and cellulose fiber. The tile is approximately 37 in. long and 24 in. wide. Thickness is just over 1/4 in. Coverage is 24 tiles per roof square at an installed weight of 380 lb a roof square.

Concrete tile is manufactured from Portland cement in the same manner as clay tile is manufactured. Concrete tile, also clay tile, may be colored with the addition of specially formulated colorants. The colorants must be formulated to withstand the alkalies in the cement and not fade.

Cement tile may contain wood fiber as a filler. This, combined with a top coat of perlite, Portland cement, and iron oxide, makes the tile fire-resistant and as workable as wood. One brand is DuraTrend™. It is available in a shake design, and is designed to re-place wood shake. The tile is 22 × 12, with 5 in. or 7 in. widths, and has a 10 in. weather exposure. The shape is much the same as wood shake, the thickness of the tail end is 3/8 in. and the butt end is 7/8 in. thick. The installed weight is 650 lb per roof square dry, 750 lb per roof square wet. Fasteners used are 2-1/2 in. staples or 8D nails.

Metal and plastic tiles are formed similarly to clay and cement tiles. The sheet material is placed on a female die and a male die is brought into contact under pressure. Sheet plastic is usually heated to its glass state before being formed. The glass state is that condition where the plastic is pliable and starts to turn translucent in color.

Metal tiles are generally formed and then coated, but need not be. Many manufacturers of sheet metals coat the sheets with paints, oxides, and other chemicals (Fig. 14–1). Some manufacturers will supply metal sheets that are clad with coppers, aluminums, plastics, or other materials. These coated or clad metals may be formed directly into finished tile without secondary manufacturing techniques.

Metal tiles vary in size, but most are 45 in. long and 15-1/2 in. wide. You will need about 23 tiles of this size to cover a square. Weight per square is light compared to other tiles, with the average weight per square being 140 lb.

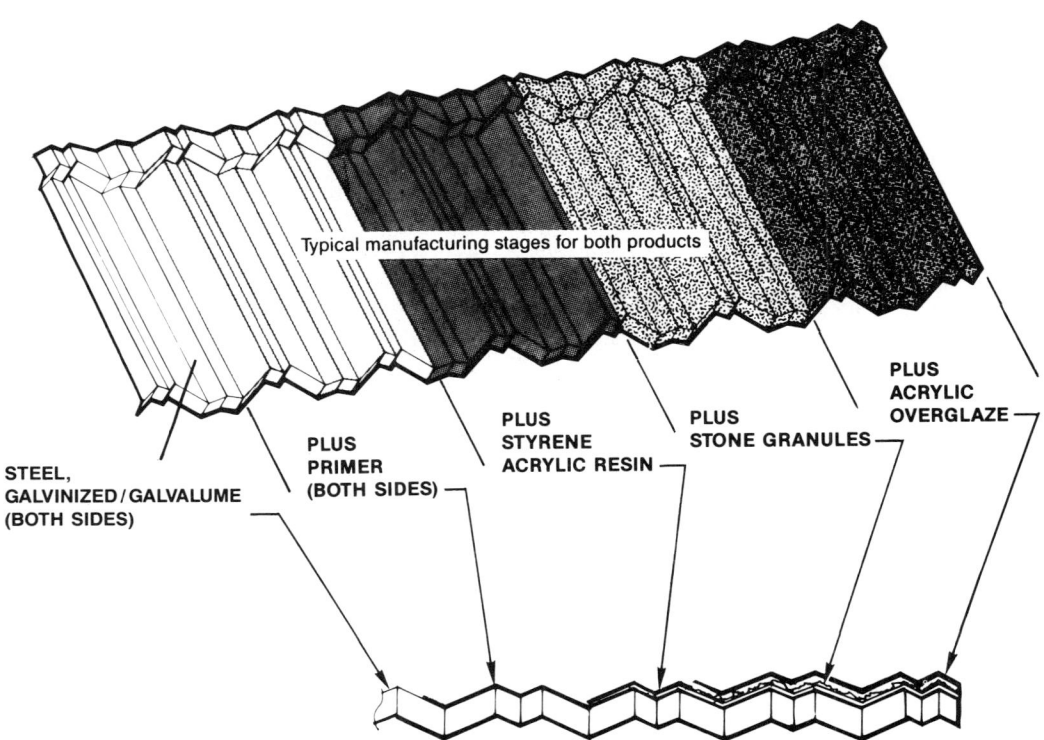

Figure 14–1 Steps used to make a metal shingle (Reprinted with permission of Gerard Roofing Technologies)

Some plastic tiles are manufactured by either the rotational or the injection molded processes. Injection molding is where a liquefied plastic is injected under extreme pressure into a fixed two-part mold. When the plastic cools, it solidifies and the mold is opened, releasing the tile.

The rotational process is a process in which a solid piece of plastic is inserted into a heated ball. The interior of the ball contains from one to six forms. The ball is then rotated at high speed in all directions. This rotation flings the heated, now liquid, plastic into the forms by centrifugal force and creates the tile. The plastic is allowed to cool and solidify and the tiles are removed. The rotational mold process is used to mold plastics into everything from soda bottles to complete truck bodies.

Slate tile is quarried sedimentary stone. Sedimentary stone took nature thousands of years to make. Mud, sand, minerals, and silt washed into lakes and oceans and settled in thin layers. As the weight of the layers and water pressed down, the mud and silt compacted. Added to this were dissolved cements and calcium deposits that acted as binders. Slate obtains its colors from the minerals contained within.

The sedimentary layers, under pressure, became rock. During great earthquakes, land uplifts, and settling of the earth's crust, the layers compressed, lifted at an angle, and stratified. It is this stratification that differentiates slate from shale and sandstone. Slate used for roofing is arqillite; it is less grainy than sandstone and not as oily as shale. The stratification of arqillite allows it to be easily split into thin sheets. These sheets are then sawn into slate roofing tile.

The main area of the United States where slate is quarried is the Northeast—Pennsylvania, Vermont, Maine, New York. Slate is quarried in the Southeast in Virginia, Georgia, and Maryland. Wales and France supply most of the slate used in the European countries. There are dozens of imitation slates on the market. Some are made from fortified and modified compressed clays and closely resemble the real slate. Others are made from stamped metals or compressed cements and clays. From a distance, they closely resemble real slate.

Natural slate tile is cut to the size of 20 in. long and 3/16 in. thick with varying widths. Natural slate tile is designed for a weather exposure of 7-1/2 in. One brand of manufactured slate tile is 16 in. long, 9-1/2 in. wide, and 1/4 in. thick. This manufactured slate tile is designed for a weather exposure of 7 in. Another brand is 24 × 12 in. and is 3/16 in. thick, the exposure is 10 in. The weight of slate tile is 800 to 1,000 lb per square for natural slate, about 450 lb for manufactured slate.

EXPECTED LIFE OF MATERIAL

Tile is a long-lasting material. Many buildings, built hundreds of years ago, used slate tile for their roof covering. The slate is still as good as new. Most of the slate tile used in the United States in recent years is imitation slate. Real slate, quarried stone, is much too expensive for the average person.

Metal and clay tiles will last 40 or more years, if they receive proper care. They may bleach out or fade somewhat, but they will remain waterproof if properly installed and maintained. Plastic tile is expected to last from 15 to 25 years. It will craze and fade in color long before that.

Tile Can Be Damaged

All tile is durable, but not indestructible. Figure 14–2 shows damage caused by neighborhood children throwing rocks on an otherwise nicely installed roof. Clay tile is not the strongest roofing material. Items thrown on it, people walking on it, and large hail stones can crack or break it.

Metal tile can be damaged in the same way. It will not break, but it will dent and bend. Most metal tiles are light-gauge steel or aluminum with a coating of binder and granules, not a very rugged surface.

The metal tiles are also subject to cleaning damage, only this time by alkyds or strong caustic solutions. Metal tiles are colored and protected with paint or mineral granules. Strong acids and caustics will dissolve many paints used to coat the tiles.

Clay and ceramic tiles can be damaged by physical abuse and by improper cleaning. The use of acids, usually hydrochloric, can and will eat the protective glaze coating. This allows more moisture absorption and eventual decomposition of the tiles.

Plastic tile, used for decorative slopes of 21:12 or more, can be broken by hard, sharp blows or by thrown rocks. Plastic tile may not be used on the roof of a normally occupied residential building. The tile is frequently used for commercial building Mansard roof

Figure 14–2   Damaged tile on roof

fronts and siding. Plastic does get brittle with cold, and when cold can be easily damaged. Plastic tile can be damaged by sunlight and by many petroleum-based chemicals.

Metal, slate, plastic, cement, and clay tiles can be damaged by people walking on them. They do not support heavy weights.

TOOLS REQUIRED

To install slate, clay, or cement tile, you will need the following tools:

- A circular saw with a carbide blade for cutting sheathing and lumber
- A masonary blade for cutting the tile
- Eye protection (essential).
- If you are doing much tile work, invest in a good cut-off saw. Be sure the saw has a means of supplying cooling water to the blade. The saw should have miter box capability so you can cut angles.
- You will need an assortment of trowels for applying cement. You will be cementing tile in place at the ridge boards and flashing areas.
- You will need a cement mixer or mixing pan and a bucket or two. A mortar board may be helpful.
- You will need a good quality drill with an assortment of bits. The bits should be 5% to 10% larger in diameter than the nails used. Purchase masonry or carbide-tipped bits. Standard bits will wear out very fast.
- You will need a quality extension cord. I recommend a 100 ft of 20 ampere, 3-wire plus ground, cord. If you make your own extension, use #10 or #12 wire. Wire thinner than #12 will cause current and voltage loss that may burn up your equipment.
- You will need a hammer or an automatic nailer/stapler. A nail puller is recommended since there will always be a mistake or two.
- Standard tools such as chalk line, screwdriver assortment, and a few hand saws will come in handy.
- For flashing you will require a good pair of metal shears. You will require a method of bending light-gauge metal.

Sheathing for Tile Roof

Tile roofs should be sheathed with solid sheathing. If reroofing over a roof that had wood shingles, and those shingles are removed, you should change the spaced sheathing to solid sheathing. If using slate, clay, or cement tile, then I recommend 5/8 in. or thicker sheathing be used, due to the weight involved. The sheathing can be 3/8 in. or 1/2 in. if using plastic or metal tiles and local code permits.

It is permitted, by code, to use spaced sheathing with this exception. In snow areas, you must use solid sheathing with a minimum of 1 ply of #15 felt.

Minimum Roof Slope for Tile

A tile roof is not a solid, watertight, membrane-like medium. It may not be used on slopes of less than 3:12 without the building inspector's specific permission.

Felt underlayment is required under all tile. The underlayment helps prevent water entry into the sheathing. The method of installing the felt will vary depending on type of tile and the tile manufacturer's recommendations. There are minimum requirements for felt underlayment for different types of tile.

Felt Underlayment

Roof tile does not prevent water from entering under it. It is used to direct water off the roof and into the gutters or onto the ground. Wind-blown water can and will enter under the tiles from time to time.

The roof sheathing must be protected from this water entry or be severely damaged. The use of felt underlayment is required by the building code. See Appendix B, table 16.

Some manufacturers of pantile recommended #30 felt under their tile for a Class B fire rating. They recommend a #75 felt for a Class A fire rating. I recommend that you decide the fire rating required and consult with the tile manufacturer and your local building inspection department. Do this before you spend much money and time installing what may be the wrong felt underlayment for the application.

I also recommend that, where you have a choice of 1-#30 ot 2-#15 felts, you use 1-#30 felt. It is difficult to lay 2-#15 felts without getting wrinkles and blisters in the felt layers. The thicker #30 felt will lay smoother and be easier to install. Plus, two #15 felts will tend to slide upon each other and make walking on them a hazard.

Some tile manufacturers recommend that you sprinkle mop the felt underlayment to the roof decking when the roof slope is between 3:12 and 5:12. Other manufacturers recommend solid mopping, and still others recommend standard nailing. I recommend you follow local building code and if applicable, the manufacturer's directions. They know the codes, the products, and the applications. On slopes of 5:12 and over, all recommended standard nailing and lapping procedures should be followed.

The felt used for flat tile is shingled the same as for wood shake. Use a 36 in. wide felt at the eaves, rakes, hips, and ridges. Then shingle in 18 in. wide felts between each tile course.

The felt used for curved tile, plastic, and metal tile, is 36 in. wide and is applied to the sheathing. Shingling is not required or possible.

Batten Strips

See Fig. 14–3. Edge nailed steel, aluminum or plastic shingle tiles require the installation of batten strips. These are often 1×2 or 1×3's. The distance from batten to batten is critical. If the distance is too short, the tile will not lay flat. If the distance is too far, the nailing edge of the shingles will be damaged. Batten to batten distance tolerance is 1/16 in. for most tiles. I highly recommend that the batten be installed as the tile is placed, batten, tile, batten, tile,

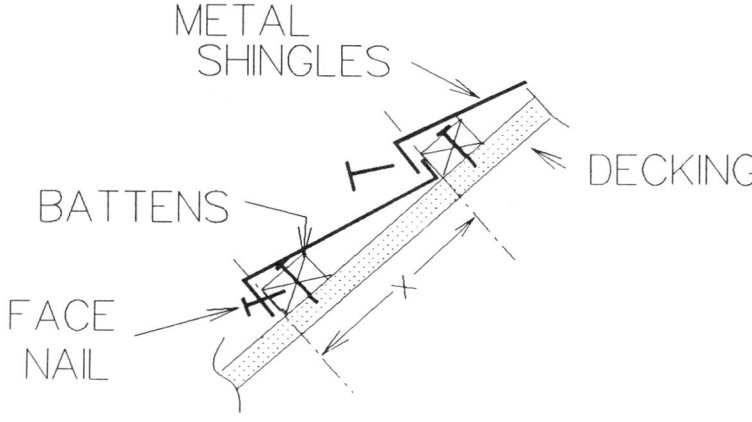

'X' DISTANCE IS CRITICAL

Figure 14–3 Nailing metal tile to battens

etc. This way, proper spacing can be maintained. Take care to maintain correct horizontal alignment across the roof.

Rake Treatment Using Metal Tile Shingles

See Fig. 14–4. Metal shingle tile used along the rake should be bent at the rake edge. The 225 degree bend forms a 45 degree V with the face of the tile. This edge now acts as a channel to direct water runoff down the roof to the gutter or eave. Chipped or exposed metal should be touchup painted to prevent future rusting or poor visual effects. A drip edge may be installed over the bent-up channel if desired. See the chapter on sheet metal roofing.

Riser Strips

When installing asphalt shingles, you start the first shingle at the eave and over a starter strip. This starter strip prevents water from entering through the tabs. Also it builds up the eave edge so that the first shingle lies even with the rest of the field shingles. Installing roof tile works very much the same. You must elevate the leading edge of the first course of eave tiles if you want the tiles to lie properly.

Figure 14–5 illustrates three courses of flat cement tile on a shed roof. See the two problems? The one problem is obvious, that of the ridge tile not overlapping the upper two field tiles. The second problem is at the eaves. There should have been a riser strip under the starter tile. A riser strip installed there will keep the slope of the tiles constant.

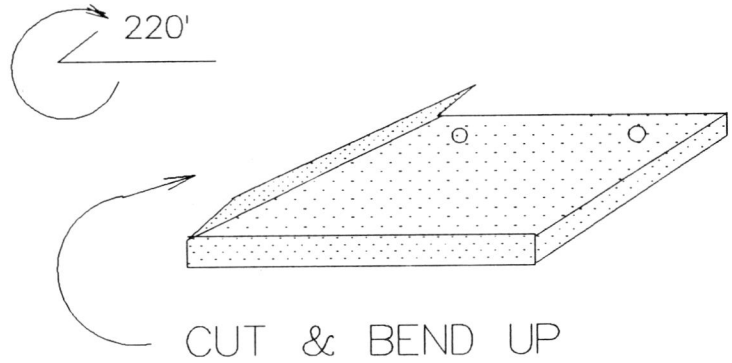

220'

CUT & BEND UP

METAL SHINGLE FORMS WATER
TROUGH AT RAKE EDGES

Figure 14–4   Bend up rake tile to form water runoff channel

Leak Sources

Like any other roof, leaks spring up when you least expect them. Most leaks are created by the improper installation of the roofing or by someone else doing work on the roof.

Many leaks start after you have completed your perfect roofing job. Shown in Fig. 14–6 is a commercial tile roof, bright blue, with outdoor lights installed. Each light fixture is a potential water leak source since the fixture's mounting base penetrates the tile. It is

Figure 14–5   Flat tile installation, not correctly installed

374

Figure 14–6 Light fixture becomes leak source

recommended that all standoff type penetrations be designed with an integral umbrella flashing and the penetration point be sealed with cement or other compatible sealer.

Bird Stop

Installed curved tile, pantile, will be open along the eave edge of the building. This opening will be a home to birds and small animals if you do not close the opening. There is a special flashing for this purpose, called bird stop. You will require a different size and shape bird stop for each different type of manufactured tile. Order sufficient bird stop when you order your tile. Figure that you must install bird stop along all eaves. Order an extra 5% or 10% to take care of the overlaps, installation mistakes, and estimate miscalculations.

Weight Distribution on Roof

Clay and cement tiles have considerable weight. When loading a roof, keep the weight evenly distributed across the roof (Fig. 14–7). Use care in stacking the tile. Stack only what is needed by the installer at that location of the roof. When using two-piece barrel tile, be sure that there are an even number of each piece available to the installer. Don't forget to supply ample fasteners and bird stop.

Breakage and A Way Off the Roof

This is a problem. The installer works from the eaves to the ridge, only to find that he does not have a way off the roof other than by walking on the tile. Walk he does, and as he does, he breaks 20 or more tiles. Be sure you have a way down that does not require walking on the tiles or be sure you have a walkway built into your roof. A walkway on a tile roof is where you have added extra support under the tiles. This walkway can be lumber bracing

Figure 14–7   Tile being loaded onto roof

or concrete underlayment. Mark this walkway on your contract. This way, both you and the building's owner will know where it is. This is in case someone has to work on the roof in the future, which is always a good possibility. People must get to antennas, fireplace chimneys, and vents.

Flashing Used for Tile Installation

The hips and ridges on tile roofs generally do not require metal flashing. The use of flashing is per the manufacturer's installation instructions. Most cement, slate tile, and clay tile are cemented in place along the hips and ridges. Plastic and metal tile may require flashing, depending on the tile design.

The valleys are to be flashed with W-valley flashing, having a splash diverter rib of 1 in. in height. The metal is to be a minimum of #28 gauge and an underlayment consisting of a single layer of #15 felt is to be used. The width of the W-flashing is to be a minimum of 22-1/2 in. The code states that the valley must extend 11 in. outward from both sides of the splash diverter rib. The exception is for metal tiles.

W-flashing for metal sheet or tile is to be a minimum of #28 gauge and have an underlayment of a single layer of #15 felt. The splash diverter is to be a minimum of 3/4 in. high. The W-flashing width is to be a minimum of 16-1/2 in., with 8 in. extending outward from both sides of the splash diverter.

Wood strips are required along the full length of both sides of the W-flashing whenever batten boards intersect with the flashing. The wood strips keep the battens from crushing the turned up edge of the W-flashing (Fig. 14–8).

The rake and eave treatments will vary depending on type of tile and tile installation

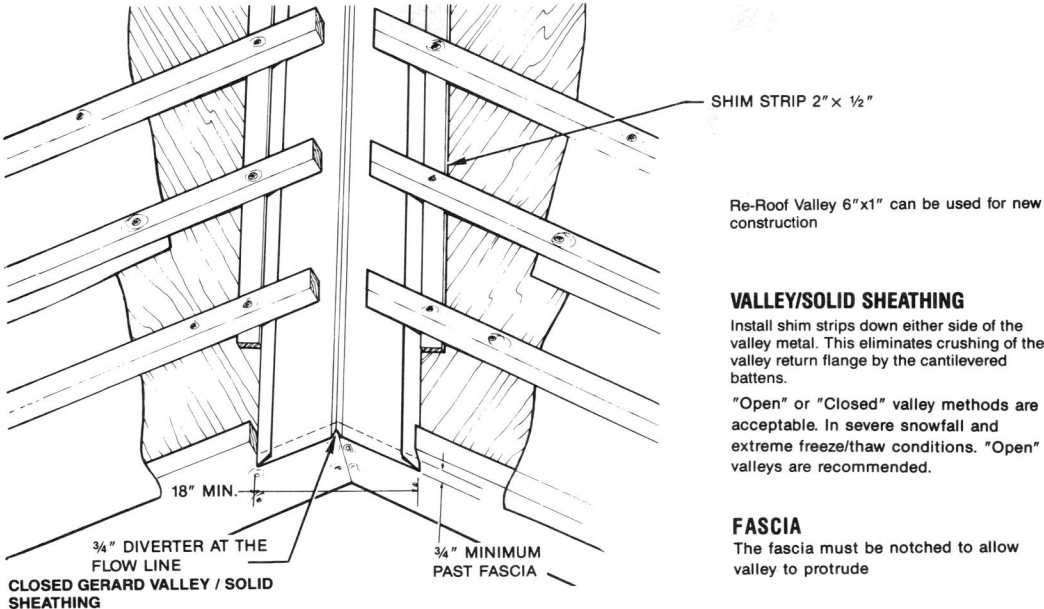

SHIM STRIP 2″ × ½″

Re-Roof Valley 6″x1″ can be used for new construction

VALLEY/SOLID SHEATHING

Install shim strips down either side of the valley metal. This eliminates crushing of the valley return flange by the cantilevered battens.

"Open" or "Closed" valley methods are acceptable. In severe snowfall and extreme freeze/thaw conditions. "Open" valleys are recommended.

FASCIA

The fascia must be notched to allow valley to protrude

18″ MIN.

¾″ DIVERTER AT THE FLOW LINE
CLOSED GERARD VALLEY / SOLID SHEATHING

¾″ MINIMUM PAST FASCIA

Figure 14–8 Use shims near valley (Reprinted with permission of Gerard Roofing Technologies)

used. If the rakes or eaves are not to be covered with tile, then drip edge flashing is recommended.

Whenever possible, chimneys, skylights, and dormers should be flashed as normal, using a base, head, and step flashing. For '~', 'U', or 'UUU' shaped tile, use techniques shown in Figures 14–19 and 14–20. Vent pipes require standard flashing or special flashing techniques to make them leak-free. Standard vent pipe flashing is used in most instances.

The remainder of this chapter will be on the preparation and installation of common roof tiles. Several examples will be shown.

ROOF PREPARATION

Figure 14–9 shows a roof being prepared for mission tile. The sheathing is covered with felt, the vent flashing is installed. The nailers and the hip and ridge boards are in place. The fascia boards have been painted and the valley flashing installed (Fig. 14–10).

The underlayment felt goes on the sheathing and under the battens. The felt should be brought over the edge of the eaves by 1/2 in. so that water on the felt will drain into the rain gutters (Fig. 14–11).

If metal tile is used, the eaves should be raised by placing 2 × 2 in. riser battens along the eave. This then requires than a cant strip be installed. The felt goes over the cant strip and drains into the gutter.

Figure 14–9   Felt and battens installed

Attaching Tile to Sheathing

Metal tile will normally require the installation of 2×2 in. battens. The tile is held in place by nailing through the face edge of the tile to the battens, using galvanized 8D nails to hold the tile in place.

How do we hold mission, Spanish, or two-piece barrel tile to the roof sheathing? One

Figure 14–10   Valley metal and hip risers installed

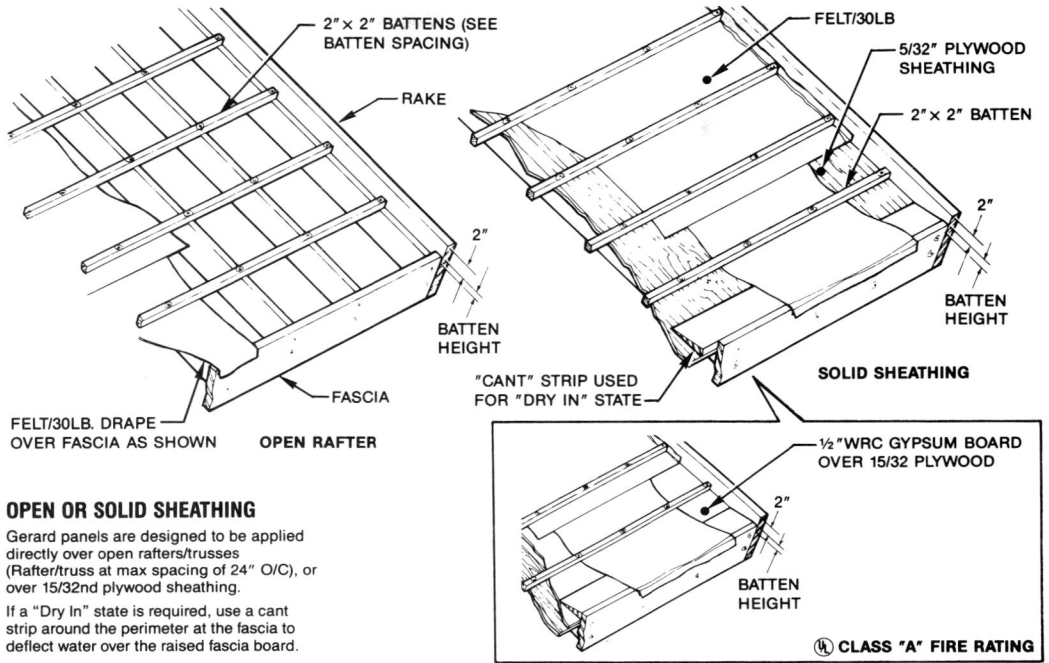

2″ × 2″ BATTENS (SEE BATTEN SPACING)

RAKE

FELT/30LB

5/32″ PLYWOOD SHEATHING

2″ × 2″ BATTEN

2″

BATTEN HEIGHT

2″

BATTEN HEIGHT

FASCIA

"CANT" STRIP USED FOR "DRY IN" STATE

SOLID SHEATHING

FELT/30LB. DRAPE OVER FASCIA AS SHOWN **OPEN RAFTER**

½″WRC GYPSUM BOARD OVER 15/32 PLYWOOD

2″

BATTEN HEIGHT

Ⓤ **CLASS "A" FIRE RATING**

OPEN OR SOLID SHEATHING

Gerard panels are designed to be applied directly over open rafters/trusses (Rafter/truss at max spacing of 24″ O/C), or over 15/32nd plywood sheathing.

If a "Dry In" state is required, use a cant strip around the perimeter at the fascia to deflect water over the raised fascia board.

Figure 14–11 Sheathing for metal tiles (Reprinted with permission of Gerard Roofing Technologies)

acceptable method is with wire ties. Figure 14–12 shows specially made wire chains draped down the roof sheathing. These wire chains are spaced to match the tie-in, nailing holes in the tile in which they are installed. This distance is 10-1/2 in. for most tiles. Figure 6–24 shows the attachment of the tiles to the wire chain via small lengths of #14 gauge tie wire.

Eave Tile Installation

Installation of mission tile at the eaves is as follows. A 1 × 2 in. riser is installed first (Fig. 14–13). The roofing felt is brought up and over this riser block and the tiles are placed. This riser is needed to angle or tilt the starter row of tile at the same angle as the field tile. It also becomes the nailer strip for the bird stop.

Figure 14–14 shows a two-piece eave tile installation. A 2 × 3 in. or 2 × 4 in. nailer is used per the tile manufacturer's recommendations. The roof felt completely covers the nailer. Field tiles are placed so that a water drain channel is created. The outer edge tile is then nailed in place. Finally, the cap tiles are installed. The outer edge tile may require the drilling of a second nailing hole to keep it from tilting. Instead of the second nail, the tile may be cemented in place. If a second exposed nail is used, it should be painted to match the tile color.

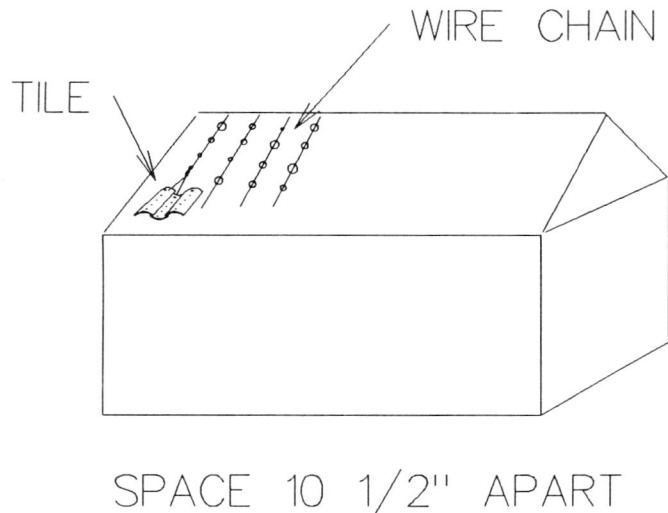

Figure 14–12   Wire chains used for securing clay tile

RAKE TILE INSTALLATION

Rake tile is formed by the manufacturer of the field tiles and should be purchased when purchasing the field tiles. Be sure that the color batch codes match up. You should order 5% extra to be on the safe side and to account for any possible breakage.

Metal tile may require that the rakes be built up high enough to be just above the top of a bent-up metal tile flange, approximately 1-1/2 in. The tile is cut and bent up 1-1/2 in.

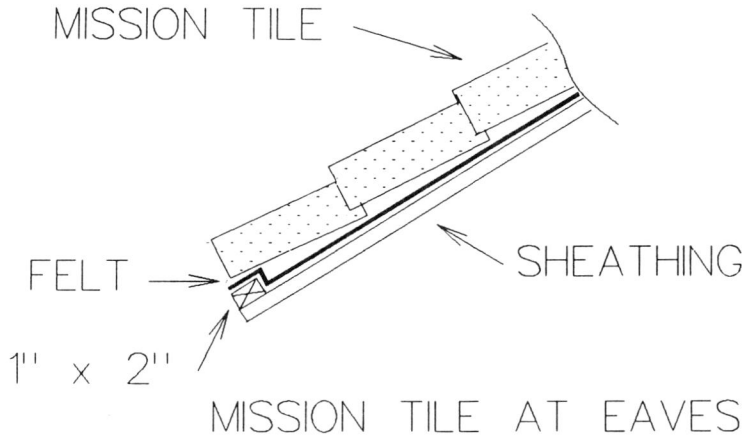

Figure 14–13   Correct method of starting tile at the eave line

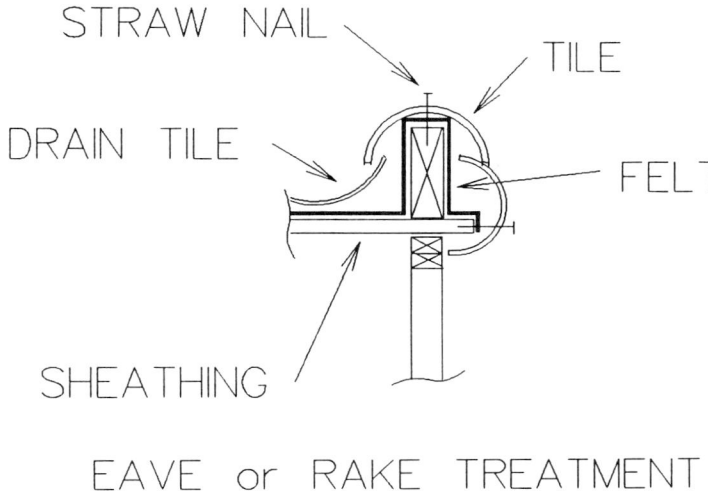

Figure 14–14   Using two-piece tile at eaves and rakes

to form a water barrier or flow trough along the rake. The bent-up section is nailed to a 2 × 2 in. batten installed along the length of the rake. To this, a special rake edge tile is installed. This tile covers the bent-up field tile edge.

Figure 14–15 shows a typical installation of mission tile at the rakes. The first step is to install 2 × 3 in. nailers the full length of the rake. Second, the roofing felt is brought from the roof up the side of the nailer. Do *not* completely cover the nailer as the cement fill must adhere to bare wood for best results. Nails driven in to half their length or stapled on wire mesh will aid in keeping the cement secured. The third step is to mound the cement over the

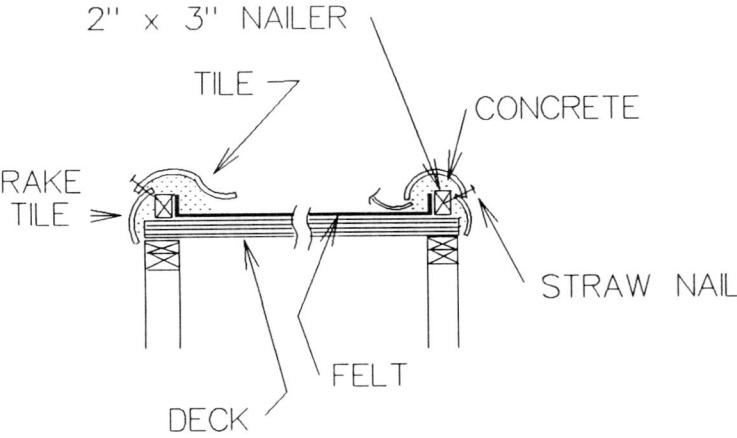

Figure 14–15   Build up rakes with a nailer

nailer strip, then press the tile into the cement and nail. NOTE: The left and right rakes are different. This difference is caused by the tile design and the way it interlocks from tile to tile.

Figure 14–16 illustrates the rake end of a clay tile roof. See the problem? The rake tiles are nailed directly to the rake fascia. The roof tile is at its low point under the rake tile. This will cause water damage problems to the fascia by allowing water to seep between the rake tile and the fascia.

VENT FLASHING

When using flat tiles, the vents should be flashed using standard vent flashing. Metal tile uses a slightly different technique. The vent is installed on the roof sheathing and a second piece of sheathing slipped over the vent pipe. A metal tile is then notched or cut to fit over the pipe and slipped down into the pipe flashing. A second but pliable flashing is then slipped onto the pipe and formed to the tile surface. Over this, a second metal tile is inserted. This second metal tile is close cut to the pipe and sealed with an elastic type sealer. This vent flashing method will result in a watertight vent seal that matches the remainder of the tile on the roof (Fig. 14–17).

SIDE WALL INTERSECTIONS

Metal tile can be installed to a side wall using standard counterflashing method or by using standard Z-bar flashing. The tile is cut and bent up at its edge to form its own flashing. The Z-bar or the counterflashing then covers the tile's bent-up edge. You can install pan flashing at the wall using standard counterflashing or Z-bar to cover the edge of the pan flashing. The field tile is brought to, and ends over the top of the pan, emptying into it.

Figure 14–16   Improperly installed tile at rake

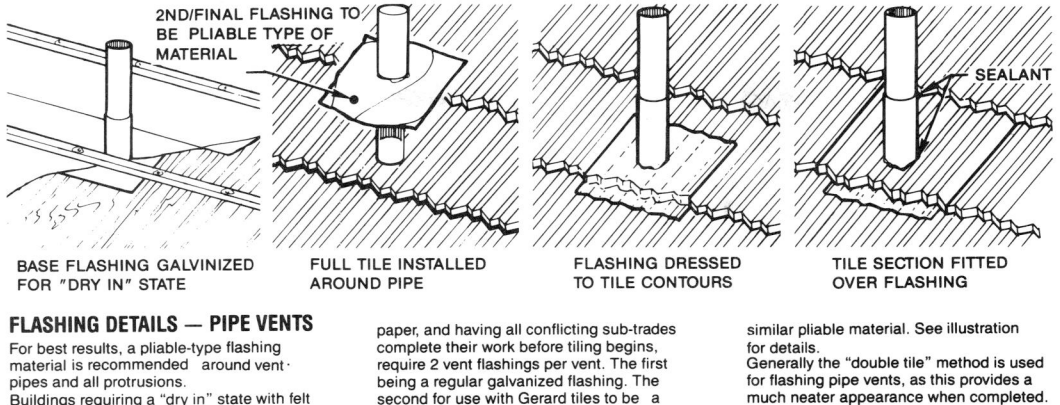

BASE FLASHING GALVINIZED
FOR "DRY IN" STATE

FULL TILE INSTALLED
AROUND PIPE

FLASHING DRESSED
TO TILE CONTOURS

TILE SECTION FITTED
OVER FLASHING

FLASHING DETAILS — PIPE VENTS

For best results, a pliable-type flashing material is recommended around vent-pipes and all protrusions. Buildings requiring a "dry in" state with felt paper, and having all conflicting sub-trades complete their work before tiling begins, require 2 vent flashings per vent. The first being a regular galvanized flashing. The second for use with Gerard tiles to be a similar pliable material. See illustration for details.

Generally the "double tile" method is used for flashing pipe vents, as this provides a much neater appearance when completed.

Figure 14–17 Installing metal flashing at vent pipe (Reprinted with permission of Gerard Roofing Technologies)

Figure 14–18 shows the installation of mission tile at parapet, dormer, or other side wall surfaces. The wall flashing must be installed while the wall is being constructed. Do NOT nail the drain pan to the decking. It should be nailed to the side wall. Nail the pan about 1/2 in. below its top edge. The wall flashing should be nailed as shown (Fig. 14–18). This area will be under the applied siding. NOTE: The left and right channel details are different due to the way the tiles lay and interlock.

Shown in Figure 14–19 is the installation of mission tile below a vertical wall, as at the front of a dormer. The reglet flashing is an architectural trim piece that doubles as a

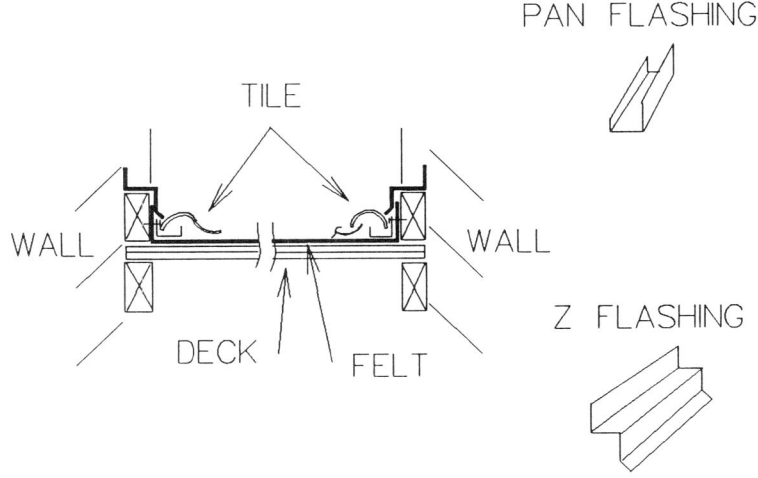

Figure 14–18 Use of pan and Z-flashing at side walls

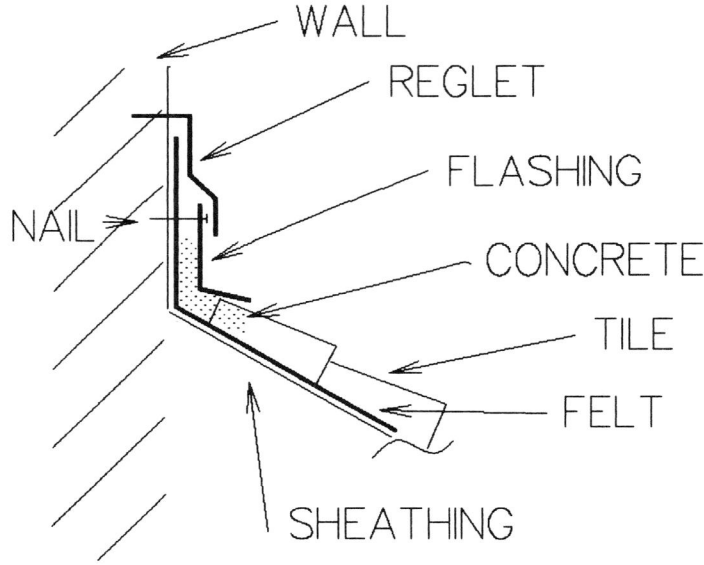

Figure 14–19   Correct tile to vertical wall intersection

counterflashing. Roofing felt is brought from the roof up onto the wall surface. The tile is brought up to within an inch or two of the wall and then filled with concrete. The V flashing is then embedded in the wet cement and nailed to the wall as shown. The reglet is then installed so that it cover the nails of the V flashing. The reglet is nailed to the wall and the siding is completed.

Tile Used as Drain Pan

If you do not want to use metal drain pans along the walls, you may use inverted tile (Fig. 14–20). The roofing felt is continued from the roof up the wall. Concrete is then mounded along the roof deck to wall intersection. The inverted tile is pressed into the concrete; no nails are used. Be sure that predrilled holes in the tile are filled with cement. Be sure the inverted tiles overlap in the same manner as the field tiles. The wall flashing and field tiles are then installed.

RIDGE CAP TILE INSTALLATION

The field tile will usually double as ridge tile. Use the same weather exposure as used on the field tile's weather exposure. Be sure to order sufficient tile for the ridge and then some extra for miscalculations and possible breakage. Some tile manufacturers have specially designed ridge tile for their product installations. If they do, then I suggest you purchase it.

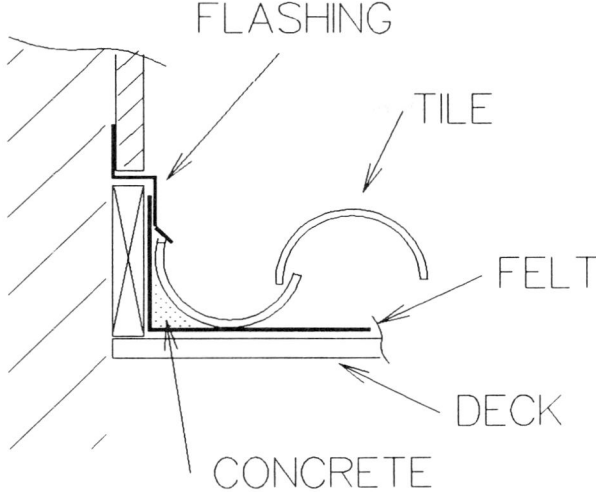

Figure 14–20 Using tile as a rain runoff pan

Metal tile roofs will have a manufacturer's supplied ridge tile. Be sure to order correct ridge tile for the application. You must build up the ridge peak with 2×2 in. battens. Most use 2 battens stacked vertically for the ridges, 2 battens side by side for the hip ridges.

The ridge cap tile of a Spanish or mission tile roof requires concrete fill (Fig. 14–21). This fill keeps water from entering the building under the ridge tile. It keeps water from entering under the field tiles. The field tiles should be partly filled to the point of being blocked. The concrete is then mounded up to the top of the tile nailer. Additional cement is placed in the tile and the tile is then inverted and nailed in place. Excess cement is cleaned off the joints and used for filling the next row of field tile.

**TYPICAL MISSION TILE
RIDGE TREATMENT**

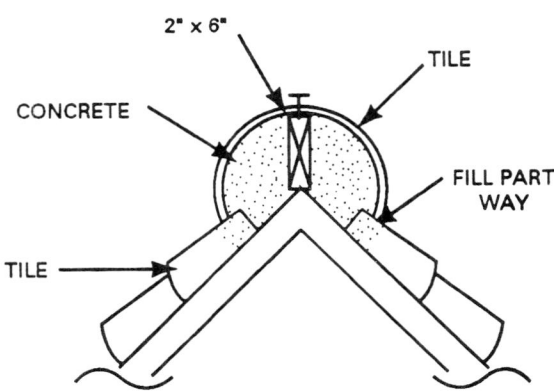

Figure 14–21 Ridge treatment for mission tile, nailers not used under field tiles if tile is wired in place

Keep the cement moist and stiff for best results. Mixing small batches as you proceed is better than mixing one big batch that may dry up too quickly. If the day is hot or if the cap tile fails to adhere properly, then dip the cap tile in water before applying cement. This helps keep the tile from absorbing the water out of the cement too fast. Water absorbed too fast will leave dusty, crumbly, and nonadhering cement at the bonding junction. In other words, the cement will not stick to the tile.

Recommended Mortar Mix

The cement or mortar mix I recommend is Type S mortar. Type S mortar has a high bond strength and is for extreme wind and weather exterior work. The mix is 1 part Portland cement, 1/4 part hydrated lime, and 3-1/4 parts clean mortar sand. Mix with just enough water to make a workable paste. NOTE: Do NOT use ocean or river sand. Ocean sand has been rounded by the wave action and does not form a good grip with the cements. Quarried sand is irregular in shape and does form a good bond.

Flat to Slope Roof Intersection

A flat roof to slope roof intersection is typical of a Mansard or Cape Cod roof. Metal tile can be bent at the intersection to make the slope change. If the tile ends at slope change intersection, then a 2 × 2 in. batten should be used at the intersection. Metal tile(s) requires a batten for proper securement of the tile(s).

Many roof designs call for a flat upper roof section and a sloped lower roof section. The upper roof, being flat, is a built-up roof that may or may not be covered with gravel. Figure 14–22 shows the proper method of intersecting a flat roof to a sloped mission tile roof. The roof felt should be continuous from roof to roof. Often this is not practical. If each roof has its own felt system, then bring one piece of felt from the upper roof and lap it over the lower roof. The U flashing is then nailed to the wall riser. Mission tile is brought to within a few inches of the riser and packed with concrete. The concrete covers the U flashing nails. The gravel stop is then placed over the U flashing. The upper roof is now installed.

Mansard Roof Pitch Change

A mansard roof changes pitch or slope on its side walls. You must bend the metal tile around this change if the slope change is in the middle of a tile (Fig. 14–23, left). To do so, you must V notch the edge of the tile and then seal the cut with solder or mastic. If you or the building framer had planned everything properly, this could have been prevented. The slope change should be at an even multiple of the metal tile's length so that the change is at the butt edge of the tile (Fig. 14–23, right).

Head Flashing for Chimneys

Metal tile is generally 40 or more inches wide. Most chimneys are not as wide as this. You would normally need to fabricate a cricket to divert water around a chimney, but using

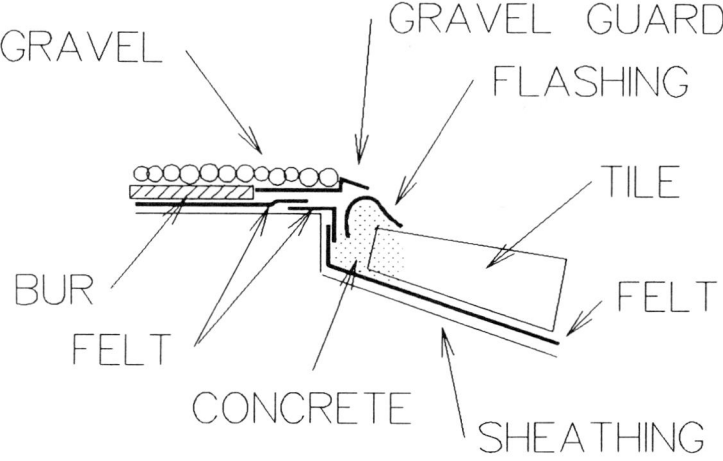

Figure 14–22   Flat roof to tile roof intersection

metal tile eliminates this need. Cut the front, center edge of the tile, perpendicular to the front edge. The cut should extend 3 to 5 in. into and toward the tail of the tile. Now cut a shallow V from left end to right end of the tile, starting 2 in. in from each end. The point of the V should end 1-1/2 in. before the end of the perpendicular cut. Fold the metal up 1-1/2 in. from end to end. This forms a V shaped splash diverter. The split at the apex of the V is then flashed with a small 1-1/2 × 6 in. piece of metal cut from the discarded V. Bend this piece to fit the apex and secure with two sheet metal screws (see Fig. 14–24).

If the chimney is wider than the tile, you must make a standard cricket and flash it accordingly. Not doing so will result in damage to the roof sheathing and building's contents.

FRAMING DETAILS — MANSARD/FACADES

Gerard roof panels require a batten structure to be affixed to. The battens are 2″ x 2″ nominal. A vertical wall or steep angled Mansard is installed in the same manner as for lower pitched structures. When the roof line changes pitch drastically as in a "Cape Cod" roof design, a batten should be installed at the pitch changes. See illustration. When an equal number of full panel courses cannot be accommodated at the pitch change, a full panel can be bent to suit. See illustration.

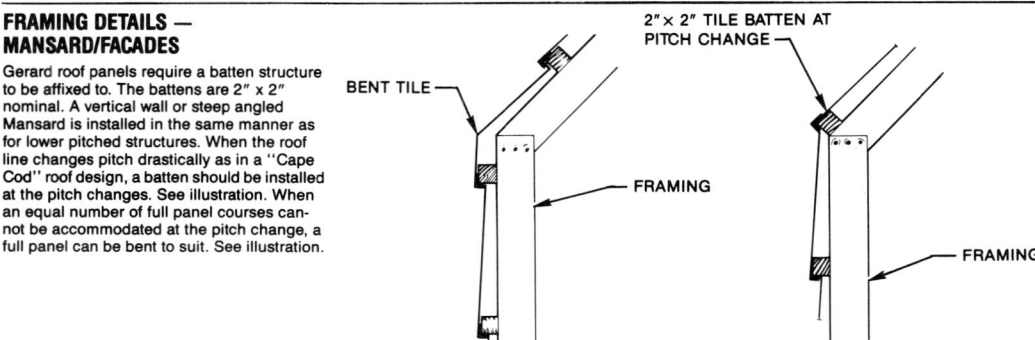

Figure 14–23   Framing detals—mansard/facades (Reprinted with permission of Gerard Roofing Technologies)

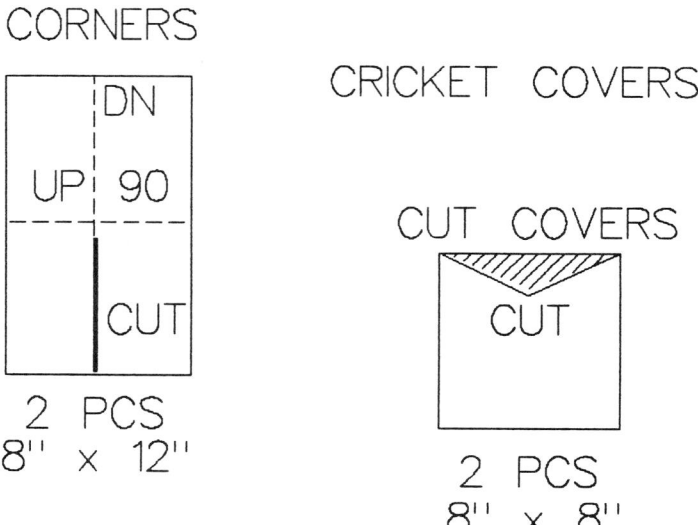

Figure 14–24   Cricket covers

Figure 14–25   Poor job of making a rake edge

Cementing Tile in Place

Figure 14–25 illustrates a clay tile roof with a different style of tile joint. There was probably a reason for this pile up of concrete. Repairs like this one are generally used to fix another problem; in this instance, the poor connection between the ridge tiles and the hip tiles.

Extra Income for You

Figure 14–26 shows a clay tile roof with an easily cured problem. The valley needs a cleaning. The willow tree, front left, has left a mess. Its droppings are quickly becoming a home for insects and fungus. The chemicals in the droppings, the leaves, will leach and discolor the tile if the leaves are allowed to remain.

This is spare change to you the roofer. With proper equipment you can clean this in a few minutes and earn \$20–\$30. You need a ladder, an extension broom, and an extension water hose. Do *not* climb on the roof, as you may break a tile or two. Then you are liable for the repairs. From the ladder sweep the gutter and then using the extension hose from top of the valley down, hose off the remaining loose particles. Clean up and collect your money. While you are cleaning the valleys and gutters, do a roof inspection. Give the homeowner a written report. They will thank you and recommend or use your services when they do require roofing work.

Replacing Tile

Tile replacement can become a nightmare. There have been literally hundreds of tile companies over the years. Each has their molds and formulas for texture and color. As in all businesses, many of these companies are no longer in existence. Sometimes the molds are purchased by other companies, but most of the time the molds end up in someone's garage. Your chances of finding a direct replacement for an older tile are slim to none.

Figure 14–26   Make yourself an extra dollar or two; inspect roof while you're at it

The solution is to repair the existing broken tile(s). This can be done by removing the broken pieces and reassembling them with a plastic or cement backer. The curved tiles can be filled with mortar and replaced. The flat tiles take a different technique. Use auto body filler putty and auto body fiberglass reinforcement tape on the back side of the tile to hold the pieces together. Instead of the auto body putty, you can use cold applied asphalt adhesive and roofing tape.

Selling Point of Interest

Slate, clay, and cement tile have an advantage that few other tiles have, that of heat retention. During the day the tiles absorb the sun's heat and tend to keep the interior of the building cool. At night the tiles give up this absorbed heat and help keep the interior of the building warm. This is the basic principle behind solar heating.

15

_____

Estimating

ESTIMATING THE JOB

One difficult item on any job is not the work itself, but the pricing of that work so that you can pay your bills and still make a profit. This is termed estimating and is a leading cause of business failure in the construction industry. You must understand your cost of doing business. You must understand that you are the expert, people call you because they feel you will do the job correctly. For this expertise you charge, for your overhead, your materials, your labor, your employees, your taxes, etc. And, most importantly, you add in a profit.

PROFITS FROM YOUR BUSINESS

What is a good profit? Is it 5%, 10%, 20%? Well, let's look at a necessary evil, taxes. If you are in the 28% tax bracket, you will pay $0.28 for every dollar you earn above cost. Therefore, if your net pay on line 31 of your 1040 is $30,000 and you have only 1 deduction, you will pay $4,316.00 (1990 figures).

This is a sizable chunk of money. As a self-employed person, you also have to pay some 15% extra, about $3,600, for Social Security, which totals $7,316.00. This leaves you with $22,684 to live on. I didn't include sales, property, or state income taxes, which will further lower your take-home pay considerably. In all, that $30,000 is taxed about 42% or more for the average American. The after-tax take-home pay is about $17,400 for a person earning $30,000.

You can put $10,000 in a CD (Certificate of Deposit) or savings account and receive a return of from 5.5% to 8.5%. This is with very little risk of losing your money. Going into business presents a high risk of losing your money. Therefore, your after-tax profit should at least be equal to or exceed the amount you would make in a savings account.

What is your profit on a job? The Gross Profit before taxes should be about 54% to make going into business worthwhile. This includes your salary. If you consider profit as above your salary, then a Gross Profit of from 18% to 24% is correct.

ESTIMATING TIME

To make money in roof contracting, you must estimate time, materials, overhead, and profit. The major item here is time, as it is usually the biggest dollar item involved. On a new, first-time roof, you can be pretty exact on time and materials. Redoing an old roof is where you can get into trouble. It's ten o'clock at night and you get the call, "Come give me an estimate on this blank-blank roof, it's been leaking ever since I owned the place!" You get the person's name and address and set up an appointment for 9 AM in the morning. You arrive at 9 and find out that the roof is about 15 years old and leaks in the master bathroom. Out onto the roof you go and you find a vent pipe coming through the roof over where the master bath area should be. You climb down and inform the customer that it will cost him $25.00 to fix the leak and he accepts. Back you go with your can of magic putty and you seal the vent pipe. You put your tools away and collect your money. The next day it pours cats and dogs, and at 11 PM your phone rings. "You, stupid so and so, I paid you yesterday to fix this leak and it's pouring in my bath, did you do anything or just collect my money." You have a callback on your hands and a not so happy customer. Anything you do for this person at this point will be suspect. Back on the roof you go. Looking around you notice something, the shingles over the master bedroom and the master bath are a different shade of color than the rest of the house. You climb down and ask the customer if he had this area reshingled before. "No, that area is an addition we put on about 3 years ago." How long has it leaked? "It's leaked right from the first year." You have just cost yourself a bundle of money and a customer. The tie-in from the old to the new roof was not properly installed. You must now rip out a 12 ft long section of shingles to fix the real source of the leak.

You now have several options open to you. You can give the customer back his money and leave, but he'll not be very happy about that. You can take his money and leave, and he'll surely not be happy about that. You can ask him for more money to fix the problem, but he'll not be happy about that either. You can spend the next 6 hours fixing the problem for the $25.00 you received, and you'll not be happy about that. Would it not have been better to do a good estimate in the first place?

What is a good estimate? A good estimate is where you play detective. You find out what the problem is, where it is showing up, how it is being created, and how you are going to fix it. Walk the building at ground level, look for tell tale signs of water damage, poor construction, etc. Then walk the roof inspecting every inch of it, and keep notes. Walk the interior of the building and check for signs of water leaks. Climb into the attic or crawl area and look for signs of water or water stains on the underside of the sheathing. Now, go home or back to your office and prepare a good estimate that takes all findings and problems into account.

ESTIMATING MATERIALS, NEW WORK

When you estimate materials for new work, new construction, you include everything right to the last nail. To do this, you must keep accurate purchase and pricing records. You must know your suppliers and anticipate their increases.

You must anticipate problems such as: the roofing framers not getting their job completed on schedule, inclement weather, or the chance that the supplier will be out-of-stock. There are hundreds of little items that can quickly kill your chance of doing a good job and getting paid a profit.

Take your time and factor in everything. Sit on the estimate for a day or two and then review it. You will find additional items that you forgot to include. Include them and type your final bid.

ESTIMATING MATERIALS, OLD WORK

On old work, a reroof job, you have to figure that you will be doing a new roof plus the possibility of a tear-off. Code generally restricts a building to three roof coverings. This is because the weight exceeds the load limit weight limitation of the trusses and wall studs. If a tear-off is not required, then you must see how much of the existing roof material can be reused. Be careful because, though it can be reused, it may require additional steps to make it suitable for reuse or match in color to the new roofing.

A classic example is happening as I write this. Three houses down the street from me, there is a single-story home. The clay Spanish tile roof was apparently leaking. The roofing contractor decided to reuse the existing tile. He tore off the tile and stacked it on the roof. First he made a stacking mistake, he laid the tile in piles as it came off. He soon found that the tile was in his way when he decided to install the new roofing felt. So, he moved the stacks of tile. He then installed the felt on one side of the roof and moved all the tile to that side. When it became time to felt the second side, he had to move half the tile back. During all this, he broke several tiles. He also broke all the tiles used for the ridge caps, which were cemented in place. Now he had to search for a supplier who could match the tiles. This is easier said than done.

The existing tiles have faded in color over the past years. Also, the manufacturer of the tiles went out of business. Even if the manufacturer had not gone out of business, the chances are that he would have changed design somewhat. It is now going on 3 weeks and this neighbor's roof is still not half completed. Rain is on the way and plastic tarpaulins cover the roof. Did this contractor make money from this 'cheap' reroof job?

REPLACE ALL OR NOT

I personally think that he will not only not make money, but lose both money and a customer. The roof tiles will never match in color and the job looks like a patchwork quilt.

I think this roofer should have approached this job with a bid that covered full replacement of tiles. He could have given a discount for tiles removed and used them on another, but smaller roofing job. Say $0.10 per tile discount, resold for another job at $1.25 per tile. Doing that, he would have stood a chance of making money.

It is your decision whether to replace a roof or not. I can only give you a few guidelines.

- If the roof is asphalt shingled and about 20 years old, consider roofing over the old roof. Use shingles of the same type, three-tab over three-tab, etc.
- If the roof is shake, consider tearing it off, putting on new sheathing, and reroofing with another, more fire-resistant material.
- If the roof is cement, clay, or slate tile, consider repairs only.
- If over 40 years old, then tear off and start anew.
- If the roof is built-up asphalt, then check the weight problem. If within limits, then roof over it. If not within the weight limitations of the framing, then tear off the roof.
- If the roof is young, under 15 years old, then try to patch or repair it.
- If the roof is severely damaged from accident or fire, then consider a tear-off and a new roof.
- If the roof is steel and very rusted, replace it. If the rust is only superficial, sand and paint.

Procedures for these repairs and replacements are explained in other chapters.

Replace Only One or Two Shingles

If you are only replacing one or two or a few shingles, then be careful of your time. The job is small, but may take upwards of a full day after you do your shopping, travel, setup, and the repair and money collection. These little jobs will get you every time.

Replace Flashings

If the flashing is rusted through, estimate a full replacement of the roofing.

READING BLUEPRINTS, NEW WORK

In new construction, one item that you have to face is not having a building to face, as the building may not yet be built. So how do you do your bid? You do your bid from the blueprints. This take-off, as it is called, requires you to think things through on paper.

You want to see the roof plan view, the front, back, and side elevation views, and the architect's sketch. You also should ask to see the material list. From these drawings you can find the roof line and the problems involved. Remember that drawings are two-dimensional. You must think in three dimensions—roofs have height, depth, and width.

We will estimate the amount of roofing needed (see Fig. 15–1). Since you are looking at the drawing, you can measure directly. Measure from one eave to the peak. This will give you a roof surface width. Multiply that by the length, and that answer by two and you will have the square footage of this roof. In our example this measurement is 17 ft and $17 \times 60 \times 2 = 2040$. We need to cover 2,040 sq ft of roof. Remember always to round up, so figure 21 squares. You must add additional shingles for the starter strip and the roof peak cap.

The starter strip is one layer of shingles along each eave or a starter roll. So we have 60 ft $\times$ 2 (sides) $\times$ 1 ft (width of shingle) or 120 sq ft of starter shingle, another 1.2 squares. The roof peak is 60 ft long, $60 \times 1 = 60$ sq ft. Add that to the 1.2 squares and we have 1.8 squares. Rounding up for packaging and waste and we have 2 squares. Add that to our 21 squares and we need 24 squares to complete this roof. There will be some extra, which is acceptable since there will always be damaged shingles. Plus, nine out of ten times you must cut some shingles for some reason or other. The extra square is your security blanket, after all it's only $27.95. Far cheaper than having to stop work to purchase one or two more bundles.

Lack of Information

In the chapters on problems and water problems, we covered most of what can go wrong with a roof and how to fix these problems. I will not rehash those chapters here. I will say

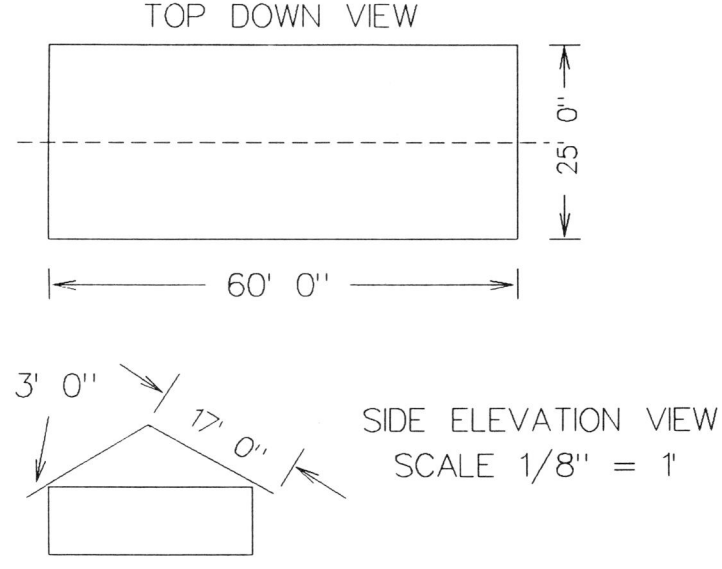

Figure 15–1 Estimate example

that it is the items that you don't look for or find that will cost you dearly during a roof job. SINEE (Suspect, INspect, and Estimate Everything) is your key to a profitable job.

VARIABLES IN ESTIMATING

People are a variable in that not all do the same task the same way or in the same amount of time. If you have hired helpers, then keep careful track of their job performance. You may have a good worker who can shingle a square in 20 min., but another worker who takes 40 min. If you plan your job bid using the first worker and then use the second worker, you will lose money.

A beginner will take two or three times as long to do a job as an experienced person. Put a beginner with an experienced person and the experienced person will take 20% longer than normal. He has to slow down and answer questions or has to fix the inexperienced person's mistakes.

You may have the problem of two or more people working together. Who does what and when? Make sure that everyone understands what their job functions and responsibilities are, as you don't want misunderstandings to slow the work. Worse, you don't want someone getting so ticked off that there is a fight and a work stoppage. Besides, a fight at the job site between employees is poor for customer relations.

TASK ASSIGNMENTS

Assign each employee a task or group of similar tasks. Rotate this employee to another group of tasks during the next job. Very shortly, you will have fully trained employees who understand everyone's job functions. Watch each employee as he or she works. You will find that there will be one who is better at a particular task than another. This will help you place the best worker for a task on that task and speed up your production.

Task groups might be:

- Installing flashing
- Installing sheathing
- Installing felt underlayment
- Installing shingles or tile
- Prep and painting
- Cleanup and safety
- Precutting and shaping
- Hauling and stacking
- Installing hips and ridges
- Installing valleys

You will get a feel for the task groups and who can do what best.

THE SCHEDULE

Make a schedule of what has to be done, when, how, and by whom it will be done. Do this even if you are a one-man company. Too often, you start a job without an inkling of what the job entails or how it is to be done. You then wonder why you can't make money, though you are working very hard.

Your schedule should include each step of the operation that must be completed. A start and a finish date and time is assigned for each step. The person who will do the task is assigned to each step and informed of his/her task. Reserve a place on the schedule for notes and problems.

Materials

Schedule the materials to arrive one day before the date that they must be installed or prepared. Materials are money. You have to take money out of the bank and spend it on the materials you are to install. Order everything simultaneously to receive the biggest discount you can, but, have the materials delivered as needed and not before. There are several reasons for this. First, you should be paying for the materials upon delivery or on time payments. The longer you keep your money in the bank, the more interest you will accumulate. Second, the longer you delay receiving the materials, the less interest charge you have to pay to the supplier. The system is called JIT or Just-In-Time purchasing.

The reason for JIT is to cut your losses. Materials that are left lying around the construction site will evaporate. Evaporate into thin air, i.e., find its way to someone's garage or basement. This is costly as you now have to spend profits replacing the missing goods.

The next reason for JIT is material damage control. The less time the material is in storage, the less chance there is of it being damaged. Damage comes in many forms; it can be deliberate, by accident or an act of weather or God. It does not matter how the material becomes damaged, it is your profits that must be used to replace the items.

Weather Delays

Roofing is a great job on a warm sunny day. You are in the sunshine, breathing fresh air and getting a little exercise. On a cold rainy day the job can be pure misery or worse, nonexistent. Use the rainy days to catch up on your bookkeeping or shopping. Roofing in the rain is unsafe and not recommended.

You must plan for these weather breaks in advance. You don't want ten workers showing up to work and expecting to be paid, when all they can do is sit around and play cards.

The second part of planning for the weather is the payments to your suppliers and yourself. Don't plan a job so tight that you have commitments to meet and find yourself behind the eight ball because of a missed day's work.

You must have a contingency plan in your sales contract, something that explains to your customer that weather delays are beyond your control and, therefore, you cannot be held responsible. If you do not have this in writing, you can be sued for nonperformance of contract.

Weather delays are not always because of rain. Snow, wind, severe storms, or too much heat can interrupt work as well. Ever try to work a roof in 110 degree heat? The shingles melt under your feet. You get tired easily and your normally fast working pace slows to a crawl.

Injury or Health Delays

No one wants to get sick or injured and few will do so on purpose. But, it does happen and can cause problems. It can slow down the job to the point that you must hire temporary help. Temporary help is great, providing you find the right person; otherwise, this can be costly and time consuming. Plan for delays in every job and have an alternative plan of action ready.

Again, have a clause in your contract about what is done if there is a delay because of illness or injury. This is essential.

You should keep accurate records on your employees, who got sick or injured and when. Problem-prone employees will cost you money and should be spoken to or dismissed. Accidents must be reported to the proper authorities and this too can cause days of delays.

Material Delays

There will be those days when the delivery truck is running late or not at all. Have an alternative plan of work for each day, for example, if the shingles don't arrive, you can do the fascia painting and so forth.

The second material delay is when the material does arrive but is the wrong material. This will take care of a day or two of work. You not only have to return the wrong material, but then attempt to convince the supplier to send you the correct items and send them now. Easier said than done.

The third material delay is when the materials arrive damaged. Of course, we have the "out-of-stock" days and the "I can send you half a load" days. Then there are the "But, the rest is back ordered, be here sometime in August," days, and it's now May.

The point is, business does have its little trials and tribulations. You can get angry at your suppliers or your employees or your customers, but anger will accomplish very little, while good estimating and planning will accomplish a lot.

The Materials Used

You have been using brand A shingles for a year or two. Suddenly your supplier tells you that brand A is no longer available, he ships you brand B. You do your estimate as you always do and then lose money. The reason, brand B was different from brand A. Even a small difference can make a big difference when you are accustomed to doing something one way and have to make a change. It is called the learning curve. Employees must relearn how to do what they have been doing all along. Your installation time jumps about 90% for the first few days and then starts to decline and level off. When all the employees master the new product, the job is complete. This learning curve holds true for the situation where different types of roofs are being installed. You and your employees may have installed a hundred tiled roofs. But, change to asphalt shingles for a month and go back to installing tile, and you will find the first few jobs losing money. Everyone is relearning how to install tile (Fig. 15–2).

Job Fatigue

This is the opposite of the learning curve. The employees have learned their job and are going at it strong. The job is within 2 or 3 days of being completed. The job fatigue curve kicks in and the time to do a task goes up by 40%. You finish the job a day late and at a loss. Job fatigue is normal and should be factored into your estimates. It is not a big factor on a short-duration job, but can cost you plenty on a long-duration job.

Material Coverage Factor

Another material factor to consider is the coverage of the material being used. It will take fewer shingles and, therefore, less time to install metric sized shingles than it will to install standard sized shingles. A square of metric shingles can be installed about 8% faster than a square of standard shingles. Up to a limit, the larger the material, the faster the job can be

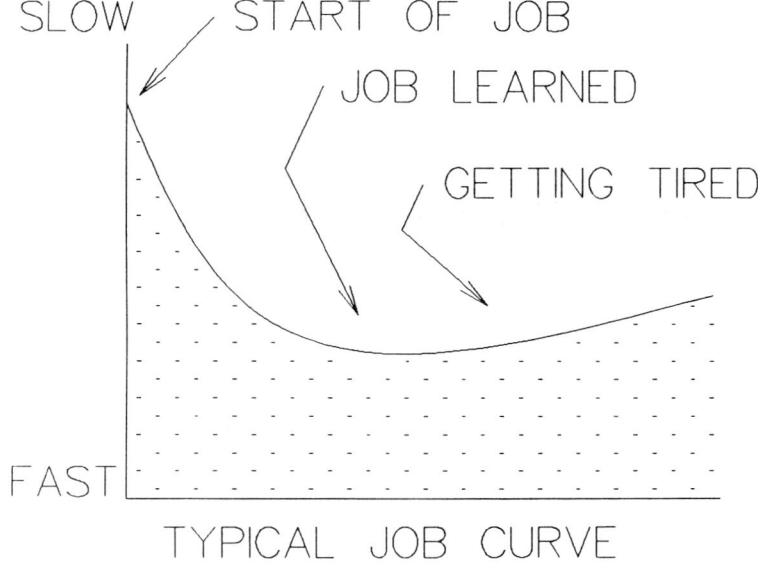

Figure 15–2 Learning curve

completed. The limit is when the material is so large that it takes two or more people to install it.

Materials Bought on Sale

This is a tricky one. Do you charge your customer the regular price of the item or the sales price of the item? I'll leave that decision up to you. I will remind you that the item may be advertised where your customer will learn of it.

      The thing to be careful about is record keeping for future estimates. You may have purchased the item on sale for the current job, but will the item be on sale when you are bidding future jobs? The probability is that it will not be on sale. An item like asphalt shingles can get very expensive very fast if you price the job at a defunct sales price. Example: Sale price of a bundle of asphalt shingle is $7.95. The regular price is $9.95. You are doing a job that requires 20 squares of shingle. That is 20 squares times 3 bundles per square times the $2.00 difference or $120.00 difference between the sales and the regular price.

      This cost delta or difference can happen when you change suppliers or product brands. It can happen when you change your purchase quantities. For example: For job A you purchased 150 bundles of shingle and were charged a quantity discount price of $7.95. For job B you purchased 60 bundles of shingle and are charged $9.95 a bundle.

      The price delta can come into effect when purchasing the same brand from the same supplier. Suppliers can and do frequently change their pricing. They may not have received a discount on their order to their supplier.

      The price delta can come into effect from county to county. For example: Materials in Orange county, New York (at this writing) are several percent higher in price than in Ulster County, New York. The counties border each other and share contractors. The difference? Orange County is experiencing a building boom, Ulster County is not.

Job Site Variations

Does the job look difficult or easy? A single-family house, sitting alone in a field, will appear to be an easier job than a single-family house sitting on a tree-studded hillside. The jobs are the same and should take the same amount of time to complete. But, the house on the hillside looks more difficult. It will take about 5% more time to complete than the roofing job done on the house sitting in the open field.

      You will have real variations as well. A new home or building may not be landscaped and the yard may be full of mud. This will slow the work by 5% to 10%.

      The steepness of the roof being roofed will have an effect on time to complete. A steeper roof takes more time to shingle for several reasons:

- The materials are harder to load onto the roof.
- The work becomes less safe and more precautions are required.
- The workers do not have as firm a foothold and are more careful.

- Scaffolding will be required and erection and disassembly time must be factored into the estimate.
- Shingle tabs must be cemented in place, the self-sealing strips will not work properly on steep slopes.
- Valleys will be longer and may require more splicing of the flashings.
- There is a constant fear of falling that slows work.
- Tools must be secured, else they will roll off the roof.

Add at least 25% to 40% to your normal roofing time estimates for steep roofs, those over 8:12 slope. Keep careful watch of the time and problems so that your next estimate will be more accurate.

Roof Shape

Odd-shaped roofs will take more time to roof than a plain rectangular roof. Odd shapes include U, T, L, O, Y, H, F, E, or just plain irregular. Anytime you have a break in a roof's surface, you will be adding time. This break is for the hips, valleys, and intersections that accompany an odd-shaped roof, places where extra care and extra flashing are required. More cutting and trimming are required. Add 10% to 15% to your time estimate for each direction change. This is compared to a straight rectangular roof.

Roofs that change slope on the roof's field also will increase the installation time. You should add 5% for this circumstance, i.e., Dutch hip or Mansard roofs.

Reroofing

Reroofing over existing roof can increase your labor by 10% to 50% or more. You must fix the old roof before you can install the new shingles. You must tear off the hip and ridge shingles and maybe tear out the valley flashing.

If you are reroofing with tile or metal, you must build up the existing roof with the addition of battens and counterbattens, which increases labor and materials. Add 20% to 40% to the labor.

If you are tearing off the old roofing, add 30% to 50% to the labor. You have the tear off, roof sheathing repair, and cleanup. You also may have dumpster rental and hauling charges to consider. If the weather is poor, you will have the problem of keeping the materials and the building's interior dry. A few very large and often expensive tarpaulins are required.

Delivery Charges

The supplier may offer free delivery but you do pay for it; the cost is built into his price. Some discount suppliers will add a delivery cost. Other suppliers will only add a delivery charge if there is equipment involved. The equipment can be a forklift or a conveyer for getting the materials to the roof.

Delivery charges can be charged to your customer or absorbed in your overhead. Delivery charges are a cost of doing business and are tax deductible.

Sales Taxes

Sales taxes are tax deductible if you pay them without collecting them from your customer. If you collect the money from the customer, the tax is not deductible as a cost of doing business. Sales tax money must be sent to the state at least once every three months.

Sales tax is overlooked by many estimators. Do not fail to include it in your estimate. For example: $2,000 worth of materials at 6% tax = $120.00. Can you afford to give $120 away, just because you forgot to add it into your estimate? No? Then you better add the cost of the sales tax paid into your bid without showing it as a separate item.

Travel Time

It takes time to get from one place to another. You should account for this time if possible. Remember that you had to go shopping for materials. You had to go to the building inspectors for the permits. You had to make a trip or two during the estimate and sales presentation.

You also have added miles to a vehicle. Miles cost money for wear and tear, fuel, oil, repair service, vehicle taxes, loan interest, and license plate fees. This must be recovered somewhere. The current industry standard charge is 28 to 40 cents per mile. To remain competitive, you may consider free mileage if the customer is within 10 miles and charge for all miles over 20 miles each day. The alternative is to bury the mileage in the overhead factor. Mileage is a tax deductible item.

Overhead

Overhead is where you account for all dollars spent that are not directly related to a particular job. This can be for office help, accounting services, equipment cost, advertising cost, building rent, and the like. Overhead can add 40% or more to a job cost. All overhead expenditures should be tax deductible. You must include overhead expense in your estimates. It can be included as a separate line item or as part of the labor total. Most contractors include overhead as a portion of the labor total. The exception is when you are working on a government project. Government procurement people want to see overhead and profit as separate line items on the bids and billings.

Supervision of employees is considered an overhead expense. If you have a working supervisor, then his/her time should be separated into productive and nonproductive time for accounting purposes. A working supervisor will spend about 20% to 25% of his or her time supervising, and this is nonproductive time.

Storage of Materials and Tools

You must find a safe haven for your materials and tools. Protection of these valuable items will cost you in both time and dollars. You may need a fenced off area, some skids, and some tarpaulins. Or, you may need a fenced off area and a storage shed. You can rent chain link fencing for your fenced area. Skids can be purchased or fabricated. Tarpaulins and sheds should be purchased. Good locks will be required and a security guard may be required. There will be extra time involved in setting up this security and extra time involved whenever an employee wants to remove an item from the secured area. Set up and tear down of the security area will take a day. Add 2% to the employee labor times for getting in and out of the secured area.

Lunch, Breaks, and Other

Employees should receive two 15 min. breaks and a 30 min. lunch period each day. There will be unscheduled bathroom, smoking, and talk breaks. The average employee works 6-1/2 hours out of every 8 hour day. This 1-1/2 hour deduction of production must be considered. There will be time spent looking for tools and materials, there will be time spent sharpening tools. There will be an occasional telephone call or interruption. All are to be factored into your labor rates and hours of production.

Cost of Labor

You must know your labor cost and include the proper cost in your estimates. Your labor cost includes the salary paid, your portion of the employee taxes paid, the insurances paid, and all other benefits paid to an employee. It is not uncommon for an employee to receive a benefit package of 28% or more. Use the true cost of the labor, not the salary cost. Example: John gets paid $10.00 per hour plus full benefits. John's real salary is $10.00 times 1.28 (salary plus benefits) or $12.80 an hour.

If you use my figure of 6-1/2 hours productive time in an 8 hour day, you will find John's cost to be $15.25 per hour (12.80 × 19.2% loss time). You charge your customer for John's time at $10.00 an hour and you will lose $5.25 per hour, or $42.00 a day.

To remain competitive you may have to calculate each employee's true cost and charge your customer accordingly. Example: John is the roofer and costs $15.25 per hour. Bill is a helper and costs $8.75 per hour. For the current job, John will work 16 hours and Bill will work 8 hours. We charge the customer (15.25 * 16) + (8.75 * 8) = (244) + (70) = 314 or $314.00 for labor. (The asterisk (*) means multiply.)

Total It Up

Let's assume that you did all your calculations and know the labor and material involved for a roofing job. We can now total everything and find a sales price that will pay the bills and earn a profit. Here is what we are going to do:

- Materials times a markup of 15% for handling (Material * 0.15)
- Sales tax paid on the materials (Mtl's. * tax of 6% or Mtl's * 0.06)
- Travel expense at our cost (Miles * 28 cents per mile)
- Labor cost for each employee (Labor hours * true labor cost)

- Overhead cost factor on labor (Labor dollars * OH cost or 0.32)
- Subtotal all the above
- Profit added (subtotal * profit of 18% or subtotal * 1.18)

For example: We are bidding on a residential asphalt roof that requires 20 squares of shingle. The job will take 3 days to complete. John will work all 3 days and Bill will work 2 days. The address is 32 miles one way from the business. The shingles cost $25 a square and the state sales tax is 6%. John earns $15.25 per hour and Bill earns $8.75 per hour. What is our bid?

- Materials: 20 square * $25.00 a square = $500.00
- Material handling costs: $500.00 * 0.15 = $ 75.00
- Material sales tax cost: $500.00 * 0.06 = $ 30.00
- Travel is 3 days@ 64 miles: 3 * 64 * .28 = $ 53.76
- John's labor is 24 hours: 24 * $15.25 = $366.00
- Bill's labor is 16 hours: 16 * $ 8.75 = $140.00
- Overhead on labor: ($366 + $140) * 0.32 = $161.92
- Subtotal: (500 + 75 + 30 + 53.76 + 366 + 140 + 161.92) = $1,326.68
- Total with profit added: $1,326.68 * 1.18 = $1,565.48

Bid should be for $1,565 + 5%. We do not include the odd cents in a bid. The courts will hold you to the bid price if odd cents are included. The odd cents are to be included in the final customer's bill. In our example, we can charge this customer from $1,486.75 to $1,643.25 and legally be within 5% plus or minus bid (rounded, $1,487 to $1,643).

TIME AND STEPS

How closely do you make your estimates? Should you be so exact that you count every nail? The answer probably is yes. It depends on the scope of the job being estimated. A small job should not take days to estimate. A very large job should take days to weeks. The reasons are as follows:

- A small job will use small quantities of materials. It doesn't really matter if you use 300 nails or 340 nails—the cost delta is small.
- A large job uses a vast amount of materials and material count becomes critical. You figure on using 50 lb of nails and then ended up using 600 lb of nails and you have lost money.
- A small job can be completed in a somewhat short period. You miscalculate the man hours involved and you lose a few dollars.

- A large job will require many man hours, sometimes into the hundreds or thousands. You miscalculate by 10% and you will have lost most if not all your profit on the job.
- Make a mistake on a small job and you lose a few hours or a day correcting it.
- Make a mistake on a large job and you may lose weeks of work. A small mistake can be repeated hundreds or thousands of times and, therefore, must be corrected the same number of times.

What about the medium-size jobs? It is up to you to determine how accurate you wish to be. You must balance the labor cost of doing the estimate to the cost of doing the job. Here are a few examples:

- Job price of $100 or less, the estimate will take up to 25% of the total price charged.
- Jobs priced from $101 to $500. The estimate should not take more than 15%.
- Jobs priced from $501 to $5,000. The estimate should not take more than 8% of the total.
- Jobs priced over $5,001. The estimate should not take more than 5% of the total.

Cost of Sales

Remember that your estimate includes the cost of sales. This is the time used to travel to and sell the job to your customer. It includes the measuring of the roof, the troubleshooting of the roof problems, and the time used to prepare the estimate. The time also includes obtaining price quotes from your suppliers and writing up the bid proposal.

Man Hour Tables

There are many companies producing manuals of man hours required to do various steps of various jobs. The man hour tables in these manuals are based on someone's experience and time studies. The manuals will give you a good idea of what should be completed in a given period. The key word is should. My recommendation is to create your estimating manual using your experience and knowledge.

In creating the manual, you will learn about your job and all that it entails. For example: Let's say you make an entry of time and materials used to install flashing around a dormer. What is required?

Set up ladder and climb onto roof.	5 minutes
Measure the width and length of the dormer.	3 minutes
Climb down off the roof.	1 minute
Gather material and move to work area.	12 minutes
Gather tools.	4 minutes
Measure and cut the head flashing.	15 minutes

Measure and cut the base flashing.	15 minutes
Measure and cut the step flashing.	20 minutes
Prime coat all flashing.	12 minutes
Wait time for prime to dry.	20 minutes
Paint all the flashing.	12 minutes
Wait time for paint to dry.	45 minutes
Bring all material to the roof.	5 minutes
Install the base flashing and shingles.	16 minutes
Install the step flashing and shingles.	20 minutes
Install the head flashing and shingles.	14 minutes
Caulk as needed.	5 minutes
Touch up paint and do final trim.	5 minutes
**** Total labor =	144 minutes plus
	65 min. waiting

This comes to a total of 229 minutes.

The materials:

Flashing metal, 12 in. wide galvanized steel	18 linear feet
Nails, 7/16 in. roofing	80 each
Caulking, silicon RTV, 14 ounce tube	1/2 tube
Metal primer	1 pint
Semi-gloss latex enamel, light brown	1 pint
Paint brush, disposable	1 each

We do the quantity times cost multiplications and find that the total cost of material is $38.00 plus 6% tax = $40.28. We did not count the shingles as they are in another part of the estimate. The labor is figured at $15.00 per hour and totals $45.00. We apply 15% materials handling and 32% labor overhead to find the material cost of $46.32 and labor plus overhead (OH) cost of $59.40. Adding material and labor gives a job cost of $105.72. We add in 18% profit to find a selling price of $124.74 for installing flashing and shingles at this dormer.

Can we use this figure, $124.72, for our future estimates? Yes, if the cost of labor and materials remain the same. To be on the safe side, use a figure of $130.00 for this roofing step. Review these costs a minimum of every six months.

We now have a building block cost figure, the $130.00 for flashing and shingling around a dormer. You should do this cost exercise for each of the following:

- Installing valleys
- Installing chimney flashing

- Installing DWV vents
- Installing flue-type vents
- Installing ridge vents
- Installing other air vents
- Installing hip and ridge shingles
- Installing soffit and fascia
- Installing starter strips
- Installing a skylight
- Installing a square of roofing

You should use your figures and material prices. The $130.00 in the example may NOT be correct for your location or use. Do several cost breakdown sheets for items that require variations. Example: Valleys can be open, half-lace, full-lace, or W-flashed, and a separate sheet is required for each.

When you complete this exercise, your estimating will be a snap. To estimate a job, just copy the appropriate estimate pages and assemble them into a job binder. Add up the total from the sheets and you have your estimate. By using this block or unit estimating method, you can estimate most jobs in 30 minutes or less.

If you have a computer, type these estimate pages into a spreadsheet program like Lotus 1,2,3™. Spreadsheet programs will allow you to change pricing with a few key-strokes.

TIME SAVERS

There are three ways to do a job: the easy way, the hard way, and your way. Think things through before doing. Here are time and money savers you should consider:

- Purchase factory formed hip and ridge shingles rather than fabricating them from shingles.
- Purchase starter strip in rolls rather than making starters from shingles.
- Purchase preformed flashing rather then site forming them from sheet metal.
- Use a power stapler or nailer rather than a hammer or roofer's hatchet.
- Use a power caulking gun rather than a hand trigger-pull gun.
- Install more than one hose on your air compressor, this allows you to keep two air tools running without changing back and forth.
- Use a chain saw for cutting thick lumber rather than making several cuts with a hand saw or circular saw.
- Organize your tools into kits used for a particular function.
- Standardize your material usage as much as possible.

Workmanship and on-the-roof procedures that save time:

- Stack your roofing materials across the roof and in stacks that can be reached by the roofer. Stack only what is needed for that area being roofed. Do this after the felt is in place.
- Have a helper do all the lugging and material handling. Why pay a roofer $15.00 an hour to haul shingles when a helper costs $7.50?
- Use blue chalk for chalk lines. Red chalk stains the materials and is difficult to remove.
- Lay out your shingles and allow them to stabilize to the ambient temperature. They will be much easier to install.
- Tie air hoses and power cords to the roof. Keeps them from falling off and your having to climb off the roof to retrieve them.
- Use very flexible air lines rather than the hoses that are supplied with most air compressors. Flexible hoses are much easier to handle.
- Keep a good supply of nails, staples, and caulking on the roof until the job is completed. Be sure to have all the various types and sizes needed.
- Store all accessory materials and tools at the ridge. That way, they don't get in your way.
- Preheat mastic and sealer in a bucket of warm water. Application is faster and easier.
- Use jigs and guides wherever possible. Makes for a better roofing job and cuts down on mistakes.
- If possible, preform and prepaint everything on the ground. Working on a flat level surface is easier than working on a steep roof surface.
- Check the work and the workmanship frequently. Catch mistakes before they become major problems.

Before-the-job money savers:

- Purchase materials in the largest quantities you can use.
- Combine material purchases for several jobs into one order for your best discounts. Balance this against the paperwork required to split out the materials to the correct job cost sheet.
- Keep abreast of the trade. Subscribe to and read all you can about new products, procedures, and technology.
- Spend some time shopping around from time to time. The lowest price you found for an item in January may be the highest price in June.
- Specialize in the type of roofing you are good at doing. Why take a job you know you do not understand?
- Keep yourself informed about labor, safety, and tax laws. Make a scrap book of the articles you come across that are pertinent.

- Keep accurate job records and use these records for the basis of estimating future jobs.
- See to it that your employees are trained and kept informed of the latest roofing and safety methods.
- Plan all jobs as if your livelihood depended on them; it does.

16

Specifications and Tests

INTRODUCTION

Codes, agencies, and regulations are the subjects of this chapter. I'll cover building codes as they pertain to roofing, and the various agencies that oversee the roofing industry. Also included are the specifications used for compliance testing on the various roofing products and the regulations that apply.

Please note: Testing agencies are generally profit-oriented agencies. They generally charge the manufacturer of a product for each sample submitted to them and for each test made on those samples. Thus, not all tests will be paid for and done on all products from a manufacturer. Nonprofit agencies such as Underwriters Laboratories do charge; they, like everyone else, have their bills to pay.

Each testing agency may have their specific rules for a test. A product approved by one agency may not gain approval from another agency for the same named test.

Each state, county, and city may have their particular specifications that a product must meet before it can be installed in a structure in those localities. What passes Los Angeles City code may not pass New York City code and vice versa.

The result is that you, the roofer, must decide if the materials you are selling or installing are approved for the location, building type, and building code. Add to this the requirement that the material should meet the building designer's/owner's intent and the challenge of the weather in your area and you could have problems. Fortunately, most suppliers of roofing materials have already checked things out and are selling products approved for your area.

Not meeting code, either with the installation procedures or the product, can mean heavy fines, rework or loss of the job, lawsuits, and loss of your contractor's license.

AGENCIES

The following is a listing of the agencies involved in testing or approving roofing products. From the looks of it, roofing products are probably the most tested products of all building products. Then I suppose they should be, since the products have to last 15 or more years in normal service. Roofing products have to protect the contents of the building from sunlight, rain, snow, ice, wind, and falling objects. They have to be fire-resistant and not generate a condition that will kill someone in a fire. They have to be made from readily available materials and sold at an affordable price. Here is the list of agencies.

APA	Trademark of the American Plywood Association
ASTM	American Society for Testing & Materials
ARMS	Asphalt Roofing Manufacturers' Association
CFM	California fire marshal
CSI	Construction Specifications Institute
Metro Dade	Fire department of Dade County, Florida
F.M.	Factory Mutual System, sometimes called F.M.S.
ICBO	International Conference of Building Officials
LA City	Los Angeles city fire department
NERCA	Northeast Roofing Contractors' Association
NRCA	National Roofing Contractors' Association
MRCA	Midwest Roofing Contractors' Association
RIEI	Roofing Industry Educational Institute
RIC/TIMA	Thermal Insulation Manufacturers' Association
SBCCI	Southern Building Code Congress International
SPRI	Single Ply Roofing Institute
UL	Underwriters Laboratories

This list is not inclusive, there are several more testing agencies involved. Many roofing products will have approval stamps from dozens of agencies from around the world.

I note with curiosity that there are no consumer groups, to my knowledge, involved in roofing products or application. Perhaps the industry is doing an acceptable job of policing its own.

CODES AND SPECIFICATIONS

Next I'll summarize some codes and specifications involved. Note that most material specifications are from the ASTM and most installation specifications are from the UBC (Unified Building Code, ICBO). These two organizations do control most roofing product specifications.

ASTM B-117 Mechanical fasteners in exposure to salt fog. This is for corrosion resistance. Especially needed in sea coast communities.

ASTM B-368 Mechanical fasteners in exposure to copper-acidified salt spray.

ASTM C-165 The compression resistance of perlite and other foam insulation products.

ASTM C208-72 covers the thermal criteria for wood fiber products.

ASTM C-209 Weight and water absorption of perlite. C-209 additionally covers, laminar tensile strength and the internal bond strength of perlite.

ASTM C-355 covers the vapor permeability of perlite.

ASTM C-409 covers the internal bond or tensile strength of perlite.

ASTM-C-512 Test the K factor of insulated vinyl curb shields.

ASTM C-581 Test the C value of perlite and other foam roof insulation.

ASTM D-312 Type I, II, III, IV specifies the heating and mopping temperature of hot asphalt. Also covered under federal spec SS-A-666-3.

ASTM D-412 Test the tensile strength and elongation percent of insulated vinyl curb shields and roofing membrane elongation at 77 degrees Fahrenheit.

ASTM D-573 covers heat aging of roofing membranes.

ASTM D-1621 covers compression resistance of polyisocyanurate insulation sheets.

ASTM D-1622 covers the density of polyisocyanurate insulation sheet.

ASTM D-1668 covers the glass fabric used in roofing systems.

ASTM D-1863 covers the aggregate surfacing material used for cap sheets.

ASTM D-2178 and federal spec SS-R-620B cover glass sheets.

ASTM D-2178-85a Type IV covers the ply sheets used in making asphalt shingles.

ASTM D-2523 checks the tensile strength at 0 degrees Fahrenheit of modified bitumen cap sheets.

ASTM D-2525 tests the elongation percent at 0 degrees Fahrenheit of modified bitumen cap sheets.

ASTM D-2626 and federal spec SS-R-501D cover asphalt coated base sheets

ASTM D-2822 and federal spec SS-C-153 cover the plastic roofing cements used.

ASTM D-2842 covers roofing membrane water absorption by percent of weight.

ASTM D-3018-82 covers type 1 asphalt shingles.

ASTM D-3161-81 covers type 1 asphalt shingles.

ASTM D-3462-83 asphalt shingles.

ASTM D-36-86 specifies the softening point ranges of asphalts.

ASTM D-41 and federal spec SS-A-701b cover asphalt primers.

ASTM D-92-85 covers the minimum flash point of asphalt.

ASTM E-84 Flame spread tunnel test for polyisocyanurate insulation.

ASTM E-96 Water vapor transmission through a roofing membrane.

ASTM E-108 Fire resistance of membrane materials.

FM 1-28 covers the minimum thickness requirements for various roof insulation.

FM I-60 and FM I-90 cover the nailing patterns for securing foam type insulation to roof sheathing. Basic wind storm ratings.

Federal Specification LLL-1-535A, B, Class C covers the C and R values of wood fiber products.

NIOSH Publication DHEW 78-107 covers the "Criteria for a recommended standard. Occupational Exposure to coal tar products," asphalt.

PS1-83 standard by the APA covers the types and grades of plywood used for roof sheathing.

UBC 25-12 covers the fabrication, design, and identification of build-up wood members including plywood. Code refers you to UBC 25-18.

UBC 25-17 Roofing nails and corrosion resistance of same.

UBC 25-C-1 covers makeup of laminated beams used for roof and floor rafters.

UBC 25-S-1 & UBC 25-S-2 Plywood sheathing for roofs. UBC 25-S-1 gives the allowable spans for plywood roof sheathing over two or more spans and with face grain of plywood perpendicular to the supports. UBC 25-S-2 is the same except the face grain is parallel to the supports.

UBC 25-R-1 & UBC 25-R-2 Lumber sheathing for roofs. Table UBC 25-R-1 gives the allowable spans and Table UBC 25-R-2 gives the minimum board grades for sheathing.

UBC 25-S-3 Particleboard sheathing for roofs. This table gives the allowable loads.

UBC 25-U T&G planking used as sheathing for roofs. Gives the allowable spans and loads.

UBC 25-U-R-1 to -14 give the allowable spans for different slopes of rafters under various live load conditions.

UBC 27-9 covers the spans of rafters and beams and their end connections.

UBC 32-A Chart of maximum weather exposure allowed for wood shingles and shakes.

UBC 32-B Chart giving the roofing material, roof slope, and application methods of built-up roofs and shingled roofs.

UBC Section 32-08 gives specifications for flashing materials.

UBC 32-2 Built-up roofing using asphalt for a hot cement and mopping coat.

UBC 32-3 Installation of asphalt shingles and sheet or roll roofing products.

UBC 32-4 covers the installation of metal roofing sheets.

UBC 32-5 specifies mineral roofing aggregate with a weight not exceeding 60 pounds per cubic foot.

UBC 32-6 The standard for wire used to attach clay, concrete, and slate shingles to a roof.

UBC 32-7 gives specifications on fire ratings of roofing.

UBC 32-8 Standards set down for wood shakes, slate shingles, and wood shingles.

UBC 32-9 The installation of asbestos cement shingles.

UBC 32-10 covers the installation of slate shingles.

UBC 32-12 The installation of clay tile and concrete tile shingles.

UL 723 and ASTM E-84 covers flame spread, fuel contribution, and smoke developed for perlite roof insulation and build-up panels.

UNI 8202 norms, the European standard for bitumen roofing products.

CLASSIFICATION RATINGS

UL Class A, best fire rating.

UL Class B, acceptable fire rating.

UL Class C, least acceptable fire rating.

TERMS EXPLAINED

Compression

Compression refers to the amount of pressure that can be put on an object before it permanently deforms or fails to return to its original shape. Most roof insulation is made from some sort of plastic foam. The foam should withstand the pressure of someone walking on it and of the roof covering and snow loads it will be subject to, without deforming.

C-Value

The C-value is the amount of conductance of heat or cold through a material or insulator. The lower the C-number the less conductance it has and the better the material is. $C = 1/R$-value.

Deflection

Deflection is the measurement of how far from a given plane a material will bend when subject to a load. For instance, one would not want a floor that deflected 1 in. when walked on. Nor would one want a roof to deflect, bend inward, with a foot of snow on it. Most sheathing materials will bend, even of their own weight if allowed to lie flat and supported in the air only by their edges. Codes for maximum allowance are an attempt to minimize this deflection. Code is 1/360 of span deflection for flooring and some decking, 1/240 of span for roofing, as installed. Deflections of up to 1/4 in. are generally acceptable, over 1/4 in. are not. To correct the problem, more support is required in the form of beams, joist or rafters. (See tables 4 and 6 in Appendix B.)

Flash Point

Most materials give off some vapor at some time during their life span. This vapor is sometimes flammable if exposed to an open flame. The flash point is the lowest temperature at which the vapors will ignite when exposed to an open flame.

Density

Density refers to how solid a material is. Stone is much more dense than foam rubber. In roofing, density is usually used to measure the solidness of the foam insulation used. There is a point where the foam insulation can be too dense and not contain enough trapped air cells, which results in a lower insulation value. There is a point where foam insulation is not dense enough and will not support a load. The ideal material has a density that will support normal roof loads and still provide a good degree of insulation value.

Dimensional Stability

Dimensional stability is the measurement of a material subject to prolonged heat or cold. Most materials will expand or contract to some extent. In roofing, you want a material that is compatible with the dimensional stability of the other roofing members. In other words, the rafters, sheathing, and roofing should all expand and contract at or near the same rate to prevent failure from pulling apart, delamination, or cracking.

Elongation

If you suspend an asphalt shingle by its edge and measure the length and width of the shingle, you will find that over a period the measurements will increase. The shingle has elongated or gotten longer. The effect is caused by gravitational pull on the material. Temperature and weight of the material will increase this elongation. The warmer and heavier the material is, the more elongation it will incur. The steeper the roof slope that the shingles are mounted on, the more elongation present.

Flame Spread

Flame spread is measurement of how much and how fast a material burns. The flame spread of a sheet of gypsum board is slow, whereas the flame spread of a piece of dried pine is fast. The lower the flame spread the better the material.

Flexibility

The material's resistance to damage at low temperatures is called its flexibility. The roofer who works in very cold weather will appreciate roofing materials that bend and will not crack during installation.

Fuel Contribution

Fuel contribution is the amount of material that will burn or add to a fire condition. The lower the fuel contribution the better the material is. A concrete or steel roof will not contribute to a fire as much as a shake or wood shingle roof will.

Heat Aging

Heat aging and heat cycling is where a material is subjected to controlled but extreme variations in temperature over a fixed period. Heat aging is used to represent the natural weathering of materials when in actual use. The aging or cycling is a sped-up process where years of weathering are simulated in days or hours. Heat aging may or may not be a true representation of actual conditions.

Mopping Temperature

Mopping temperature is the temperature at which the asphalt or tar is applied or mopped onto the roof. Too low a temperature and the ply and cap sheets will not bond properly, thus resulting in delamination of the roof. Too hot a temperature and drips, runs, and the possibility of melting the plastic ply sheets occurs. Too hot a temperature can result in a fire during application.

Melting Temperature

Melting temperature is the temperature at which a material will melt and no longer hold its shape. The material enters its plastic or liquid state. Most materials have a melting temperature.

Permeability

Permeability in roofing materials is related to the amount of water moisture that will pass through the material in a given period. The more vapor that passes through, defuses in, or penetrates a material, the more permeability the material has. Steel has low permeability to water vapor, a sponge has a high permeability.

Pot or Kettle Temperature

Pot temperature is the temperature of the tar or asphalt being heated in the kettle. Too hot a temperature may result in a fire. Too low a temperature will cause the mopping or spreading temperature to be incorrect. Various grades of asphalt roofing require different pot temperatures. See Appendix B, table 20.

R-Value

The most common measurement of the quality of insulation is the R-value. It is the resistance to the flow of heat or cold through a material. The higher the R-number, the more resistance there is and the better the material. Most construction has floors insulated to R-19, walls to R-11, and ceilings to R-25. Newer construction is R-19 in the floors and walls, R-38 in the ceilings. Some building codes, in colder areas of the country, require R-25 in the floors, R-19 in the walls, and R-49 in the ceilings.

To figure out the R-value of fiberglass insulation, measure its thickness and multiply by 3.23. The result is the R-value.

See the section on DOE for additional recommendations.

Smoke Developed

This is a measurement of the amount and density of smoke produced when a material burns. The lower the smoke development the better the material is. Many materials burn somewhat smoke-free, propane gas is an example. Some materials produce great amounts of smoke during combustion, some foam rubber for example. Smoke development is important for several reasons.

First is the ability of people to see their way through a fire to safety. The less smoke generated, the easier it is to see. Second is the amount of toxins produced by the smoke. Smoke produces toxins, poisons, that can kill. Third, most smoke is the result of incomplete combustion. This smoke contains gases and particles of matter that can burn explosively when heated enough, a very dangerous situation.

Span

Span is the distance between two supporting members, i.e., rafters or beams. Span is measured in inches. Normal spans for construction are 16, 24, and 48 in. UBC tables give allowable spans for floors, walls, ceilings, and roofs. Since most beams and rafters are used to support sheathing, the span is important. The sheathing must not bridge a span so large that the sheathing cannot hold its weight or the weight of people, snow, drywall, or roofing materials without serious deflection or failure. See Appendix B, tables 4, 6, and 8.

Tensile Strength

Tensile strength is the force required to pull or tear a material apart. This force is usually measured in P.S.I. or pounds per square inch.

Water Absorption

Water absorption is the measurement of how much water enters a material. The lower the amount of water absorption, the better the material generally is. Sponges are an exception.

Included in this test is the amount of water released, by weight, from the material. You don't want a roofing material to dry out completely and become brittle.

Wind Rating

The amount of upward force required before the roofing material blows away or the roof trusses break loose from the wall's top plates. Most areas that are subject to high wind conditions require, at the minimum, that a roof withstand 100 mile per hour winds.

There are some basic rules that you must follow to comply with the UBC. This is not an all-inclusive listing.

UNIFORM BUILDING CODE SIMPLIFIED

Asbestos Felt

Asbestos felt used for roofing shall have glass fibers spaced 1/4 in. apart running for the full length of the sheet. The sheet thickness must be greater than 0.022 in. and the sheet weight must be greater than 12 lb per 100 sq ft. Asbestos felt must contain a water- and fire-resistant binder and must not be saturated with bitumus. When used under wood shingles it must be lapped 2 in. or more at all joints. When used under wood shake it not only has to be applied to the sheathing but also each layer of shake must have an 18 in. wide strip of asbestos felt shingled between it. The felt may not be exposed to the weather.

Asphalt Shingles

Asphalt shingles of 18 in. or less width shall be fastened to solid sheathing with no less than two nails. Asphalt shingles with a width of 18 in. to 39 in. shall be fastened to solid sheathing with no less than four nails. Type 15 or greater underlayment felt is required, except if reshingling over an existing roof of 4:12 slope or less.

Minimum Slope

Minimum slope is 3:12 for flat or curved interlocking or noninterlocking tile and metal shingles. Exception: Tile and metal shingles used below 3:12 if interlocking and two layers of Type 15 underlayment felt are used. Felt must be solid mopped with approved roofing cement and applied in shingle fashion.

Minimum slope is 4:12 for asphalt shingles, slate shingles, and for wood shingles or shakes. Exception is by building inspector who may allow use of wood shakes on a 3:12 slope. Exception 2: Asphalt shingles laid in double coverage and over two layers of Type 15 underlayment felt. Exception applies if shingles are hand sealed or approved as self-sealing. They can then be laid on a 2:12 slope.

Minimum slope is 5:12 for asbestos cement shingles. Exception: A 3:12 slope is permitted when there are two layers of Type 15 underlayment felt overlapped in a shingle fashion. Minimum slope is 0:12 for built-up roofs (0:12 is dead flat).

Maximum Slope

Maximum slope for built-up roofs with slag or gravel is 3:12. Built-up roofs greater than 3:12 shall not be coated with slag or gravel but with an approved coating. Plies shall be laid parallel to the slope of the roof deck.

Roof Valley Flashing

Roof valley flashing is to be no thinner than 28 gauge. It shall be of galvanized or approved corrosion-resistant metal or material. It should be installed over Type 15 or greater felt. It shall have a 1 in. high water diverter rib in its center (W type flashing). (3/4 in. water diverter permitted if using metal shingles).

Flashing shall extend a minimum of 8 in. from center line to edge (of the shingle) when using asphalt, metal, or wood shingles, and 11 in. from center when using wood shakes. Overlapping sections of flashing shall overlap from top down over lower flashing by 4 in.

Valley flashing may be made from full-laced shingles or from 90 lb mineral surfaced cap sheet. Upper sheet must be a minimum of 24 in. wide and have the mineral granules exposed. To this upper sheet you must have a bottom sheet cemented. Bottom sheet is to be a minimum of 12 in. wide and the mineral granules are to face down toward the roof sheathing. There shall be Type 15 or greater felt used as an underlayment. Underlayment must extend a minimum of 18 in. from both sides of the valley's center line.

Vapor Barriers

Vapor barriers shall be installed between the roof deck and the roof insulation of built-up roofs. This is where the average January temperature is below 45 degrees Fahrenheit or where excessive internal building moisture is expected.

Wood Shakes

Wood shakes shall be doubled at the starter or eave line. They shall be nailed with two nails placed 2 in. above the weather exposure line and in 1 in. from the shakes edges on both sides. The spacing between shakes can be no more than 5/8 in. and no less than 3/8 in. (3/8 in. and 1/4 in. for shakes treated with preservative). The joint side lap between adjacent courses shall be no less than 1-1/2 in. There is to be 18 in. wide Type 30 underlayment felt shingled between each course. Underlayment may not be exposed to the weather. Weather exposures must comply with UBC 32-A.

In areas where there is wind-blown snow, the roof sheathing must be solid. The sheathing must be covered with a minimum of Type 15 felt.

TESTING AND TEST METHODS

Burning Brand Test

Brand in this sense is not a name or trademark, it is a piece of material that burns. The test subjects the roofing material to a flame and measures the amount of material that burns and compares this to a standard.

Dimensional Stability Test

Material is heated to 176 degrees Fahrenheit for 3 days and then cooled for 24 hours. Overall length, width, or thickness should show little or no change.

Drip Test

Drip test or drip flame spread test is where a material is suspended in a vertical plane and then heated to the point it melts or catches fire. The drips produced are an indication of the ability or inability to spread that fire. If the drips have an absence of flame, then they are permitted. If the drips carry the flame with them, then the material is subject to more regulations. These regulations might include coating the material with a nonflammable substance, limiting the size of the material that can be used in building construction, etc. This test is called running point testing by some manufacturers.

Flexibility Test

This is a test to see how well a material bends when cooled to some very low temperature, usually in the teens. The material is cooled to 14 degrees Fahrenheit for several hours and then bent 180 degrees around a 20 millimeter diameter steel rod in 5 seconds. If the material does not split or crack, it passes.

Flying Brand Test

Used to test wood or shake shingles to find the amount of flaming ash that is released into the air. See intermittent flame test for setup.

Fuel Contribution Test

See spread of flames test. The material is weighed before and after the test. The difference in weight, expressed as a percentage of material lost, is the fuel contribution. The lower the percentage the better the material is.

Hot Water Soak Test or 60 Day Test

Material is placed in 140 degree Fahrenheit water for 60 days. There should be no evidence of cracks, delamination, shrinking, or swelling. Test is for rain followed by blazing hot sun that produces steam during the evaporation process.

Intermittent Flame Test

Material is mounted to a 3-1/3 by 4-1/3 ft roof deck that is protected around the sides and its bottom. The deck can be sloped no more than 5 in. for built-up roofing material, no less than 5 in. for shingle roofing. A 12 mile per hour air current is then driven from the lower to the upper edge. A 1,400 degree Fahrenheit luminous flame is applied across the bottom edge and over the material. This flame is cycled on and off over a set period and then turned off. A comparison of the material to a standard is then made to determine the Class A or Class B fire resistance. If the material fails the test, a new sample is used with a 1,300 degree Fahrenheit flame to determine if the material meets Class C fire rating. If not, then the material is classed as nonusable for roofing.

Resistance to Aging Test or Aging Test

Material is subjected to 1,000 hours of sun, temperature changes, and intermittent water spray. See weatherometer test.

Running Point Test or Melting Point Test

Material is hung vertically and heated to the highest temperature it can withstand without liquefying and running or dripping.

Slippage or Slipping Test

This is the movement or sag of the material caused by temperature and its weight. You do not want the roofing material to slip or slide down a roof during hot days. Test is usually at 190+ degrees Fahrenheit. The material is laid on or attached by one edge to an incline of up to 90 degrees in slope, and then heated to the test temperature for several days. The amount of slippage is then measured. A good material will not slip more than a few millimeters.

Slippage Test, 7 Day

Built-up roofing material or membrane is installed to a 90 degree inclined roof surface and heated to 194 degrees Fahrenheit for 7 days. Slippage is to be less than 1 millimeter.

Spread of Flames Test

See intermittent flame test for setup. The measurement in this instance is the amount of material burnt and its contribution to the flame. Speed of the burning is considered.

Stress or Tensile Test

Material is clamped on one edge and then pulled from the opposite edge. The point where the material snaps or breaks is the tensile strength in PSI.

Temperature Cycling Test

The weather can change drastically from night to day, hour to hour, day to night. This tests for signs of delamination, cracking, splitting, crazing, and for fastener failure during extreme temperature changes. Material is cycled for 25 hours from −40 degrees Fahrenheit to 70 degrees Fahrenheit to 180 degrees Fahrenheit and to 0 degrees Fahrenheit. Material must sit for one hour at room temperature before cycling begins, then be subjected to a minimum of 3 test cycles during the 25 hour period.

Water Permeability Test

To test for standing water damage over a period, the material is subjected to a 2 in. column of water standing on it for 24 hours. No water should pass through or into the material.

Water Resistance Test

Material is weighed and then placed under water for 28 days, removed and reweighed. Difference in weight is measured in percentage. Material should not gain or lose weight.

Weatherometer Test

This is used to test for exposure to U/V radiation for evidence of flaking, chalking, color fading, blistering, crazing, or cracking. Test exposes material to a carbon arc light for 2,000 hours, thus duplicating the effects of 6 months of weather exposure.

Wet/Dry or 50 Cycle Test

This test is used to find the adverse effects of rain and sun on roofing materials. The material is cycled 50 times between being soaked and completely dry. There should be no evidence of cracking, shrinking, or other defects.

Wind Uplift Test

FM I-52 test. Material is fastened to a substrate or deck and then pulled upward from the substrate until it breaks loose. The measurement in pounds is taken at that point. Good measurements are between 100 to 200 lb. FM I-90 requires material to hold tight at 134 lb uplift for 60 seconds.

This short chapter was just an overview of the main agencies, specifications and tests. It did not cover the agencies of other countries; they too have volumes of specifications covering roofing products.

_____ Appendix A

Abbreviations

ABBREVIATIONS USED IN ROOFING

A . . . .

AB	Anchor bolt
ABS	Acrylonitrile butadiene styrene
ABV	Above
ACR	Acrylic plastic
AD	Area drain
ADH	Adhesive
ADJT	Adjustable
AGG	Aggregate
AL	Aluminum
ALT	Alternate
ANOD	Anodized
ANT	Antenna
AP	Access panel

APP	Atactic polypropylene
APX	Approximate
ARCH	Architect
ASB	Asbestos
ASC	Above suspended ceiling
ASPH	Asphalt
A/C	Air conditioning
A/R	As required

B . . . .

BD	Board
BD FT	Board foot
BDL	Bundle
BIT	Bituminous
BKT	Bracket
BLDG	Building
BLK	Block
BLKG	Blocking
BPL	Bearing plate
BTU	British thermal unit
BUR	Built-up roof
BVL	Beveled

C . . . .

C	Centigrade
CD	Cadmium
CEM	Cement
CER	Ceramic
CFL	Counterflashing
CFM	Cubic feet per minute
CHT	Ceiling height
CIPC	Cast-in-place concrete
CIR	Circle

CIRC	Circumference
CJT	Control joint
CK	Calk or calking
CLG	Ceiling
CLR	Clear or clearance
CLS	Closure
CMU	Concrete masonry unit
COMP	Compress or compressed
COMPO	Composition
CONC	Concrete
CONST	Construction
CONTR	Contract or contractor
CORR	Corrugated
CRG	Cross grain
CPE	Chlorinated polyethylene
CPR	Copper
CPVC	Chlorinated polyvinyl chloride
CRS	Course or courses
CRS	Cold rolled stell
CSPE	Chloro sulphinated polyethylene (Hapalon)
CTR	Center
CTSK	Countersunk screw
CWF	Clear wood finish
CX	Connection

D . . . .

D	Drain
DIAG	Diagonal
DIAM	Diameter
DIM	Dimension
DL	Dead load
DP	Damp proofing
DS	Downspout

E . . . .

E	East
EB	Expansion bolt
EL	Elevation
ENC	Enclose or enclosure
EPDM	Ethylene propylene diene monomer
EPS	Expanded polystyrene
EQ	Equal
EQP	Equipment
EST	Estimate
EXG	Existing
EXH	Exhaust
EXP	Exposed
EXS	Extra strong
EXT	Exterior

F . . . .

F	Fahrenheit
FAS	Fastener or fasten
FBD	Fiberboard
FBO	Furnished by others
FBRK	Fire brick
FE	Fire extinguisher
FF	Factory finish
FGL	Fiberglass
FHMS	Flathead machining screw
FHWS	Flathead wood screw
FJT	Flush joint
FLG	Flashing
FLX	Flexible
FP	Fireproof
FR	Frame or framing

FRC	Fire-resistant coating
FRP	Fiberglass reinforced plastic
FRT	Fire-retardant
FS	Full size
FT	Foot
FUR	Furring or furred
FUT	Future

G . . . .

GA	Gage or guage
GC	General contract
GI	Galvanized iron
GKT	Gasket
GLF	Glass fiber
GP	Galvanized pipe
GPM	Gallons per minute
GPPL	Gypsum plaster
GRN	Granite
GR WT	Gross weight
GSS	Galvanized sheet steel
GV	Galvanized
GVL	Gravel

H . . . .

HBD	Hardboard
HD	Heavy duty
HDR	Header
HES	High early strength cement
HOR	Horizontal
HR	Hour
HT	Height
HV	High voltage
HVAC	Heating/ventilating/air conditioning
HWD	Hardwood
HX	Hexagon

I . . . .

ID	Inside diameter
IN	Inch or inches
INCL	Include or included
INCR	Increase
INS	Insulate or insulation
INSC	Insulating concrete
INSF	Insulating fill
INSP	Inspect or inspector
INT	Interior

J . . . .

J	Joist
JCT	Junction
JF	Joint filler
JT	Joint

K . . . .

| KD | Kiln dried |
| KO | Knockout |

L . . . .

L	Length
LAB	Laboratory
LAD	Ladder
LAM	Lamination or laminate
LB	Lag bolt or pound
LBS	Pounds
LE	Leading edge
LH	Left hand

LL	Live load
LMS	Limestone
LP	Liquefied petroleum
LP	Liquefied propane
LPT	Low point
LTL	Lintel
LW	Lightweight
LWC	Lightweight concrete

M . . . .

M	Meter
MAS	Masonry
MAX	Maximum
MB	Machine bolt
MBF	1,000 board foot
MBR	Member
MECH	Mechanical
MED	Medium
MET	Metal
MFR	Manufacturer
MI	Malleable iron
MIN	Minimum
MISC	Miscellaneous
MLD	Moulding or molding
MMB	Membrane
MOD	Modular or modify
MOV	Movable
MP	Melting point
MRB	Marble
MRD	Metal roof decking
MSDS	Manufacturer safety data sheet
MT	Mount or mounting
MTL	Material

N . . . .

N	North
NAT	Natural
NFVA	Net free vent area
NIC	Not in contract
NL	Nailable
NMT	Nonmetallic
NO	Number
NOM	Nominal
NR	Noise reduction or not required
NTS	Not to scale

O . . . .

OBS	Oriented structural board
OBS	Obsolete
OC	On center
OD	Outside diameter
OJ	Open web joist
OP	Opaque
OPG	Opening
OPP	Opposite
OPS	Opposite surface
OZ	Ounce or ounces

P . . . .

PAR	Parallel
PBD	Particle board
PCC	Precast concrete
PCF	Pounds per cubic foot
PCS	Pieces
PERF	Perforated
PERI	Perimeter

PERP	Perpendicular
PFB	Prefabricated or prefabricate
PFL	Pounds per lineal foot
PLAS	Plastic or plaster
PNT	Paint
PRELIM	Preliminary
PRF	Preformed
PSC	Prestressed concrete
PSDN	Product safety disclosure notice
PSF	Pounds per square foot
PSI	Pounds per square inch
PTC	Post tensioned concrete
PTN	Partition
PU	Pickup
PVC	Polyvinyl chloride
PWD	Plywood
PWR	Power

Q . . . .

QT	Quart
QTY	Quantity
QUAD	Quadrant

R . . . .

R	Riser
RAD	Radius
RB	Rubber base
RBT	Rabbet (cut)
RD	Roof drain
RE	Reinforced
REF	Reference
REM	Remove
REPL	Replace

RET	Return
REV	Revision
RFG	Roofing
RFH	Roof hatch
RFL	Reflective
RH	Right hand
RO	Rough opening
RTV	Room temperature vulcanizing
RVT	Rivet
RWC	Rainwater conductor
RWD	Redwood

S . . . .

S	South
SBS	Styrene butyl styrene
SCH	Schedule
SCN	Screen
SD	Storm drain
SEC	Section
SHT	Sheet
SHTH	Sheathing
SIM	Similar
SKL	Skylight
SNT	Sealant
SP	Soundproof
SPEC	Specification
SQ	Square
SST	Stainless steel
ST	Steel
STD	Standard
STOR	Storage or store
STR	Structural
SUPT	Supervisor
SUR	Surface

SUS	Suspended
SYS	System

T . . . .

TEL	Telephone
THK	Thickness or thick
TOL	Tolerance
TSL	Top of slab
TST	Top of steel
TV	Television
TW	Top of wall
TYP	Typical
T&G	Tongue and groove

U . . . .

U/V	Ultraviolet
UNF	Unfinished

V . . . .

VB	Vapor barrier
VERT	Vertical
VG	Vertical grain
VJ	V-joint
VRM	Vermiculite

W . . . .

W	West or width
WD	Wood
WIN	Window
WO	Without
WP	Waterproof or waterproofing

WR	Water repellent
WS	Water stop
WTW	Wall to Wall

X . . . .

X	Unknown quantity

Y, Z . . . .

Z	Zone

Note: Some abbreviations may be written in lower case letter(s) and some may be spaced with periods (.).

SYMBOLS . . . .

′	Foot or feet or minutes
″	Inch or inches or seconds
× or ×	By or times by
/	Over (fraction)
<	Less than
>	Greater than
=	Equal to
< or ◁	Angle
℄	Centerline
D	Penny
⊥	Perpendicular
∅	Diameter or round
* or ×	Times by or multiply
+	Addition
−	Subtraction
/	Divide by
~	Approximate
&	And
%	Percent or percentage
#	Number or pounds

@	At
:	Ratio to or of
°	Degrees (angle or temperature
^{n}	To the power of, number times itself n times
^{2}	Square, number times itself once

_____ Appendix B

Tables

TABLE 1   ASPHALT SHINGLES

Width (inches)	Length (inches)	Strips/square	Nails/square	Notes
12	36	80	320	Standard
12	37	78	312	
13 1/4	38 3/4	66	264	Metric
13 1/4	39 3/8	65	260	Metric
17	40	48	288	

TABLE 2   SHEET INSULATION

Board thickness (inches)	R-value ranges
1/2	1 to 3 1/2
3/4	2 to 5 1/2
1	2 1/2 to 7 1/4
1 1/2	4 to 10 7/8
2	5 1/2 to 14 1/2
2 1/2	6 3/4 to 16 1/2
3	8 1/3 to 20
3 1/3	22 to 25
4	11 1/8 to 26 1/8

R-value depends on the age of the material, the moisture content, the foam type used, and the facing materials.

TABLE 3 T & G NAIING

Lumber size	Recommended nail locations			
	Tongue	Face	Edge	
2 × 4	1	1	—	Number
2 × 6	1	1	—	at
2 × 8	1	2	—	each
2 × 10	1	2	—	rafter
4 × 4	1	1	1	inter-
4 × 6	1	1	1	section

TABLE 4 PLYWOOD SHEATHING

Maximum span edge blocked	Maximum span non-blocked	Thickness of plywood sheet	ID index marked	Live load	Total load
24"	16"	3/8"	24/0	#50	#65
24"	24"	1/2"	24/0	#50	#65
30"	26"	5/8"	30/12	#50	#70
32"	28"	1/2"	32/16	#40	#55
32"	28"	5/8"	32/16	#40	#55
36"	30"	3/4"	36/16	#50	#55
42"	32"	5/8"	42/20	#35*	#40*
42"	32"	3/4"	42/20	#35*	#40*
42"	32"	7/8"	42/20	#35*	#40*
48"	36"	3/4"	48/24	#35*	#40*
48"	36"	7/8"	48/24	#35	#40

ID index first number is roof sheathing, second number is flooring.

*Decrease span 13% for 40/55 load or use next higher index number. Table is for C-C, C-D, and Structural I or II plywood.

Note: Table is for 4 ply plywood, 5 plywood can handle 15 more pounds of live load, 20 more pounds of total load. Loading is 1/240 of span deflection under live load, 1/180 inch deflection under total load.

TABLE 5 SHEATHING/SQUARE

Sheathing type	Sheathing size	Number required
Particle	4′ × 8′	3 1/8 Sheets
Plywood	4′ × 8′	3 1/8 Sheets
T&G Plank	4" × 8′	44 1/8 Planks
T&G Plank	6" × 8′	27 1/4 Planks
T&G Plank	8" × 8′	20 Planks

TABLE 6   T & G MAXIMUM SPANS

Minimum T & G thickness	Maximum rafter span
3/4"	16"
1 1/2"	24"
2"	32"
3"	48"

TABLE 7   WOOD GRADES

Grade recommended		Roofing usage
#2 or Better	*	Rafters
#3 or better	*	Roof boards/decking
B or better		Exterior trim
C-C, C-D, CDX		Plywood sheathing
M.D.O.		Paintable eave plywood
Type II		Particle board sheathing

*West coast Hemlock or Douglas Fir required. The U.B.C. allows other lumber species per their tables 25-G and 25–17.

TABLE 8   SPAN DEFLECTIONS

	Rafter spans			
Deflection	16"	24"	30"	48"
1/180	.089"	.133"	.167"	.267"
1/240	.067"	.100"	.125"	.200"
1/360	.045"	.067"	.083"	.133"

The 1/360 of span is for flooring and is recommended by some BUR manufacturers for flat roof decking.

TABLE 9   NAIL USAGE

Required connection	Nails used	'D' number	Nailing method
Rafter to top plate	2	8D	Toenail
Rafter board to rafter	3	10D	End
Rafter to valley rafter	3	10D	Toenail
Rafter to hip rafter	3	10D	Toenail
Rafter to ceiling joist	5	10D	Face
Rafter to 1" collar beam	2	8D	Face
Rafter to 2" collar beam	2	10D	Face
Ceiling joist to plate	3	8D	Toenail
Ceiling joist to blocking	2	10D	Face

+ Plywood sheathing, 3/8" thick or less. Face nail with 6 D nails along edges, 12" along intermediates.
+ Plywood sheathing, 1/2" to 1" thick. Face nail with 8D along edges, 12" along intermediates.
+ T&G, 2" × 4". One 8D toenailed through tongue, one 8 D face nailed through flat at each rafter intersection.
+ Particle board sheathing is the same as plywood. Edge nailing must be 1/2" minimum to 3/4" maximum from the edges.

TABLE 10   NAIL SIZES

Nail length (inches)	'D' number
0.875 (7/8)	Roofing Nail
1.00	2D
1.25	3D
1.50	4D
1.75	5D
2.00	6D
2.25	7D
2.50	8D
3.00	10D
3.50	16D

TABLE 11   AL. NAILS PER POUND

Used for	Nail length (inches)	Number/pound
Asbestos shingles	1 1/4	785
	1 1/2	659
	1 3/4	544
Asphalt shingles	7/8	663
	1	605
	1 1/4	491
	1 1/2	417
	1 3/4	368
	2	336
	2 1/2	274
Cedar shake	1 1/4	1300
	1 3/4	724
	1 1/2 (3D)	1480
Corrugated	1 3/4	318
Aluminum, plastic.	2	285
(With washers)	2 1/2	242

NOTE: Aluminum nails are made for the product being installed and the thickness and shape of the nails vary accordingly.

TABLE 12   READY MIX CONCRETE

Compression strength	Ready mix from truck
2000 PSI	5.0 Sack mix
3000 PSI	6.0 Sack mix
4000 PSI	7.1 Sack mix
5000 PSI	8.5 Sack mix

TABLE 13 FLASHING METALS

Metal type and symbol
Aluminum (AL)
Brass (copper and zinc alloy) (Cu + Zn)
Copper (Cu)
Lead (Pb)
Monel * (Mn + Si + Ni + Cu)
Steel, galvanized, iron (Fe) coated with zinc (Zn)
Steel, stainless, type #304
Terne metal or ternplate**
Tin (Sn)

*Monel is a replacement for stainless steel (S/S). Monel contains iron, manganese, silicon, nickel (66%), and copper (30%). It is a very corrosion-resistant metal.

**Terne is steel hot dipped in tin-lead alloy. Terne is used for painted metal roofing panels and flashings.

TABLE 14 MATERIAL WEIGHTS

Material type	Weight per square foot
Aluminum	.4 to .6 pounds
Aluminum, corrugated	.5
Asphalt shingles	2.15 to 3.00
BUR, 3 ply w/gravel	5.0
BUR, 5 ply w/gravel	6.5
BUR, 3 ply w/slag	4.5
BUR, 5 ply w/slag	5.5
Copper sheet flashing	1.0
Gypsum per inch thickness	8.0
Iron, steel, corrugated	2.0 to 3.0
Lead sheet flashing	4.0 to 8.0
Plywood, 1/4 in. thick	.70
Plywood, 5/16 in. thick	1.0
Plywood, 3/8 in. thick	1.1
Wood shingles	2 to 3
Wood shake	2 to 3
Tin sheet flashing	1.0
Zinc sheet flashing	1 to 2

TABLE 15 CRS WEIGHTS

Gauge number	Thickness (inches)	Weight/sq ft (pounds)
18	1/20	2.00
20	3/80	1.50
22	1/32	1.25
24	1/40	1.00
26*	3/160	0.75
28*	1/64	0.62
30	1/80	0.50

*Standard gauges for roof flashing. CRS = cold rolled steel.

TABLE 16 TILE MINIMUM SLOPES

Minimum allowed slope	Tile type	Underlayment	Number of ply required
5:12	Asphalt-asbestos	#15	1
3:12	Metal	#30	1
4:12	Slate	#15	2
4:12	Slate	#30	1
3:12	Flat clay or cement	#15	2
3:12	Flat clay or cement	#30	1
3:12	Curved clay or cement	#15	2
3:12	Curved clay or cement	#30	1

TABLE 17 NAILING TILE(S)

Tile type	Recommended nails to use
+ Asphalt-asbestos, or aluminum	1 1/4, 1 1/2, or 1 3/4 in.
+ Curved clay or cement	8D aluminum or glavanized
+ Flat clay or cement	8D aluminum or galvanized
+ Metal	8D aluminum or galvanized
+ Slate	1 1/4 in. copper

TABLE 18 BUR SLOPES

Slope range	Materials to use
0:12 to 1:12	Base sheets, felt, or glass fiber felt
0:12 to 3:12	Slag, 300 #/sq. or gravel, 400#/sq.
1/2:12 to 1:12	Cap sheets

TABLE 19   RESIDENTIAL BUR

Decking material	#15 Felt Ply	#30 Felt Ply	#90 Cap Sht.	Mop Coats	Gravel Weight
Wood	2	1	—	3	#400
Concrete	3	—	—	3	#400
Gypsum	3	—	—	3	#400
Wood	2	—	1	2	—
Wood	4	—	—	4	#400
Concrete	4	—	1 or gravel	5	#400
Gypsum	4	—	—	5	#400

TABLE 20   BUR ASPHALT TEMP.

A.S.T.M. type	Kettle temperature (°F)	Mopping temperature (°F)	Slope inch/foot
Type I	425	300 to 375	1/2
Type II	450	350 to 425	1/2 to 1
Type III	500	400 to 450	1 to 3
Type IV	500	400 to 475	3 and up

TABLE 21   HENRY™ COLD BUR

Henry product to use	Amount to use
+ Fiberglass base sheet	3 plies
+ Henry #203 cold application	2 coats
+ Henry #107 asphalt emulsion	1 coat
+ Henry #120 aluminum reflective	1 coat
or	
+ Henry #184 rufon polyester fabric	2 plies
+ Henry #106 asphalt emulision	3 coats
+ Henry #120 aluminum reflective	1 coat

These are for residential built-up roofing applications.

TABLE 22 AGGREGATE

Aggregate type	Weight pounds per cubic yard
Dormite	225 to 275
Gravel	225 to 275
Firebrick (crushed)	375 to 425
Lava	180 to 240
Marble	240 to 260
Porcelain, white (crushed)	190 to 210
Rock, desert	250 to 270
Rock, desert bronze	250 to 270
Rock, igneous, desert green	240 to 260
Slag	165 to 210
Tile, arctic white (crushed)	210 to 230
Tile, canyon red (crushed)	230 to 250

TABLE 23 FORMULAS

Item	Type	Formula
Square	Area	Length ^{2} ($^{\wedge}$ = to the power of)
Cube	Volume	Length × width × height (L × W × H)
Rectangle	Area	Length × width (L × W)
Rectangle	Volume	Length × width × height (L × W × H)
Triangle	Area	(Base × height)/2
Pyramid	Volume	(Area of base × height)/3 ((A × H)/3)
Cylinder	Area	Circumference × height + 2 × end area
Cylinder	Volume	Radius of base ^{2} × 3.1416 × Height)
Cone	Area	Base circumference × height)/3
Cone	Volume	(Radius of base ^{2} × 3.1416 × height)/3
Sphere	Area	Diameter × circumference
Sphere	Volume	(Radius ^{3} × 3.1416 × 4)/3
Circle	Area	Circumference ^{2} × .07958
Circle	Area	Diameter ^{2} × .7854
Circle	Circum	Diameter × 3.1416 or radius × 6.2832
Circle	Diameter	Circumference × .3183
Circle	Radius	Circumference × .15915

TABLE 24 COMMON DECIMALS

Fraction	Decimal	Fraction	0.625	Fraction	Decimal
1/8	0.125	5/8	0.750	15/16	0.938
1/4	0.250	3/4	0.781	8/8	1.000
3/8	0.375	25/32	0.781	9/8	1.125
1/2	0.500	7/8	0.875	5/4	1.250

TABLE 25 LUMBER MOISTURE

Lumber type	Moisture content (%)	
Plywood	<15	
Some KD lumber	<15	
Most KD lumber	<12	
S-dry	<19	
S-grn	>19	<28
Wet	>28	

TABLE 26 INSULATION

R#	C#	Thickness (inches)	Use batts (inches)
R-11	0.091	3.41	3 1/2
R-13	0.077	4.02	3 5/8
R-19	0.053	5.88	6
R-22	0.045	6.81	6 1/2
R-25	0.040	7.74	3 5/8 + 3 5/8
R-30	0.033	9.29	6 + 3 1/2
R-33	0.030	10.22	6 1/2 + 3 5/8
R-38	0.026	11.76	6 + 6
R-49	0.020	15.17	6 + 6 + 3 1/2

See D.O.E. recommendations for use.

1/2" Plywood	R-0.63	Above R-values are for
5/8" Plywood	R-0.79	fiberglass insulation.
Asphalt shingles	R-0.44	Approximately ½" less required
Wood shingles	R-0.87	if using mineral wool.
Vapor barrier	R-0.00	
>1" Air space	R-1.2~	

TABLE 27   U.B.C.

U.B.C. Section:	Covers:	{1985 edition, I.C.B.O}
1701	Weather protection	
1709	Parapet walls	
1710	Roof projections, overhangs	
1712	Foam plastic insulation	
1713	Insulation	
1806	Type I fire resistive buildings	
1906	Type II one hour fire resistive buildings	
2005	Type III one hour fire resistance building	
2106	Roof decking for heavy timber construction	
2305	Roof design and loading	
2306	Roof design and loading	
23-C (table)	Minimum roof live loads	
23-D (table)	Maximum deflection for structual members	
24-A	Mortar mix ratios	
2511	Laminated beams and ponding	
2516	Draft and fire stops	
2517	Roof and ceiling framing, general	
25-Q (table)	Nailing schedule	
25-R-1 (table)	Lumber sheathing spans	
25-R-2 (table)	Board grades of sheathing lumber	
25-S-1 (table)	Plywood sheathing spans	
25-S-2 (table)	Plywood loading	
25-S-3 (table)	Particleboard loading	
25-U (table)	T&G, 2", allowed spans	
25-U-R-n (tables)	Rafter spans for various slopes	
3202	Roof construction and materials	
3203	Roof coverings	
3204	Roof insulation	
3205	Draft and fire stops	
3206	Venting	
3207	Drainage	
32-A (table)	Wood shingle/shake weather exposure	
32-B (table)	Roof covering application	
3209	Re-roofing	
3702	Chimney specifications	
4305	Fire protection	
5206	Plastic roof panels	
5207	Skylights	
2318 (appendix)	Roof projection loading	
4901	Patio covers	

Glossary

Acetylene Welding gas, colorless hydrocarbon gas made from calcium carbide and water.

Across the supports Grain direction of plywood or lumber as it crosses, or spans, the rafters or joist. The allowable span varies with lumber grade, thickness, and grain direction, consult U.B.C.

Acrylic coatings Any number of paints, coatings, and sealers manufactured from methyl methacrylate. Coatings are breathable, flexible, have good adhesion, and are generally water based.

Adhesive a) Any of the glues or binders that are used to bind two items together into a unistructure. b) May be an elastomeric adhesive. Construction grade adhesive meeting AFC-01 specifications and used to attach plywood to joist or rafters. Adhesives can be used in below freezing temperatures.

Aggregate Small stone or crushed rock or slag used for a roof coating. See Appendix B, table 22.

Aging resistance (See retained R-value, aged)

Alignment offset The offset of alignment slots from one shingle to another from course to course.

Aluminum flashing   Flashing manufactured from aluminum sheet. May be extruded but is more likely roll formed.

Aluminum paint and aluminum coatings   a) A paint that contains a high content of aluminum. Used for its reflective qualities to reflect light and heat away from the item painted. Frequently used for factory and mobile home roofs. b) Water-based asphalt emulsion containing aluminum. Spreading rate approximately 3 gallons per roofing square.

Ambient temperature   The current air temperature in a room or outside area.

APP   APP is the abbreviation for atactic polypropylene. Plastic used as a modifier to bitumen products. APP gives low temperature flexibility and crack resistance to asphalt and asphalt-based felts and membranes.

Application rate   a) Refers to the speed of applying a material to another material. b) The controlled quantity of a material applied to another material.

Application temperature   a) The temperature of the material being applied to another material, such as tar to a roof. b) The temperature of the material being applied to. c) Refers to the ambient or surrounding air temperature during application of one material to another.

Approved adhesive   An adhesive approved, or recommended, by the manufacturer of a product or by the adhesive manufacturer. Not using the approved adhesive can void the warrantees of one or both material manufacturers involved.

Asbestos felts   Roofing felt made from matted asbestos fiber and a binder. Not impregnated or saturated with bituminous or asphalts. Used under wood shingles or wood shake. May no longer be available due to the asbestos scare of the 1980s.

Asphalt   The residue remaining from the distillation of petroleum. Asphalts are selected for their various melting points, which enables them to be used on various sloped roofs. They are further selected for their flexibility and their ability to penetrate substrates.

Asphalt felts or asphalt rag felts   Felts that are saturated with asphalt and used as underlayment on roofs. The felt is a combination of wood fibers, paper fibers, and rag fibers. Weight is 15 lb per 108 sq ft. Sold in rolls of 4 squares, 432 sq ft, weight 60 lb. Also produced in a 30 lb weight, rolls of 2 squares weighing 60 lb.

Asphalt primer and concrete primer   Cutback asphalt formulated to adhere to concrete and concrete decking.

Asphalt shingles   Standard strip, no tab, two-tab, or three-tab shingles manufactured in layers. Usually, with a layer of asphalt, a web or mat, a layer of asphalt, and a layer of mineral granules.

Attic   The open space between the underside of the roof sheathing and the upper side of the ceiling directly below the roof. Access is sometimes provided. Utilities are sometimes installed in an attic.

Attic vent   A vent or vent system used to expel hot, moist air from the attic. All attics should be vented.

Attic walkway   A board or network of boards placed across the ceiling joist in an attic and used as a foot path. The walkway prevents one from falling through the ceiling.

Backing down   Method of aligning shingles from one side of a dormer to another. One shingles up one side and then down the other.

Balanced ventilation   a) Design of roof venting system so that proper venting occurs. b) Combination of soffit and ridge peak vents that provide proper air flow through the attic area.

Ballast   Top coating of rock or slag needed for its weight to hold the roofing plies in place. Usually doubles as the reflective coating.

Band sticks   Protective boards placed on both sides of a bundle of shake or wood shingles. This protects the shingles from damage from the banding straps.

Bare spots   Places that are missing granules on roll roofing or shingles. May be caused by someone walking on the roof.

Barge rafter or barge board   The end rafter or the fascia board covering the end rafter of a gable.

Barrel tile   Clay tile shaped like a U or semicircle.

Base ply   (See Base sheet)

Base flashing apron   The lower horizontal mounted flashing for a chimney, skylight, or dormer.

Base sheet   The first sheet of felt or roofing applied to a roof. Base sheets are to be covered with other plies.

Battens and counterbattens 2×2 in. boards nailed from rake to rake and used as a nailer strip for installing roof tile. Counterbattens are 1×4 boards nailed from eaves to ridge. The battens nail to the counterbattens. Recommended lumber is #2 or better construction grade Douglas Fir.

Beadboard   Insulation sheet made from white styrofoam beads, heat and steam expanded in a mold.

Bearing area   a) The point on a wall that bears the weight of the item(s) above it. b) Top sill plate of a wall that supports the roof.

Bed molding   a) The trim molding directly below a wooden gutter and attached to the fascia. b) Trim molding that finishes off the junction of the soffit and the frieze molding. c) A molding found at the bottom of a deep hollow. d) The trim molding used at the junction of the underside of the roof sheathing and the building's siding in an open cornice design.

Bellows   The 1/4 or 1/2 in. round flashing used at roof seams or roof to wall intersections. The round or bellows shape allows for expansion and contraction flexibility of the intersecting joint (see Joint shield).

Below grade waterproofing   Water barrier placed against the foundation and below the ground level.

Bevel heel gusset   (See Heel Gusset)

Bevel square   Square that has a tongue adjustable to various angles.

Bird holes   Screened vent holes drilled in the blocking lumber used between the rafters at the eaves.

Bird stop   Wood, metal, or plastic strip shaped to fit the eave edge of a corrugated or nonflat tile roof. Used to dress off the eaves and to prevent insects and birds from nesting between the sheathing and the roofing.

Bitumen applicator   a) The person who is applying asphalt to a roof. b) The tool used to apply asphalt to a roof. c) A broom, mop, brush, or spray outfit used to apply asphalt materials to a roof or the roof plies.

Bitumen drippage   a) The dripping of asphalt onto the roof where it is not wanted. b) The unwanted dripping of asphalt off the roof onto items below the roof.

Bitumus or Bituminous or Bitumen   These are common names for various hydrocarbon distillation byproducts. Tar and asphalt are bitumus products.

Blind nailing   Shingles are nailed in such a location that when the next shingle is applied, the nails of the first shingle do not show.

Blind valley   A valley that normally is not seen from the ground level.

Blisters   Bubbles formed under a roofing membrane by the expansion of trapped heated water.

Blowing temperature   The temperature of the asphalt when air is blown through it. Air is blown at high pressure into the asphalt to lower its viscosity and give it better flow characteristics.

Blue board   (See Polystyrene board)

Body of square   The wider portion of a metal carpenter's square.

Bond line   a) Term given to the pattern generated by the cutouts of tab shingles. b) The seam between two items glued or bonded together.

Bondability   a) The condition of a roof or of the roof installation that allows it to qualify for bonding (see bonded roof). b) Bondability is the ability of two or more materials to be fastened together.

Bonded roof or surety bonded roof   A roof, usually a builtup roof, that is bonded. Bonding means that a dollar amount bond is purchased to cover the roof and any repairs that may be needed over the life expectancy of the roof. Bonding allows the manufacturer of the roofing products legal access to inspect the roof installation. Bonding also allows the roof owner legal right to hire qualified contractors, other than the installer, to make repairs, if required.

Bonding agent   a) A chemical that will act as a primer in that it prepares the existing surface for application of another paint or chemical. b) A chemical added to a material to help it bond tightly to another material. c) A person who sells surety bonding for roofs.

Boot, rubber   A rubber or plastic shield used to seal a vent pipe to the vent flashing.

Bored hole   Hole drilled in joist or rafter or other framing member for passage of pipes, wiring, or for air venting. By code, hole diameter must not exceed 40% of the width of the lumber the hole is drilled in for load bearing lumber. It must not exceed 60% for nonload bearing lumber, i.e., partition wall studs.

Braces and cross- or X-bracing   a) Diagonal wood or metal straps used to stabilize a structure against racking or tilting or falling over. b) X-bracing is two thin strips of metal or lumber used in an X pattern between joist or rafters (see bridging).

Bridging or X-bracing   Short diagonally placed boards or metal strips between joist or rafters. Bridging aids in keeping the rafters or joist from warping and twisting.

Bright white roof coating   A highly reflective roof coating used on mobile homes and factory buildings. The coating reflects the sun's rays and thus the sun's heat. This in turn reduces A/C cost and cuts down on roof repairs or replacements.

Brooming in   a) Using a broom or squeegee to apply a thin glaze coating of asphalt to a roof ply. b) Spreading the still hot liquid asphalt with a broom or squeegee. Brooming in helps to prevent blisters caused by trapped water vapor. To be done any time a roof ply is to be left uncovered overnight.

Brown coat, stucco   The brown coat is the second coating of stucco on a wall. This is the coating that determines the wall's flatness.

Bubble, in shingle   A bubble or blister in or under an individual or group of shingles. This can be caused by trapped, heated water, the shingle drying out, or an object left under the shingle during installation.

Building code   A body of rules. These govern how a building is to be constructed so that it is long lasting and will afford maximum protection to its occupants and their possessions (see U.B.C.).

Building paper   The paper or material used under the siding to aid in prevention of air leaks into the building. Being replaced with polyethylene building wrap on newer buildings, Tyvex™ or similar material.

Bull   Usually refers to a trowel or male cow. In roofing it refers to plastic cement. Term is not in common use.

Bull paddle   A wood stick used to apply plastic cement (see Bull).

Bundle   A package of shingles as sold. Normally 3 bundles of asphalt shingles make up one roofing square (100 sq ft).

Bundles per square   The number of packages of shingles sold that will properly cover 100 sq ft of roof area.

Bungalow siding   Siding made from beveled lumber. Lumber may or may not be rabbet cut.

BUR   Abbreviation for Built-Up Roofing system.

Butt edge, shingle   The butt edge is the bottom edge. This term usually refers to the thicker end of an item, such as the bottom or thick end of a shake shingle. It also refers to items being placed end to end, such as the side to side joints of asphalt shingles.

Butterfly roof   A roof that starts high at the eaves and slopes toward the center in a V shape.

Butting up   The placing of new shingles over old shingles. Done in such a manner that the top edge of the new shingle contacts the bottom or butt edge of the old shingle.

Butt joint   The side to side joints created when two shingles are installed horizontally across the roof (also see Butting up).

Butt nailing   Nailing the shingle through the butt or weather exposed lower edge of the shingle. Not normally recommended.

Butyl sealant   Synthetic rubber sealer used to fill cracks and holes in roofing and roof sheathing. Recommended sealer for isolation of the intersection of metal flashing to redwood.

Callback   Returning to a job to repair or correct a problem. Callbacks are usually at the roofer's expense.

Cantilever or cantilever roof   Load bearing joist or rafter or truss that extends beyond the load bearing walls into space.

Cant strip   A wood V or wedge-shaped strip used along the intersection of a flat roof to vertical wall. Used to prevent the roofing from splitting or pulling loose.

Cap flashing   The metal flashing that caps off the top of a wall or parapet.

Cap sheet   Roll roofing material. A sheet of felt or fibers that is coated or saturated with asphalt or bituminous and coated with mineral granules, talc, slag, or other similar material.

Capping in   Refers to putting a weather-resistant cap on the building. This cap is the roofing felt applied over the bare sheathing.

Carpenters' square   (See Square.)

Caulking gun   A trigger pulled dispenser of caulking. Caulking is supplied in drop-in tubes and a plunger pushes caulking out of tube each time the trigger is pulled. May be manual or pneumatic controlled.

CCA   a) Chromated copper arsenate. A preservative used to protect wood. b) Lumber treated with CCA.

CD or CDX   The American Plywood Association's grading of plywood. The C grade in this instance is the front surface quality, the D is the back side surface quality, and the X stands for exterior use. The grades are N, A, B, C, and D, with N being the best. For some sheets the CD is omitted and Group 1 or Group 2 is stamped. Group 1 is best of the two.

Cedar or red cedar   a) A commercially grown evergreen tree valued for its timber that is naturally rot- and insect-resistant. b) The wood of a cedar tree used to make siding and roof shingles.

Ceiling joist   Horizontal supporting members of a ceiling. Usually of 2 in. by 6 in. or larger lumber placed on 16 in. centers.

Center aligning slit   Top center slit on some asphalt shingles that marks the center or midpoint of the shingle.

Centistrokes   Refers to the measurement of a liquid or semi-liquid material's resistance to flow.

Ceramic Granules   (See Granules.) Granules made from ceramic, heated clays, and used to coat the weather exposure surface of shingles or tile.

C-factor, thermal counductance   The amount of heat conducted through a material. C=BTU/HR/FT × FT/degree Fahrenheit. Conductance equals British thermal units per hour per square foot of surface per 1 degree Fahrenheit difference between sides of the material.

Chalkline   a) A line created by the snapping of a string, coated with powdered chalk, on to a surface. b) Tool used to house string and chalk. Tool is pear shaped and contains a chalk fill port and a string winder.

Channel vent   Built-up roofing term. A channel between the deck and the first or base ply. The channel is exhausted into the atmosphere to expel any built-up vapors that are under the first ply.

Channel vent sheet A specially designed base ply sheet. The sheet contains fluted ribs, used to expel moisture from under the base ply and into the atmosphere.

Cheek cut An end cut of a piece of plank lumber in which the cut is not square. Cheek cuts are used at the ends of rafters to fit them to corners, ridge beams, and for shaping the tail. Double cheek cuts are V shaped, single cheek cuts are '/' shape. A short cheek cut is the short side of a V if cut that way, the long cheek cut the opposite side.

Chimney hole A framed out hole in the roof sheathing where the chimney will be built. Location of framing is dictated by the U.B.C.

Class I deck Highest fire-resistance rating for a steel or flat roof deck. Fire spread at 800 degrees Fahrenheit is not to exceed 60 ft from fire source. Sometimes, 100 feet is permitted.

Class II deck Lowest fire-resistance rating acceptable for most steel or flat roof decks. The fire spread at 800 degrees Fahrenheit is more than 60 ft from the source. May require a fire sprinkler system, pointed up toward underside of roof and located within 5 to 6 ft of roof, to become Class I.

Class A shingle Highest fire-resistance rating for a shingle or shingle material.

Class B shingle Second-highest fire rating for a shingle or shingle material.

Class C shingle Lowest fire rating for a shingle or shingle material. Below Class C is unrated, for decorative use only.

Climatic zone chart A map of the United States broken into three or more climate regions. Each climate requires different materials and specifications for their roofing needs. There are different charts for each type of roofing and for roof and wall insulation. Climate maps for other countries are available.

Chisel Small tool with a flat sharp blade at its end. Used to cut and displace wood.

Clamping ring The ring that holds a flat roof drain in place. It is circular and has some method of spinning it tight.

Closure strip A formed strip of metal, wood, or plastic used to seal or fill between a rafter or purlin and a sheet of corrugated roofing.

Coal tar bitumen Special coal tar distillate manufactured by the Koppers Company. Coal tar bitumen is a very low fuming type product used for roofing.

Coal tar pitch products Tar residue from the production of coke. Coal tar pitch is distilled to obtain a melting point range of 145 to 155 degrees Fahrenheit. It is used on low-pitched roofs for its low melting point and self-healing abilities. Cannot be used on steep roofs. Coal tar pitch products are not compatible with asphalt or asphalt felts.

Coated roofing felts Felt sheet coated or impregnated with rubber, synthetics, asphalts, or other compounds. The coating makes a water-resistant membrane. Used between the sheathing and the top ply covering a roof.

Code approval   A material or application method that has been approved for use by the building department, an approving agency, or the U.B.C. (see U.B.C.).

Cold flex   Ability of a product or shingle to be bent 180 degrees, without splitting or cracking, when at or below a set cold temperature.

Cold process   Installing a built-up roof using asphalt or other product(s) that do not require heating for installation.

Cold mop   Applying specially formulated unheated asphalts or other roof coatings to a roof or roof plies.

Cold weather precautions   Roof installation techniques and precautions used when the temperature is at or below 45 degrees Fahrenheit. Primarily a built-up roofing term.

Collar, vent   A metal or plastic flashing that slips over a vent pipe and seals pipe to roof leaks.

Collar beam or straining piece   A horizontal beam that connects two opposing rafters together and is approximately 2 to 3 ft below the ridge beam. The collar beam keeps the rafters from spreading apart.

Combination sheet   A roofing sheet of Kraft paper laminated with glass fiber felt.

Combustible deck   A roof deck or roof sheathing that can be set on fire.

Combustible insulation   Insulation material that will burn if exposed to open flame or will generate flame when burning.

Compliances   The act of adhering to the will of the manufacturer and the laws governing roofing materials and their installation.

Composite board   Insulation sheet made of two or more materials laminated together. Example: perlite or polyisocyanurate foam.

Compressive resistance   a) The amount of force applied to an item to bend it permanently or to destroy it. b) Foam insulation sheet's resistance to deformation from an item exerting pressure on it. 16 PSI is a good rating.

Concrete bond beam   U-shaped concrete strip used on the top of a concrete or block wall. The U is then filled with concrete, plate anchors, rebar, etc. used to anchor the roof to the wall.

Concrete deck   a) A roof deck consisting of poured concrete over a wood or metal sub-deck. b) A roof deck made from preformed concrete slabs.

Concrete deck fasteners   Special fasteners designed to hold insulation or decking to concrete. Several styles are available.

Concrete nails   Nails that have been specially shaped and strengthened so that they can be used to nail a material or item to concrete.

Concrete tile   Roofing tile manufactured from concrete or a mixture of concrete and other materials.

Conditioned R-value   The R-value of insulation after the insulation has artificially weathered for 6 months. Represents 5 years of weathering.

Coping cap   Metal flashing used on top of a parapet wall (also see Cap flashing).

Copper flashing   Flashing manufactured from copper sheet. Recommended for use when in contact with redwood lumber or in areas with high year-round moisture content. Copper is a natural killer of bacteria and marine algae.

Cornice line, open, or closed   The area where the roof meets the siding. Open cornice is an overhang that does not have soffit. Closed cornice is an overhang that does have soffit.

Corrosion resistant   a) Cannot be corroded very easily. b) A metal that is not ferrous or that does not contain steel. c) A steel that has 10% or more chromium or .2% or more copper in it. d) Stainless steel, copper, brass, aluminum, lead, zinc, or galvanized metals.

Corrugated panel   A metal or plastic sheet that has ridges and valleys pressed into its full length. The combination of ridges and valleys increases the strength of the sheet.

Counterflashing   A metal flashing that extends from a wall or chimney downward and over the top of the roof shingle flashing. Counterflashing prevents water from entering behind the roof shingle flashing.

Courses   a) Courses of shingles can run horizontally, diagonally, or vertically and are sometimes termed the run of the shingle. Each run or row installed is considered a course. b) The installing of shingles one on top of each other as might be done at the eaves.

Coverage   a) The amount of a material that is required to cover an item. b) Sometimes referred to as the spreading or application rate of material. c) The area of an object that can be covered by a fixed amount of a material.

Cricket   An inverted V shaped structure used to divert water from an item. Crickets are usually constructed of wood and used on the upper roof side of chimneys.

Cricket system   a) An inverted V shaped structure built behind a chimney or other large object on a roof. Purpose is to direct runoff water around the object and prevent water or ice damming. b) The method of constructing a watertight cricket from several pieces of sheet metal or lumber. c) The method of establishing water drainage on a near flat roof using wedge-shaped roofing panels or sections.

Cripple   Short 2 by lumber section used to give added strength to a portion of the structure.

Cripple jacks   A rafter that extends from the ridge of a hip roof to a valley rafter.

CRS   Cold Rolled Steel. Metal used for most construction framing and for forming roof decking.

Curb detail, rooftop equipment   Construction or construction drawing of the vertical rise flashing used for the built-up roof to skylight or other equipment mounting.

Curb appeal   How an item or building or roof looks to people from the street or curb. Curb appeal is important to establishing image and resale values.

Curb shape flange   Flashing that properly fits a curb.

Cutback adhesive   Selected asphalts of a certain melting temperature range mixed with petroleum solvents or mineral spirits. Adhesive remains semiliquid at application temperature, but as the solvents evaporate, the adhesive solidifies into a solid. Used as cold process roofing.

Cutout, shingle, full or half   a) The side to side opening between the weather exposure surfaces of two installed asphalt shingles. Each shingle is manufactured with a half cutout, that when installed, forms a full cutout. b) A cutout is the slot in the weather exposure face of a shingle.

Cylinder dolly   A wheeled cart properly designed to secure and transport cylinders of gases. In roofing: The dolly used to transport the oxygen and acetylene gas or the L.P. (Liquefied Petroleum) tanks used as fuels for heating asphalt.

DBL coverage 'T' lock map   Map showing areas where the shingle is sold. There are maps for each type and brand of shingle.

DBL coverage 'T' lock shingle   A shingle that looks like a 'T' and has locking ears or tabs. Size is 20-13/16 × 21-1/2 in. Coverage is 100 pieces per square or 3 bundles per square. Requires 200 nails per square.

Dead level   A roof with a slope of 0:12 or less, i.e., no slope of any sort is called dead level.

Dead level asphalt   Low melting temperature asphalt used on dead level roofs. The low melting point makes the asphalt self-leveling and self-healing.

Decking screws   Specially designed screws that are used to secure items to a deck. Screws usually have drill points, fluted shanks, and heads designed not to fall out of a drill bit during installation. May include distribution plates.

Dentil molding   Trim molding that contains a pattern that resembles teeth, popular on many older buildings. Dentil molding is making a comeback in the 1990s with the arrival of low-cost plastic foam moldings.

Detail torch   Hand-held torch used for heating asphalt and impregnated membranes during installation of curbs, flanges, etc.

Dimensional stability   The ability of an object not to distort from heat, cold, or its weight.

Distribution plates   Square or round metal plates, approximately 1/2 to 2-1/2 in. in diameter, used to distribute nail or screw head pressure over a large area. Used with nails or screws to hold sheet foam insulation to a roof deck.

Divorcing layer   a) A layer of wood fiber or perlite insulation used over foam insulation to make it fire-resistant. b) A layer of material that separates two materials that may chemically react with each other.

Dolly varden   Beveled siding lumber that has been rabbet cut on its lower edge.

Dolomite   White stone used for roof coatings. Marble or limestone with a high content of magnesium carbonate or a stone containing calcium magnesium carbonate.

Dormer   A windowed structure protruding vertically from a roof surface (see Shed dormer and Gable dormer).

Double coursed   Refers to the application of wood or shake shingles on a wall or as used as siding. Two layers of shingle are applied to each row or course.

Double header   A supporting 2 by or joist or rafter that is doubled to add strength. Usually, this is to support a lookout rafter or a dormer rafter.

Double layer application   Installing two thin layers of sheet foam insulation instead of one thick piece. The second layer covers the nails and seams of the first layer. Therefore, it gives better total insulation value since heat loss, through the nail heads and the seams, is minimized.

Double trimmer   Second board nailed to a trimmer board to give extra strength.

Downspout   Gutter system component that carries the water from the gutter to the ground.

Downspout elbow   An elbow used to change direction of the downspout. Three downspout elbows are normally required in the typical installation of a downspout.

Downspout strap   Gutter system component that attaches the downspout to the building's siding.

Draft stop or fire blocking wall   A fire-resistant attic wall required if the attic area is over 3,000 sq ft. May be required between apartment or condo unit attics of any square footage. Check local fire and building codes. Usually, a double-sided wall consisting of 1/2 or 5/8 in. gypsum over 2 × 4's or a 1/2 in. thick plywood sheet will meet code requirements.

Drainage, roof   The ability of a roof to drain all water without the roof or the building being damaged. Good roof drainage is usually considered not less than 1/4 in. slope per foot. Drainage water may not enter a sanitary sewer or be allowed to flow across public property.

Dress off   a) To complete a rough carpentry or roofing job with trim pieces so that the job looks completed. b) To use trim to hide unsightly construction.

Drift down   Poor horizontal alignment of a course or row of shingles. The shingles slope downward as they extend across the roof plane from rake to rake.

Drill point screw   A screw that is self-drilling. The point is cut the same as a drill-bit point. The shank may be threaded so that the screw becomes self-tapping. Used to anchor items to light gauge metals.

Drip edge   A modified L-shaped flashing used along the eaves and rakes. The drip edge directs runoff water into the gutters or air and away from the fascia.

Drop outlet   Gutter system component used to join the gutter to the downspout.

Drying in   Same as capping in. The roof sheathing is considered to be protected from moisture, kept dry by the felt applied over it.

Drywell   A hole filled with rocks into which a rain gutter system drains. The drywell holds and evenly distributes the rain runoff water into the earth. The hole is usually 4 to 5 ft in diameter, 5 to 6 ft deep, and covered with a sheet of exterior plywood and 6 to 12 in. of earth.

Dubbed corner   The removal or clipping of the corner of a shingle used in a full- or half-lace valley. Dubbing helps prevent water from adhering to and running along the top edge of the shingle.

Ductility   The ability of a material to be molded or hammered into another shape.

Dutch lap   The course to course alignment of asphalt tab shingles. The cutouts do not align directly above each other but are instead offset by a few inches.

Dutch weave   Shake or wood roof where ever fourth or fifth shingle in a course is lowered by 1/2 to 1 in.

Durability   The quality of a material or material installation that enables it to last a long time.

Ears   Ears are small extended portions of locking type shingles. The ears or tabs are used to lock the shingle into the next shingle.

Eave   The horizontal roof overhang that extends outward and is not directly over the exterior walls or the building's interior.

Eave drip edge   Metal flashing used along the eave to direct water from the roof to the ground or into a rain gutter. The drip edge, if properly installed, keeps runoff water off the fascia boards.

Eave underlayment   A membrane of rubber or felt applied to the roof eave(s) to protect the sheathing.

Elbow joint, inside corner   Gutter system component used to join two gutter sections on an inside corner.

Elbow joint, outside corner   Gutter system component used to join two gutter sections on an outside corner.

Elongation   To extend or grow in length or width. This is not a good quality of a roofing product.

Endlap   Endlap is the lapping of felts at the end of the roll from one sheet to another. A 4 in. endlap is usually recommended; code requires a minimum of 2 in.

Energy credits   Tax credits given for home improvements that increase energy efficiency in a home, usually, added insulation.

Envelope strip   Modified L-shaped flashing used at the edges of a flat roof to prevent asphalt from dripping off the roof.

Epoxy resin cement   Thermosetting plastic cement that is very waterproof and has high bondability. Epoxy is resistant to most acids and chemicals.

EPS   EPS is the abbreviation for expanded polystyrene. Beadboard is a name for insulation sheet made from EPS.

Equipment supports   The blocks, braces, or stands used to keep permanent roof mounted equipment off of, and above the roof. Equipment is usually air handlers or air conditioning equipment.

Equipment vibration   The mechanical motion of a piece of equipment that is transferred to the roof sheathing or deck and produces cyclic, objectionable noise.

EVT (equivalent viscosity temperature or equiviscous temperature)   The temperature of a roofing asphalt when it has a viscosity of 125 centistokes. This is the ideal temperature for application of asphalt to a built-up roof.

Existing roof deck   A roof deck that already exists and may be roofed. Many manufacturers, especially of built-up roofing systems, have special requirements for roofing over an existing deck.

Expansion joint   A planned, controlled joint placed between two roof surfaces or between two sections of a built-up roof. The expansion joint allows the roof to expand without physical damage to the roof or the building.

Expansion joint shield   A waterproof cap and membrane system that fits over an expansion joint.

Expansion strap   a) A metal strap used across the expansion joint to prevent the joint from spreading too far and causing roof or flashing failure. b) A metal strap used across the building's ridge to keep the opposing rafters from separating. Essential on very steep roofs.

Exposure   The portion of a shingle not covered by the next course of shingles. This portion is exposed to the weather.

Exposure aligning slit   Small aligning slits at the edges of a shingle located just above the tabs. Use to align the shingles horizontally across the roof.

Extra steep roof   A roof with a slope of 60 degrees or more is considered extra steep. This is a roof with a rise of 21:12 or more.

Face nailing   a) Not recommended (see Butt nailing). Nails are placed in the exposed or face of the shingle. b) Nailing lumber where the nails show.

Factory edge   The edges of a shingle that are not recut by the roofer during installation.

Fascia   A wood trim board used to hide the cut ends of the roof's rafters and sheathing. Fascia is either one by or two by lumber. The gutter system is usually nailed to the fascia.

Fastener density   The number of nails, screws, or staples used per sheet of material to secure the material to the roof or roof decking.

Fastener guide   The manufacturer's suggested placement of fasteners used to secure a material to the roof or roof decking.

Fastener line   A marked or unmarked line across a shingle. This is the line on which all nails or fasteners of a course of shingles should line up. Fastener lines are located 1 to 2 in. above the weather exposure tabs and hidden by the tabs of the next course of shingles.

Feathering strip   (See Horsefeathers.)

Felt   A mat of inorganic or organic fibers saturated with a bituminous.

Felting   The act of installing felt on a roof.

Felt interlace   The act of laying a felt across a valley from one roof, then a felt across the valley, and the first felt from the second roof. This is repeated until the entire valley is covered.

FHA   Farmers Home Administration, or later, the Federal Housing Administration, a lending organization. For best resale value, homes should be constructed to meet or exceed FHA minimum specifications for construction.

Fiberboard roof sheathing   Natural fibers impregnated with a waterproofing material and compressed into a mat-like insulation board.

Fiberglass   Fiber spun from melted glass as it cools. Fiberglass is used for insulation, reinforcements, and as the base for asphalt shingles and asphalt felts. Fiberglass is a trademark name, fiber glass is not.

Fibrated aluminum roof coating   Fibrated means contains fibers. Fibers are fiberglass or polyester and add strength to the aluminum filled coating (see Aluminum coatings).

Field   a) The section of a roof that is directly over the living areas and does not include the eave or rake overhangs. b) The job site, as "we will do it in the field."

Field lamination   a) This is the act of bonding two or more items together to form a composite item. b) Act of installing several roof plies to a built-up roof thus making a monolithic or one-piece structure.

Field membrane   The ply sheet that covers the field of a built-up roof.

Field shingles   The shingles in the field of a roof.

Field tile   The tiles in the field of a roof.

Field torch   A torch that is used to heat asphalt or impregnated roofing membranes while installing the roofing on the roof field. Torches vary in design but usually have a roller and several flame heads spaced to heat the 36 in. width of a roll of roofing.

Finger wraps, vent   Circular collar that fits over a vent pipe to seal it to the flashing. Has metal fingers that spread to shape and attach to the bottom flashing.

Finish coat, stucco   The final coat of stucco applied to a surface. The finish coat containing the color and the texture.

Fire rated   Roofing or building materials test to pass three flame spread tests: intermittent flame exposure test, burning brand test, and spread of flame test. Highest ratings are Class A and Class I.

Fire-resistant or fire resistance   Material that is resistant to catching on fire when exposed to open flame or flaming ashes.

Fire-retardant   a) A spray used to make a flammable material fireproof. b) Material that resists burning.

Fish mouthed ridge   Term given to the gap under a ridge shingle caused by roofing over existing ridge shingles. Cure is to remove the buildup of shingles at the starting point of the ridge or remove all ridge shingles.

Fish mouth or fishmouth   An opening that resembles the open mouth of a fish. This occurs when a shingle buckles or warps and can be the result of the shingles drying out or becoming distorted from chemical fumes.

Five-inch pattern   The act of displacing each course of asphalt shingles by 5 in. from course to course. The cutouts of the second course are 5 in. sideways from those of the first course, etc.

Flash cones   Vent pipe flashing. Cone-shaped flashing that fits over a vent pipe.

Flashing   Metal strips used to form a watertight seal between the items butted up against the shingles. Flashing is used along walls, chimneys, and dormers. Metal is usually 28 gauge galvanized sheet metal, but may be lead, copper, tin, or aluminum.

Flashing, lead   Lead sheet that is site cut and hammered into place. Lead, being a soft metal, can be hammered to fit, and conform to, odd shapes that would be difficult or impossible to flash with other types of metal flashings.

Flashing, sheet metal   Usually 28 gauge, galvanized steel sheet.

Flashing, siding connection   (See Flashing, Z-bar)

Flashing, step   L shaped metal sheets used to flash shingles to a vertical wall. Each L is installed on the tail portion of the shingle before it and is covered with the exposed butt portion of the shingle above it.

Flashing, stucco connection   (See Flashing, Z-bar)

Flashing, W   Sheet metal valley flashing that has a 3/4 or 1 in. hump along its center line. The hump is a splash diverter and breaks up the force of the water impacting it. Flashing resembles a flattened W.

Flashing, Z-bar   Sheet metal bent to Z shape. Used at the junction of a roof and a vertical wall extending above the roof. Also used at the junction of two pieces of panel siding that are one above the other as in a two-story building.

Flashing collar   A metal collar that fits over a pipe extending through the roof.

Flashing membrane   An asphalt impregnated strip that is embedded in and then covered with asphalt, used around pipes, stands, and at parapet walls.

Flashing paper   Kraft or resin or rosin coated paper used as a separator between oil impregnated wood and asphalt felt or asphalt base products. (See Divorcing layer)

Flashing roll   Roll of fiber or metal used for flashing or for making flashing.

Flat attic vent   Small roof-mounted vent that looks like a scoop and is almost flat with the roof. Several are usually required on the average size roof.

Flat roof   A roof with a pitch of 3:12 or less is considered flat.

Flood coat   a) Built-up roofing term for a complete flooding of the roof or roof plies. Usually the last ply is flood coated with hot asphalt just before the gravel is placed. b) Asphalt coating to which the roof gravel adheres.

Float, wood   Flat block of wood with a handle used to smooth concrete.

Floor joist   Horizontal lumber that supports a floor. Usually 2 × 10 in. or 2 × 12 in. lumber.

Flute cap   A cap placed over a chimney, used to prevent water from entering the chimney.

Flute spanability or span   Refers to the distance a material can transverse unsupported across the flutes, the ridges of a corrugated steel decking material. Example: Most foam insulation sheets must contact the decking flutes every 3 to 4 in. for best foam support.

Fly-in method   The area the roofing ply is to be placed on is mopped fully with asphalt. The ply or felt is then placed immediately on the hot mopped asphalt. The roof ply is thought to be flown into place.

Four-inch pattern   The act of displacing each course of asphalt shingles by 4 in. from course to course. The cutouts of the second course are 4 in. sideways from those of the first course, etc.

Free fire zone   A secure ground area under a roof that is being torn off. The free fire zone allows you to throw items off the roof while not having to look and see if anyone is below you. Also termed Off limits zone.

Freezeback   The condition created when water is blocked from exiting a roof and then freezes. Usually caused by an ice dam.

Frieze blocks   Board(s) installed at the junction of the roof overhang and the building's siding and between the exposed rafters. Used to dress off the intersection, giving it a finished look when soffit is not used.

Frieze board   A board installed at the junction of the soffit and the siding. Used to dress off the intersection giving it a finished look.

Frieze molding   a) A frieze board that is decorative. b) A decorative trim installed on the frieze board.

Full lace   A valley in which the shingles from both roofs are brought into the valley and up the other roof in a woven pattern.

Fungi and bacteria   Minute plant or animal growth on paint and shingles. Bacteria and fungi will eventually consume the materials on which they are growing.

Fungus regions   Sections of the United States and the world where roofing is highly subject to the growth of bacteria and fungi. Usually, in the Southeastern United States, where the weather conditions are warm and humid. Coastal areas are also prime candidates for fungus and bacteria growth.

Fungus-resistant   Shingles or other material that have been chemically treated or that have a natural ability to ward off fungus growth.

Furring strips   Thin, narrow strips of rough cut lumber spaced apart and used to hold plaster or stucco in place.

Gable dormer   A dormer that has a gable type or V shaped roof.

Gable ends or gable truss   Truss that has all supporting members vertical and at 16 or 24 in. center to center spacing. The trusses make up the gable ends of a building.

Gable plate   a) Top sill plate of the wall that a gable mounts to. b) The gable plate is the top horizontal member of a snub gable.

Gable roof   A roof shaped like an inverted V.

Gable studs   The vertical 2 by lumber used to form and support the gable or gable ends of a building.

Gable vent   An attic vent placed in the gables of a building. The screened vents may be square, rectangular or inverted V shaped.

Galvanized metal   Sheet metal that has been coated with a zinc covering. The coating may be applied by dipping or by electroplating. The galvanizing helps protect the metal from rusting.

Gambrel roof line   A roof that is essentially flat on top but that has very steep angled sides.

Gas house tar   Waste product from the manufacture of illuminating gas. 1800–1900s era when street lamps were gas lights.

Gauge   a) The thickness of a metal sheet or wire. The lower the gauge number, the thicker the item is. b) The portion of a roofer's hatchet opposite the hammer portion. Used to gauge the weather exposure of a shingle during installation of the shingle. c) The act of gauging or using a gauge to find the correct weather exposure of a shingle during installation. d) The act of determining the overhang of a shingle.

Gingerbread   Decorative lattice and scroll work just below the eaves and rakes.

Glass felts   Felts that contain a web or mat of glass fibers for added dimensional stability. Sometime termed glass fiber or fiberglass felts.

Good roofing practice   a) Compliance to the manufacturer's recommendations for applying his materials. b) Compliance with the building codes and fire safety laws. c) Compliance with OSHA laws.

Grades of shingles   Shingles are graded by fire resistance, wind resistance, life expectancy, weight, size, etc.

Grain direction   a) The direction of the grain of an item. b) The long side of a sheet of plywood. c) The length of a roll of roll roofing.

Granules   Minerals, ceramics, or crushed rocks used to cover a roof or shingle surface. The granules provide the roof or shingles with coloration, protection against abrasion and weather, and protection against U/V radiation from the sun.

Gravel   Crushed stone used to cover a flat built-up roof surface. The stone can be sized from a fraction of an inch in diameter to about 1-1/2 in. diameter. Gravel is usually bright white or off-white in color and provides U/V and heat protection to the building.

Gravel guard   A flashing strip used on gravel-covered roofs to prevent the gravel from being blown or washed onto a lower roof surface or into the rain gutter system.

Gravel roof   A near flat built-up roof covered with a final coat of gravel embedded in asphalt.

Gravel stop   (See Gravel Guard.) A gravel guard used at the rakes and eaves of a built-up gravel coated roof.

Graveled in   The act of putting the final coat of asphalt on a built-up or flat roof and then applying the gravel coat.

Gross area coverage   Material coverage needed to cover the entire roof including curbs, parapet walls, and expansion joints.

Ground roughness   Scale used to decide which chart to use for obtaining wind lift factors from wind velocity charts. Ground Roughness B is for city, suburban, town, and wooded areas. Ground Roughness C is for flat, open country, and open coastal areas > 1,500 ft from coastline. Ground Roughness D is for flat open areas < 1,500 ft from a coastline.

Guarantee   A personal or company promise to fix something that is broken. Roofing lasts upwards of 20 to 30 years and the person or company may not be in existence when a repair is needed, thus roofing guarantees are not considered valid. (See Bonding.)

Gusset   a) A triangular, right angle, structural member that is used to support a building component. b) Wood or metal plate used to attach roof truss members together. c) Wood or metal plate used to attach post and beams together.

Gutter, forged   A rain gutter that is formed with a flat back, rounded bottom, and sloped front. Forged gutters are the most common in use today. Forged is a misnomer in that most forged rain gutters are roll formed.

Gutter, half round   A rain gutter that is semicircle in shape and not currently in common usage.

Gutter, wooden   Older rain gutter fabricated from milled lumber, painted, and then coated internally with tar or other waterproofing.

Gutter basket   Wire screen placed in the top or gutter end of the downspout to prevent leaves from entering the downspout and clogging it.

Gutter clip   Gutter holder shaped the same as the gutter. The gutter clip is nailed to the fascia, the gutter snapped in and a clip latched over the top.

Gutter section   a) A long, approximately 10-ft section of horizontal rain gutter trough. b) Any section or group of components of a rain gutter system.

Gutter spike   Sometimes called a spike and ferrule. Long spike that is nailed through the face side of the gutter, through the ferrule, the back inside of the gutter and into the fascia. The ferrule or spacer tube prevents the gutter from collapsing from over nailing.

Gutter support strap   A gutter system component that encircles the gutter and nails to the roof by a metal strip. Used on new construction for securement of the gutters.

Gypsum deck   A deck, usually steel, that is covered with a minimum of 2 in. thick gypsum material.

H-clip   (See Plywood support clip)

Half-lace A valley in which one side of the roof shingles are full sized and cross the valley, the other, intersecting; roof shingles are cut and end in the valley.

Half-cutout, shingle The cutout at both ends of an asphalt or fiberglass shingle. When two shingles are butted together side by side, the half cutouts form a full cutout.

Half pattern The installation pattern of two- or three-tab asphalt shingles where the next row of shingles is offset from the first row by one shingle tab.

Hand-split Shake shingles that are split by hand and frequently sawn smooth. Termed resawn.

Hatchet Refers to a roofer's hatchet. This tool has a square, flat, serrated head and an ax head. Along the ax head are gauging holes set at every 1/4 in. A one- or two-piece gauge stop is supplied. The gauge stop fits into the gauge holes, and is used for gauging the overhang of shingles at the eaves and rakes and for fast shingle pattern positioning.

Hawk (See mortar board.) Small board with a handle for handling mortar.

Head flashing apron The horizontal mounted flashing above a chimney, skylight, or shed dormer.

Header a) Supporting member or beam over a doorway, window, or open area. Usually formed with 4 by 12 in. lumber. b) Dormer rafter support board for main roof side of a shed dormer.

Headlaps The overlapping of shingles or roofing felt at their top edge. Roofing felt should be headlapped by a minimum of 2 in.

Heartwood The darker colored wood at the center of a tree. Usually refers to that of a red cedar or redwood tree. The heartwood is the insect- and rot-resistant wood of these trees.

Heat fused The joining of two or more materials with the application of heat or flame.

Heating coils for roofs Also called heating cables. These cables are constructed of resistance wire in a waterproof, heat-proof shield. They are used to melt snow and ice off of the roof in winter.

Heavyweight shingles Shingles that are heavier than 235 lb per square.

Heel gusset The gusset used to hold the ceiling joist and rafter or the truss upper cords together. Also called a bevel heel gusset because of its shape, that of an inverted V with its point cut off squarely.

Hexagon strip shingles Three-tab shingles that when installed form a hexagon pattern across the roof. A hexagon is a six-sided figure. Hexagon shingles are not readily available and you may have to cut standard shingles to pattern when doing replacement or repair work.

Hip   The ridge line that is on an angled extension portion of the roof and is not as high as the main roof ridge.

Hip pad   A rubber or other material pad worn on the roofer's hip. Hip pads provide the roofer with protection against splinters and abrasions. They also provide a certain amount of friction between the roofer and the roof, thus affording traction when lying on the roof installing shingles.

Hip jack rafters   The rafters that extend from the ridge of a hip roof to the eaves.

Hip rafters   The rafters that form the ridge of a hip roof.

Hip roof   a) A roof that extends downward on all sides from a peak or ridge. b) An angled extension roof that is lower than the main roof's ridge.

Hip shakes   The shake shingles used to cover the ridge line of a hip roof.

Hip tiles   The tile shingles used to cover the ridge line of a hip roof.

Hip/valley area factor   A numeric multiplier used to find the number of shingles required for a hip or valley area. The multiplier number will vary with shingle size, shape, and coverage.

Hop mop   To mop or apply heated tar or asphalt to a roof or to the roofing felt plies.

Horsefeathers   Thin, wedge-shaped wood strips used to create a flat roof surface when shingling over existing shakes. The thick end is butted to the butt section of the existing shakes. Also termed feathering strips or leveling shims.

Hot application   a) The act of applying heated roofing materials to a roof or to roof plies. b) Name given to a built-up roof that was installed using heated asphalt or flamed asphalt roll roofing.

Hot steep asphalt   Asphalt used for steep incline roofs. Has a high melting temperature and thus, does not liquefy or soften and flow with the sun's heat.

HUD   Housing and Urban Development Agency, a division of the U.S. Government. Aids in low-cost home and home improvement financing.

Hurricane tie   A metal plate punched with holes and bent at a 90 degree angle at its midpoint. Hurricane ties are sometimes called seismic anchors and are used to secure the roof rafters to the top sill plate. Hurricane ties are required in most earthquake and high wind prone areas. Commercially available with prepunched nailing holes for about $1.00 each.

Hydraulic cement   Cement that prevents the passage of water through it. Expands as it sets and is used to fasten bolts to masonry. Usually contains a plastic.

Ice backup   Ice behind an ice dam that enters under the roofing material.

Ice dam   A roof dam formed by water freezing into ice at the eaves or gutter. An ice dam is very harmful since it prevents melting snow or rain water from exiting the roof, and the water backs up under the shingles instead.

Identification index   Plywood marking that shows allowable center to center spacing of the supports for that piece of plywood. As an example: The marking 32/16 is 32 in. on the roof, 16 in. for a floor. Grain of plywood must run across the supports.

Imitation shake   Any number of new products designed to look like and replace shake shingles. Shake shingles have been deemed a fire hazard by many communities and can no longer be used in those communities. Imitation shakes are manufactured from fire-resistant materials such as cement, metal, clay, or other approved roofing material.

Inferior construction   Construction that does not meet code, manufacturers' recommendations, or is of poor workmanship quality.

Inferior designation   Built-up roof term. Roof has poor wind resistance to uplift. May be caused by poor design, poor decking, insufficient fasteners, too cool an asphalt application temperature, adhesive applied in the wrong direction, or combination of above items. Defect does not affect fire rating but must be corrected before new roof is installed.

Infrared roof examination   The examination of a roof using a heat detecting infrared camera system. The differences in the heat detected can be used to locate hidden bubbles, broken or split seams, and roof ply delamination. Infrared examination is becoming important for nondestructive examination testing of aging built-up roofs.

Insulating concrete deck   a) Poured concrete deck using concrete that contains lightweight aggregate, perlite, or other insulating material. b) A 1:4 or a 1:6 insulation/concrete mix that when dried weighs less than 22 lb per cubic foot. Insulated concrete should be installed over vented sheet metal where the vent openings are 1.5% of the sheet area.

Insulation   Any material that blocks heat flow through it. Fiberglass, urethane foam, perlite, and styrofoam are the most commonly used materials since they do an excellent job at blocking heat flow.

Interlayment   Felts used between each course of wood or shake or tile shingles. Minimum width of interlayment is 18 in.

Interlocking tile   Tile that has molded-in male and female joints and when installed with another tile forms a water tight joint.

Isocyanurate or polyisocyanurate   Approved, solid, plastic foam used for insulation sheets on roof decks. Covered by Federal Specification HH-I-1972/2.

Job mixed   a) Refers to mixing concrete or other cement type products at the job site. Concrete is used when installing roof tile. b) Job Mixed Insulation. Job mixed insulation is a two component foam such as polyurethane foam. Two liquids are mixed in place and a chemical reaction takes place forming a solid foam. Licensed installers are usually required.

Joint, sheathing   The interface between two or more pieces of roof sheathing. Sheathing should not have more than a 1/8 in. open, or less than 1/16 in., space between each piece.

Joint shield   (See Bellows). A bellows placed over an expansion joint in a roof to keep water out of the joint. Joint shields are usually constructed of rubber, synthetic rubber, or plastic.

Joist   a) A rafter laid horizontally from support to support. b) The 2 by framing of a floor or ceiling.

Joist hanger   A modified U-shaped piece of sheet metal used to join a joist to a beam or another, usually vertical, structure. Commercially available preformed with punched nailing holes, cost is in the $2.00 each range.

Joist span   a) The distance from one resting place, supporting member, to another that a joist must gap. b) The distance from joist to joist that another material must span.

K-value   Thermal conductivity of a material where the material is exactly 1 in. thick. K = BTU/HR/FT^/F/in. BTU = British Thermal Units: F = degrees Fahrenheit: HR = hour: FT^ = Square foot: in. = inch.

Kettle temperature   The proper temperature of asphalt as it is heated in the hot pot or kettle. Too high a temperature and the asphalt will lose its properties, burn, or flash explode. Too low a temperature and the asphalt will not form secure bonds with the roofing materials being used.

Keyhole saw   A short, about 15 in. long, rough toothed, tapered blade saw used to saw holes. A hole is first drilled, the saw inserted, and the hole cut. May be used for stroll cutting and for cutting in tight poorly accessible areas. Being replaced with electric and pneumatic powered reciprocal saws.

King post   Center vertical supporting post of a king post truss.

Knife, hook blade   Linoleum knife. Small knife with a tapered curved blade ending in a sharp hook. Linoleum knives are ideal for backside cutting of asphalt shingles.

Kraft paper   a) Paper made from boiling wood chips in a solution of sodium sulfate (See building paper). b) Kraft paper may sometimes be brown paper, paper bag paper, impregnated with oils or chemicals to make it somewhat water-resistant. Kraft paper is used as underlayment to siding to prevent air and moisture entry into a building. Being replaced by newer materials.

Kraft paper   a) Paper made from boiling wood chips in a solution of sodium sulfate (See building paper). b) Kraft paper may sometimes be brown paper, paper bag paper, impregnated with oils or chemicals to make it somewhat water-resistant. Kraft paper is used as underlayment to siding to prevent air and moisture entry into a building. Being replaced by newer materials.

Laced   Refers to the application of shingles in a valley. The shingles are woven together to form a watertight seal or surface.

Ladder loader   A mechanical or power driven hoist that attaches to a ladder. Used to transport bundles of shingles from the ground to the roof.

Laminated shingles   Shingles made up of several materials pressed or glued together. Many metal shingles are considered laminated shingles. They have the base metal, a primer, a binder, and a top coat of granules.

Lap joint strength   The ability of a lapped joint to hold under normal use.

Lath sheets   Metal screening designed to hold plaster or stucco in place.

LB/SQ   LB/SQ is the abbreviation for Pounds per Square. a) Total weight of building and roofing materials on a 100 sq ft of roof surface. b) Weight of one square of shingles or other roofing materials in pounds.

Lead flashing   Flashing, made from thin gauge lead sheeting, that can be site formed to various shapes by hammering into place. Lead flashing is not currently in widespread use. Used extensively on buildings up to 30 years ago.

Lean-to roof   Shed roof with only one roof slope.

Left end cap   Rain gutter term. The closure cap used on the left end of a section of rain gutter.

Level cut   A cut in or through a board that is level with the horizon after the board is installed.

Leveling shims   (See Horsefeathers)

Liberty guarantee   Refers to a manufacturer's guarantee for a roof. The roof must be installed according to manufacturer's specifications. The manufacturer has the right to inspect the installation before granting the guarantee.

Lightweight shingle   An asphalt shingle that weighs less than 235 lb per square.

Loading a roof   Act of stacking materials on a roof so as not to exceed the roof's load design limit.

Lookout   Also termed lookout rafter. A short board placed horizontally from the roof's rafters tail to the building's wall. The soffit is attached to the lookout rafter, if used.

Lower cord   The bottom member of a roof truss is called the lower cord.

Low pitch roof   Varies with type of roofing and manufacturer of the roofing material. Usually considered to be any roof with a rise of less than 4 in. per foot for shingles, 3 in. for built-up roofing.

Low slope   (See Low Pitch Roof)

LP gas use   Liquid petroleum used for heating a hot tar kettle or a roofing flame torch.

Main ridge   The highest and usually longest ridge on a building.

Mansard roof   (See Roof, mansard)

Marble chips   (See Granules) Granules made from crushed marble stone.

Masonry blade   A saw blade specially designed to cut masonry. The blade does not have teeth and looks like a thin grinding wheel. Masonry blades are reinforced to prevent centrifugal force breakage, but safety glasses should always be worn in case of breakage.

Mechanical equipment stand   A system of supporting mounts, usually raised above the roof line, that support air conditioners, heaters, and other plant facility equipment.

Membrane   a) A roofing sheet that forms a water- and weather-tight seal. Some may be used as the final roof covering. b) Roofing felt. c) Small strip of asphalt impregnated felt or cloth used for strengthening roofing cement during repairs or patching.

Metal strap anchor   (See Hurricane tie)

Metric shingles, west coast   Shingles sized using the metric system and used extensively in the Northwest states of Oregon and Washington. Size varies but most common is 13 1/4 × 39 3/8 in. in linear measurement.

Mineral surface roofing   Shingles or other roofing material that are coated with granules. Roll roofing is generally mineral faced and some metal tile roofing may be.

Minimum slope   The lowest slope that a shingle or roofing system can be used on and still protect the building and its occupants.

Miter saw   a) A saw used for cutting precise angles. b) A hand saw with a large rectangular shape blade used to make smooth or angle cuts in trim lumber. c) A table mounted power saw used to cut angles.

Mitered corner   A corner formed when both intersecting items are cut at angles and butted together smoothly. Usually two 45 degree angles to form a 90 degree corner.

Mitered shingle   A shingle, usually wood or shake, that is cut at an angle so that it fits tightly against another item or shingle.

Modified bitumen   (See APP) A bitumen or asphalt that has additives that change its physical properties.

Modifiers   (See APP) A filler, chemical, plastic or other substance that, when added to a material, changes the physical or chemical composition of that material.

Moldability   Something that can be molded or formed by molding. Lead flashing is moldable.

Mop and flop method   Installation technique used to apply roofing plies to a built-up roof. The heated tar or asphalt is first applied to the felt and the felt is flipped into place. It is then rolled smooth and tight to the roof sheathing or last ply installed.

Mop grade   Asphalt that when heated can be applied to a roof or roofing felts with a mop.

Mortar   A mixture of Portland cement, masonry cement, hydrated lime, sand, and water used to bond masonry or roof tiles. For roofing work, use Type N or type S mortar.

Mortar, type N   Standard exterior above ground use mortar. Ratio: 1 part masonry cement, 2-1/4 parts sand, water to make a thick paste. Or, 1 part Portland cement, 1 part hydrated lime, 5 parts sand, water to make thick paste.

Mortar, type S   Very high strength mortar for exterior use in high wind and earthquake areas. Ratio: 1 part Portland cement, 1/4 part hydrated lime, 3-1/4 parts sand, water to make a thick paste, or, 1/2 part Portland cement, 1 part masonry cement, 3-1/2 parts sand, water to make a thick paste.

Mortar board   A sheet of metal or plywood that has a handle centered on its bottom side. Mortar is mixed, placed on the board, and then carried to the work where it is used.

Mortar box   A wood or metal low-walled box in which one mixes mortar.

Mortar stop   Flashing used to form the junction between mortar or stucco and no mortar or stucco.

Nailable deck   A roof deck that can be nailed into. Most wood decks are nailable decks. Some steel and some concrete decks are not nailable decks and special fasteners must be used.

Nail area   a) The designated area or location on an object that is used for nailing the object to another object. b) The nailing locations for foam insulation. c) The area or location where nails are kept.

Nail, clinch   Nail designed to bend when it contacts a very solid surface. A V notch is cut just above the point end. This V bends and locks or clinches the items being nailed together. Used for truss assembly and for nailing sheathing or insulation to concrete decks.

Nailer board   a) A board, usually a 1 by, that is nailed to the roof sheathing. To this nailer board, one nails the tiles being installed. b) A board part way up a parapet wall to which the flashing is nailed.

Nailer header or nailing header   The board to which fascia is nailed, a fascia backer. Not always used.

Nailer plate or nailing plate   A flat metal plate prepunched with holes. Used to assemble trusses and other lumber joints.

Nailer strip   A board, usually a 1 by 2, that is nailed to the top edge of the rake(s). The rake tiles are then nailed to this board.

Nail gun A hand-held electric or pneumatic tool that drives nails into a material at the pull of a trigger. Nails are loaded into the gun via rolls, strips, or cartridges.

Nailing pattern The manufacturer's suggested nailing points for the material being nailed. Foam insulation manufacturers require a certain number or a certain type nail be used at specific places in the foam sheets. The manufacturers will include a small diagram or pattern with their products designating these locations.

Nails per square The count or weight of the nails used to install a square of roofing.

Nail punch A metal rod used to drive nails flush with or below the surface of the material into which it is nailed. The rod is about 6 in. long, is pointed on one end, and flat on the other. The flat end accepts blows from a hammer.

Nail stripper (See Stripper.)

Narrow crown staple A staple that is less than 3/4 in. wide.

NFVA Abbreviation for Net Free Vent Area. The open area of a vent or vent sheet.

Noncombustible deck A misnomer since all roof decks are combustible. Concrete and gypsum decking is considered noncombustible, incapable of catching on fire. Certain decking systems that are Factory Mutual Class 1 are considered noncombustible.

Nonnailable A deck that will not accept a nail. Special approved fasteners or glues must be used on a nonnailable deck.

Nonwoven polyester Polyester plastic cord that is randomly stacked and then pressed into sheets. Used as matting for shingles and roll roofing materials. Very dimensionally stable and inexpensive, but has a low melting point.

Nose The front edge of a roofing tile.

Nosing Flashing that fits over the lip of a rake or eave and does not have a formed drip edge. An L shaped flashing.

No-tab shingles Shingles that do not have a cutout in their weather exposure surface.

OBS Oriented Structural Board: A thin chipboard used to sheath foam insulation products. Some manufacturers require the use of certain heavy-duty shingles due to the surface irregularities of OBS.

On-the-fly method (See Fly-in Method)

Ordinary roofing Non fire rated or fire rated below Class C. Usually uncoated roll roofing, untreated wood shingle, or shake or plastic roofing. Not to be used as main roofing on dwellings or Class I or Class II buildings.

Open beam ceiling A ceiling with roof beams exposed. If roof deck is insulated, it requires a 1/2 in. fire-resistant sheathing over the insulation before the asphalt shingles or other roofing materials are installed.

Outriggers   Short pieces of supporting lumber running from two or more rafters outward the rake. Outriggers aid in holding up and strengthening the rake overhangs.

Overall economy   a) The design, the materials, and the construction that lend themselves to the best results at the lowest cost. b) A design that over time uses less heating or cooling or requires roof replacement less often.

Overlap   The condition that develops when one item is on top of another and the two items are connected.

Overlap seam   A seam that overlaps. Two items that overlap and are mechanically fastened together at the overlap. Recommended for corrugated steel decking seams. Different roofing materials require specific overlaps, check codes.

Palleting   a) Wood or metal panels that can be loaded with items and then transported by forklift. b) The loading of a pallet with shingles or tiles for storage and transport. Many manufacturers of roofing materials have restrictions on the amount of their materials that may be vertically placed or stored on a pallet.

Pan flashing   Flashing formed into a U and used as a water runoff trough.

Parapet   a) A vertical wall that extends above the roof line. Usually used to hide from view the equipment mounted on a flat roof. b) The top edge of a fire wall that extends above the roof line may be considered, by some, to be a parapet wall. This fire wall is usually seen between the fourth and fifth units of an apartment or condominium building.

Parapet tool   Small, roller torch unit used to apply flame applied edging material to a parapet wall.

Pattern   a) The application of shingles in a specific manner. b) The look created by applying shingles in a specific manner. c) The offset used between shingles during installation.

Pattern lines   The viewed lines created by the cutouts and shingle edges when shingles are installed.

Pattern, scallop   Wood shingles that have their butt edge cut in a semicircular shape. When installed, the general roof design looks scalloped. Often mixed with other patterns.

Pattern, staggered   Shingles installed so that the butt edge is not in horizontal alignment from shingle to shingle. The general alignment from rake to rake is a line with plus or minus tolerance. The staggered pattern is used to imitate a shake roof. Some shingles are manufactured with their tabs cut at different lengths to obtain the staggered look.

Pattern, straight   The standard straight cut shingle. The line from rake to rake does not vary from the horizontal.

Peak gusset   (See Heel gusset)

Penny, nail guide   a) A listing of nail lengths to penny or 'D' sizes. b) A chart by which one compares a nail, to find the penny or 'D' number.

Perimeter edge   The edge of a building or roof.

Perlite   Natural siliceous rock that, when heated, expands and forms trapped air bubbles within it. Perlite is a lightweight, fire-resistant insulation product.

Phase application or phasing in   a) Not recommended. Phase application of a built-up roof is where the roof is completed in separate sections, which can result in poor seams. The recommended method is shingle fashion. b) Phase application of a built-up roof where entire roof is covered with each ply before the next ply is installed. Not recommended since it can cause early deterioration of the roof and roof blisters.

Phenolic insulation   Phenol and aldehyde resin plastic impregnated fiber or paper.

Pin nails   Galvanized 3 or 4 D nails used to face nail shingles. The use of pin nails for face nailing is not recommended if blind nailing can be used instead.

Pine tar   Roofing tar made by distillation of pine tree wood. Not in frequent use due to its properties and cost. Pine tar should not be mixed or combined with petroleum-based tars or asphalts.

Pitch   The distance or ratio of distances of a roof is called the pitch. A roof has a pitch of 1/6 if the rise is 4 in. vertical for each 12 in. of horizontal distance from the eaves to the ridge. Pitch, expressed as a fraction, equals rise over $(2 \times \text{run})$.

Pitch dam   Flashing used around a cutout in a built-up roof. The pitch dam keeps the liquid asphalt from dripping through the opening on to people or items under the opening.

Pitch, tar   (See Pine tar.) Residue or left overs from making pine tar.

Pivotal point or center of rotation   The point on a locking shingle where one pivots or rotates the shingle to lock it into the next shingle.

Plank roof decking   A roof deck fabricated from 2 by T&G (tongue and groove), lumber.

Plastic cement   Roof sealing compound made up of various petroleum products. May have the addition of various oils such as linseed, tung, soya, and fish oil. The oils impart special waterproofing and drying qualities.

Plate or rafter plate   The top board of a wall on which the rafters or ceiling joists rest is called the plate.

Plate anchor   A bolt or strap used to anchor or secure the plate to a masonry or concrete block wall.

Plate, long cheek cut   A single cheek cut that is '/' shaped.

Plug weld   A welding technique used for attaching corrugated roof decking to metal trusses. The valleys of the corrugated decking are punched every 6 to 12 in. with an oblong-shaped hole. The weld is made by heating the metal under the hole and applying welding rod to the surface, melting it and filling the hole.

Plumb cut   a) A cut through a piece of lumber that is vertical or plumb when the lumber is installed in place.

Ply sheets   Ply sheets are the various material layers of a composite product.

Ply sheets, roofing   Built-up roofing term. The various fabric sheets, cemented with binder, that form the roof's covering.

Plywood   Sheets of laminated lumber used for sheathing. These should be 1/2 or 15/32 in. or more thick and comply with products standards PSI-83, marked APA (American Plywood Association). I.D. index 32/16 (unsanded) or Group 1 (sanded). See Appendix B, table 4.

Plywood roof decking   Roof decking consisting of 1/2 or 5/8 in. plywood over a truss or rafter network.

Plywood support clip or 'H' clip   H shaped clip approximately 1 in. long used between two unsupported sections of plywood to add support. H clips may be extruded aluminum or stamped sheet metal.

Pneumatic nailer   (See Nailers.)

Polyester mat   Matting used as a base for asphalt or other rubber or synthetic material used to make shingles and roll roofing materials.

Polyisocyanurate insulation   (See Isocyanurate insulation.)

Polymer   Chemical reaction where two or more molecules combine to make a single large molecule. Most plastics are polymers.

Polymer modified asphalt   Asphalt that has polymers or plastic added to it to change its characteristics.

Polystyrene board or blue board   a) Blue color insulating sheet popular for its low cost and high insulating value. b) Closed cell foam insulation made from polystyrene.

Ponding   Small water puddles formed on a flat roof due to indentations in the roof surface. Puddles can be from rain, air conditioners, or other condensates. Considered harmful to a roof if not drained or dissipated within 48 hours.

Portland shingles   Asphalt shingles that are 39-3/8 in. in length. Portland shingles are used extensively in the Northwest coastal states. Portland shingles are metric sized shingles and receive their name from the place used, Portland, Oregon.

Poured concrete deck   One-piece reinforced concrete poured in place over metal or other subdeck. Pour should be a minimum of 2 in. thick.

Power vent   A roof attic area vent that is comprised of a motor driven fan and a temperature controller.

Precast concrete deck or T or TT decking   A roof deck that consists of precast, pre-stressed concrete slabs. The slabs are 'T' or 'TT' shaped and form trusses from the wall to wall supports.

Prepared roofing   Roofing material that is ready to be installed as it is received from the manufacturer. Asphalt shingles, roll roofing, and tiles are considered prepared roofing. Wood shingles, shake, and built-up roofs are not.

Pressure release vents   (See Vent valve stack)

Pressurized plenum   A ceiling area that is enclosed and pressurized to above atmosphere pressure, used for clean rooms and in place of a network of heating ducts.

Primer, paint   Specially formulated paint used to aid in the bonding of the top paint coat to the surface being painted. Primers usually contain a high percentage of acid or other etchants that cut into the surface and form the bond. Primers also are used under color top coats to provide a bright white surface that aids in the color clarity of the top coat. Most primers contain waterproofing oils, binder agents, etchants, and coloration. Primers should be top coated within 90 days of application. If allowed to sit uncoated for over 90 days, one should reprime the surface.

Principal rafter   a) Main supporting rafter. b) Rafters from the collar beam to the plate on a queen post truss.

Product density   a) The weight of an item as compared to another item of the same size. If heavier, it is denser. b) Density is the compactness of cells within a product. Example: Wood is denser than polystyrene foam.

Projection of decking   The outer portions of the roof decking that are unsupported and used as rake or eave overhangs.

Projection ridge board   A ridge board that comes off the main ridge board at a 90 degree angle and forms the ridge of a roof projecting outward from the main ridge. Top down view is a T.

Projection roof   A roof that projects out from the main roof, usually at a 90 degree angle.

Pry bar   A metal bar about 1-1/2 ft long with a forked end or claw used for prying up shingles and nails.

PSDN   Abbreviation for Product Safety Disclosure Notice. Safety notice detailing use on handling of a product. Also called MSDS or Manufacturer's Safety Data Sheet.

Purlin   a) Boards laid from gable to gable on which the common rafters sit. b) Boards nailed horizontally across the rafters and used for securement and support of corrugated or other sheet roofing. Spacing from purlin to purlin is per roofing material manufacturer's instructions, but is usually 16 to 32 in. c) Gambrel roof; 2 by lumber nailed from rafter to rafter at the point the slope changes.

Putty knife   A small trowel used for applying putty or roofing cement. About 6 in. in total length and with a 1 in. wide, flat blade.

Queen post   a) Vertical posts that are between the top and bottom cords of a truss and not centered on the apex of the top cords. b) The two vertical posts of a queen post truss.

R-factor, thermal resistance   Resistance to heat flow through a material. $R = 1/C$. R is the reciprocal of C or conductance. Combined R-Value of roofing components is $Rt = R1 + R2 + R3 + R. . . .$

Rabbet cut   L shaped cut at the edge of a piece of lumber. It forms a tight seal with another piece of lumber that has a reversed rabbet cut.

Rafter tie boards   Temporary boards used to hold the rafters in alignment until the roof sheathing is applied. Rafter tie boards are usually 1 by lumber or lath.

Rafters   The lumber supports that make up the roof structure. Usually 2×10 in. or 2×12 in. lumber. The roof sheathing is nailed to the rafters.

Rafters, common   The lumber supports that transverse the gap from the building's outer walls to the roof's ridge board. May be spaced 12, 16, 24, 30, or 48 in. apart depending on sheathing and loading factors (consult codes). Rafters are usually 2×10 in. or 2×12 in. lumber.

Rafter tail   The portion of a rafter that is outward from the outer walls of the building and forms the building's overhang.

Rake   a) The overhang portion of a roof that goes from the eave to the peak. b) The exposed side of the left and right most roof rafters of a roof.

Rake drip edge   Metal flashing used to cover the exposed sheathing along the rake(s) of a building. This flashing also may provide a water runoff channel to direct water to the gutters.

Rake tiles   Tiles installed along the rake(s) of a building. These tiles hide the exposed sheathing, rake flashing, and rafter sides. The rake tiles may be installed to form water runoff valleys. Rake tiles are usually nailed, wired, or cemented in place.

Racking a joint   Mason term that refers to the act of finishing a mortar joint. A raking tool, various shapes available, is used to compress the new wet mortar into the joint, making the joint water resistant.

Random exposure   The weather exposure of shingles, shake, or tile that occurs when the butt ends are not in a straight line from rake to rake, but staggered instead.

Recurrent splitting   Recurrent is to do again. Splits that return even after repaired. Recurrent splitting is an indication of structural or other defects.

Red resin paper   An older building paper that is seldom used. Paper coated with a natural or synthetic resin, red in color.

Redwood   Tree native of the Western coast of the United States. The redwood's heartwood is valued for its ability to ward off insects and rot. Redwood is fast becoming an endangered species and substitutes should be used where possible.

Reglet   A metal flashing piece inserted or nailed to a parapet wall above the roof line. To the reglet, a metal flashing is snapped. This flashing covers the roofing material that extends up the parapet wall. Reglets are used so that the flashing is removable during reroofing, without tearing up the parapet wall.

Reinforcement   A web or mat placed into roofing cement and then covered with additional cement. This imparts a strength and dimensional stability to the roofing cement.

Resawn   a) Resawn is to saw again. b) Hand split shake shingles that are then sawn smooth.

Retained R-value   The retained R-value is the R-value of insulation after 6 months or more of exposure. The aged or stabilized R-Value is the retained R-Value after 5 years.

Reroll method   The roll roofing is rolled out and cut to fit, then rerolled. Hot asphalt is then applied in front of the roll and the roll again unrolled. Hot asphalt is continued to be mopped in front of the roll until the roll is completely unrolled. This method is recommended for applying roofing plies in weather temperatures near or below 50 degrees Fahrenheit.

Reroofing   a) The act of tearing off the old roofing and installing new roofing. b) The act of installing new shingles over old worn shingles. c) The act of installing a new built-up roof over an old built-up roof. On built-up roofs, reroofing should not be considered if there is evidence of delamination, wet roof insulation, or poor attachment of the old roof. If these conditions exist, a tear-off is required.

Ribbon courses   Ribbon courses are the double layering of shingles to accent a row of shingles visually.

Ridge   The peak or highest point of a roof, hip, or dormer.

Ridge board or ridgepole   A board or beam used to form the ridge. The rafters butt up to and attach to the ridge board from both sides of the building.

Ridge cap   Formed shingles, shake or tile, used to cover the ridge of a building.

Ridge cap vent   A special designed vent installed along the ridge of a building. Ridge vents require the use of soffit venting to operate properly.

Ridge line   a) The visual line that the ridge of a building makes when viewed from the ground. b) The ridge or peak of a building.

Ridge shakes   Shake shingles that are mitered, butted, and stapled together to form an inverted 'V' shingle used to shingle the ridge of a building.

Ridge shingle starter   The first shingle installed along the ridge. This should be installed opposite the prevailing wind end of the ridge, if possible.

Ridge tiles   Tiles used along the ridge or peak of a building. Various shapes and combination of shapes may be required depending on the manufacturer and type of tile.

Ridge vent   (See Ridge cap vent.)

Ridge, lookout   An extension of the ridge beam past the gable end of a building.

Right end cap   Rain gutter term. The closure cap used on the right end of a section of rain gutter.

Rim joist   A joist that sits on and follows the length of a sill or plate. The common joist butts to and attaches to the rim joist.

Ring shank nails   Nails that have deformed rings around their shank. The rings aid in holding the nail in place.

Rise of roof   The vertical distance from the eave to ridge of a roof if measured straight up from the eave.

Roller, roofing   A weighted, revolving cylinder with a handle. Used to compress the plies and their asphalt binders when installing a built-up roof.

Roll-in or reroll method   (See Reroll method)

Roll length   a) The square footage or coverage a roll of roofing or roofing felt will cover when properly installed. b) The linear measurement of a roll of material from one end to another along its longest axis.

Roll roofing   Granule covered, asphalt base roof covering 18 to 36 in. wide. Standard 36 in. roll covers one square when properly installed.

Roll weight   The weight of a roll of roofing felt or roll roofing material. The weight indicates the thickness of the material and, thus, its expected life span.

Roof, dutch hip   A roof that is a modified inverted V and has hip roof ends. The inverted V is steep from the ridge to about 1/3 down and then has a low slope from that point to the eaves.

Roof, flat   A roof that has a slope of 3:12 or less.

Roof, gable   A roof that has inverted V shaped rakes.

Roof, gambrel   A roof that has a low slope from the peak to about 1/3 down and then has a very high slope from that point to the eaves.

Roof, hip   A roof that has all four roof surfaces sloped downward from the peak or ridge.

Roof, mansard   A roof that is flat or near flat on top and has all four sides built on a steep slope. Top must use fire-rated materials. Sides, being steep, may be considered as decorative and not required to be fire rated. Materials such as wood shingles or shake or plastic are frequently used. Consult local building codes.

Roof, ogee   A modified gable roof in which the slope constantly changes. Resembles a bell when looking from the rake sides. Design is frequently used on barn roofs.

Roof, saddle   Same as gable roof.

Roof, shed or lean-to   A roof that slopes from a low point to a peak and does not have another downward side.

Roof board   a) Roof board is another name for roof beam. b) A plank used for sheathing.

Roof curb   (See Curb detail)

Roof deck   a) The flat portion of a roof. b) The sheathing applied to the rafters to form a deck surface.

Roof drain   a) A drain built into the deck of a flat roof. b) Gutter system at the eaves of a roof.

Roof drainage   The ability of a roof to drain off water before the roof or building is damaged (see Drainage).

Roof hips   Secondary roof ridges that are lower than the main ridge.

Roof jacks   J shaped steel bars that are temporarily nailed to the roof during shingling. A 2×4 is placed in the J hook portion and provides a foot rest for the roofer.

Roof mop   A mop used for spreading hot tar or hot asphalt on a roof during the installation of roofing plies.

Roof penetrations   Those items that pass through the roof's sheathing, example: vent pipes.

Roof plan   a) A layout drawing of a roof showing locations of holes and protrusions. b) A layout drawing of a roof showing rafter locations or sheathing locations and cuttings.

Roof plan view   A drafting term in which a drawing is made showing the property or building as seen from directly above.

Roof projection   A drawing of what the roof would look like if it is lifted off the building and then flattened. Used to find the square footage of shingles and sheathing required.

Roof rise   (See Rise of roof.)

Roof seat   A small wooden seat made so that the seat platform is horizontal to the ground. The bottom is cut to fit the slope of the roof and contains nails or spikes that dig into the roof sheathing. Roof seas are used by roofers, during the installation of shingles, to provide a degree of comfort.

Roof span   a) The distance between supporting walls on which the roof trusses. b) Distance from outer wall to opposing outer wall of a building covered with a roof.

Roof tape   Narrow fiberglass or polyester reinforced fabric used to make roof repairs and to repair failed flashings.

Roof walkway   A narrow, reinforced pathway used on flat or built-up roofs that may be damaged if walked on. Walk ways can be as simple as an extra ply of roofing or as complex as an elevated metal grill.

Roofing cement   (See Plastic cement)

Roofing hatchet   (See Hatchet)

Roofing nail   a) Any nail used to secure roofing material to the roof's sheathing or deck. b) A short, galvanized, 5/8 in. nail with a large diameter head. NOTE: Roofing nails are specified in the U.B.C. as to head diameter, length, and wire gauge.

Roofing paper   Rosin or resin impregnated paper used to isolate wood deck lumber from the base ply of a built-up roof (see divorcing layer).

Roofing plies   The layers of materials applied to a built-up or flat roof.

Roofing square   (See Bevel Square)

Run   (See Courses) A row of shingles as applied to a roof.

Run of common rafter   The run is the horizontal distance from the tail end of a rafter to a point directly below the ridge center.

Run of valley   The length of the valley from its starting to finishing point, usually from the peak to the eave. Measured horizontally from lower edge to a point vertically under the upper edge.

Rust   Metal oxide formed when steel, moisture, and oxygen are present. Rust is red in color and will stain most surfaces. Rust will eventually consume the steel so that it is no longer structurally sound.

Rustic shingles   Shingles designed to give a rough cut appearance. This appearance can be selected with texture, color, or different tab lengths or a combination of each.

Saddle flashing   An inverted V flashing used above a chimney or doorway to divert run-off water away from the chimney or doorway.

Scaffolding   A system of stands and planks used to provide a safe working platform high above the ground.

Scarifier   Scarifier is a mason term. Hand-held, rake-like tool used to create multiple grooves in wet cement or stucco. The grooves, when dry, provide locking keys for the next layer of cement or stucco.

Scarifying, stucco   The act of using a scarifier tool to create grooves in wet cement or stucco.

Scissors truss   (See Truss, scissors)

Score and break   The act of scoring an object and then bending it until it breaks. In roofing, one scores tiles, shingles, and metal flashings.

Scoring   The act of creating an indented line on an object. This line weakens the object along the line and allows the object to be bent and broken along the line.

Scratch coat, stucco   Mason term for the stucco coats that are scarified, usually the first and second coats.

Screeds, stucco   Trowel width vertical lines of stucco applied over fix depth nails. Screeds are approximately 1/4 to 1/2 in. thick. They are used as guides for even application of the remaining stucco.

Scupper flange   An opening in a wall or roof section for the release of water from the roof.

Scuttles   Covered hatchways onto a roof. Used to obtain roof access for servicing equipment on the roof.

Seal cap   A cement or mortar cap used to water seal and provide a drainage slope. Term usually refers to the cap of a chimney.

Seal down   a) To apply mastic to the tabs of a shingle. b) A shingle that will seal itself to another shingle (see self-sealing).

Sealing   a) The act of making an object resistant to water and air entry. b) Application of a sealer to an object. c) Act of applying roofing cement to a vent or roof protrusion.

Seat cut   (See Birdsmouth cut, level cut)

Self-sealing   Asphalt shingles that, when heated by the sun, bond themselves into a unified water-resistant, wind-resistant surface.

Self-sealing strip   The spots of adhesive material on the face side of fiberglass or asphalt shingles. When heated by the sun, the material becomes liquid and adheres to the shingle over it.

Selvage edge roll roofing   Roll roofing where one half its width, for the full length of the roll, is covered with mineral granules, the other half is uncovered.

Serrated   The act of forming cuts in two directions across an object. In roofing, serrated is the cross cut pattern on the head of a hammer or roofers' hatchet. The serrations help prevent the hammer blows from sliding off the nails being hit.

Serrated pattern   a) The design obtained after serrating an item. b) The pattern or layout of how an item is to be cut for serrations. c) Wood roof where the butt edge of the shingles are not in line with each other.

Shadow course   (See Ribbon course)

Shadow line   A darker colored portion of a shingle that when installed appears as shadows. The shadow line of an asphalt shingle is built in by the manufacturer with the use of colored granules.

Shake cap   Inverted V flashing used to cover shake shingles at the ridge.

Shakes   Wood shingles that are split, not sawed. Usually manufactured from redwood, red cedar, or treated pine and fir.

Sheathing   The roof covering directly over the trusses or rafters to which the roofing material is secured.

Sheathing span   The distance between rafters or trusses that the roof sheathing must transverse unsupported. This distance is covered by the U.B.C. and varies with span, sheathing thickness, sheathing material, and expected weight loading (see Plywood).

Shed dormer   A dormer that has a lean-to type roof.

Shims   Tapered pieces of lumber used to fill voids or to make alignment adjustments between two or more items.

Shingle   A flat square or rectangular sheet of material used with others of same description to cover and protect a building, its occupants, and their contents. Shingles may be manufactured from a variety of materials including tar, asphalt, fiberglass, aluminum, steel, copper, plastic, lead, and wood. The use of the material is based on looks, afforded protection, cost, and building and fire codes. The shape, size, and weight of each vary with material, type, and manufacturer.

Shingle, aluminum   Shingles manufactured from aluminum sheet stock. May be embossed with a design, colored, and coated with other materials that inject specific qualities.

Shingle, fiberglass or polyester   An asphalt shingle that uses a web or mat of fiberglass or polyester for structural strength. The web or mat holds the asphalt in place and gives the entire shingle dimensional stability with changes in ambient temperature.

Shingle, slate   Considered a tile. Slate is a natural rock that is cut and split into shingles.

Shingle molding   A trim molding placed at the intersection of the eave shingle and the fascia.

Shingle tabs   Weather exposure surface of a shingle between the cutouts.

Shingle weight   Expressed as pounds per 100 sq ft of roof coverage. Weight varies with the base material, asphalt content, granule content, and size of shingle.

Shingles, wood   Shingles manufactured from wood where the wood is sawn smooth on both sides and is of the same thickness from tail to butt. Wood shingles require air circulation on both the exposed and the bottom side.

Shiplap siding   Tongue and groove siding that has its tongue end milled concave on the exposure surface.

Shipping weight   Weight of an item including all packaging and palleting.

Short cheek cut   (See Cheek cut)

Shovel, flat bottom   A shovel used for removing shingles from a roof. The flat bottom shovel is slid under the shingles and with an up and over, lifting motion, raises the shingles, dislodging them and the nails or staples.

Sidelap   a) The offset of vertical alignment of the cutout tabs from course to course of asphalt shingles is called sidelap. b) The recommended side to side overlapping of the roofing felt by 4 in. onto the next strip of roofing felt. Code requires a minimum of 2 in. c) The recommended overlapping of roofing plies on a built-up roof.

Sign supports   Metal angle supports used to hold a sign or billboard in place on a roof or parapet wall.

Silcem slates™   Man-made slate tile imported from Europe.

Single coursed   One layer of wood, shake, or shingle applied to a surface.

Single-ply roofing   a) A roof covering that by itself forms the base and final ply. b) Single membrane roofing.

Six-inch pattern   The pattern formed when asphalt shingles are installed with a 6 in. or one-half tab offset.

Skylight   A roof window that may or may not be opened. Skylights let light into a room from above and cut down on electric light energy cost.

Slag   a) Fused leftovers from metal smelting operations. b) Volcanic lava that resembles slag.

Slag coating   Slag embedded in asphalt and used as a final roof coating. Slag may not contain over 2% water and must be free of dirt and dust. Slag is 1/4 to 5/8 in. in size. Installed at 300 to 400 lb per square into 60 lb of asphalt per square, with 50% embedment, ASTM D-1863.

Slant nailing   a) Nailing with the nail held at an angle rather than perpendicular to the surface being nailed. Not recommended on shingles. b) Slant nailing of T&G roof decking that is wider than 6 in. The lumber is face nailed at a slant through the tongue and into the groove of the next board. Recommended if T&G over 6 in. wide.

Sleeve, rubber   The rubber boot used to seal a vent pipe to the vent pipe flashing.

Slip point connection   a) A connection that slips. b) An expansion joint in a built-up roof. c) An roof connection where a vertical wall protrudes through the roof.

Slope, roof   a) The incline or angle of a roof surface from horizontal. Slope = rise:run b) An individual section of a roof.

Smooth valley   A valley that does not use W-flashing, but instead uses a sheet of smooth sheet metal or a 90 lb cap sheet. In wide use but may not meet local building codes of some areas.

Snow guards   Large V shaped flashing installed on a roof over a doorway to prevent heavy snow from falling on the persons below.

Snub gable   A gable end that is lower than the main roof ridge. A snub gable looks like an inverted V with the point of the V cut off.

Snub hip rafter   Rafter that comes from the hips to the top of a snub gable.

Soffit   A board or sheet that extends from the fascia to the building's siding and hides the bottom of an overhang. Soffit can be made from wood, vinyl plastic, sheet steel, aluminum, and other materials. Soffit may or may not contain ventilation slots depending on the attic venting system used.

Softening point range   The temperature range in degrees that a material will begin to soften and possibly flow.

Solid deck   A roof deck that contains no voids or is completely covered with sheathing.

Spaced deck   A roof deck that has voids between each sheathing member. Spaced decks are required under wood shingles and some shake shingles. The deck is usually made up of 1 by lumber strips, each spaced apart from the next by a fixed distance, the distance depending on the final roofing nailing requirements.

Spacer tube   a) Metal tube used to prevent over nailing. Used for nailing gutters to fascia and for nailing insulation to a deck. b) A ferrule is another name for a spacer tube.

Spacing of shingles   The side to side gap measurement between two shingles. A space is required for expansion and contraction of the shingle with changes in ambient temperature.

Span   a) Span is the distance between rafters or joists. b) The distance an item must transverse unsupported between two supports. c) The unsupported length of a roof truss or beam.

Spanish tile or pantile   Clay tile shaped like a lazy 'S' or '~'.

Specification pitch   Coal tar pitch specification. Specified as a range of pounds of pitch per square of roofing per ply being installed. Each roof ply has its own, usually different, amount of coal tar pitch applied to it during installation.

Splash block   A pad, usually of concrete, into which the gutter downspout empties. Splash blocks keep the emptying water from eroding the ground and help direct the runoff water on to the lawn.

Splash diverter   (See Flashing, W.)

Splice cleat   Short 1 by or plywood piece used to hold, splice two joists together over a sill plate.

Splice covers   Small covers that fit over the joints of two joint shields or flashings. Splice covers keep water out of the joint.

Spot cementing   Application of asphalt, tar, roof cements, and cold asphalt compounds to a roof or shingle or flashing in spots, rather than in continuous strips. Required to allow moisture escape paths under the items being cemented.

Spot mopping   Built-up roofing term. Refers to application of asphalt or binder in circular spots rather than in continuous strips or complete coverage. Required to allow movement of the roofing plies during expansion and contraction of the roof's components.

Spot repair   Small repairs to an item to bring it back into specifications. Use spot repairs whenever you do not wish to tear off and replace the item or roof.

Square   a) A tool used to form or create 90 degree angles, a carpenter's square. b) The placement of two objects at exactly 90 degree angles. c) A unit of measure. A square of shingles is the amount of shingles required to cover 100 sq ft of roof surface.

Square butt shingles   Standard two- or three-tab shingles where the butt or bottom edge is cut square.

Squared up   a) Placement of shingles so they are exactly 90 degrees to the roof's eave line. b) Vertical placement of shingles from eave to ridge so that the visual, vertical line formed is 90 degrees to the eave and the ridge line.

Stacked tile   Method of stacking tiles on a roof in small piles so as not to exceed the roof's load capabilities. Stacking the tile makes it easier for the roofer to reach each tile.

Stacking   Act of placing similar items on top of each other for storage.

Stagger joint   a) The offsetting of plywood sheathing by 1/2 sheet from course to course. b) The offset overlapping of two pieces of lumber when they are nailed together to form a continuous joist or a solid thick beam.

Staggered pattern   (See Serrated pattern)

Stairstepping   A method of installing shingles diagonally to a roof. First course has six shingles, then one works the second course with five shingles, then next course with four, next with three, next with two, and next with one.

Standard exposure   The weather exposure recommended for a shingle, shake, or tile, when the butt ends are in a straight line from rake to rake. (See Random Exposure).

Standing seam   a) A joint between two materials where the one material has an edge sticking straight up and the other material has an inverted J edge. The inverted J fits over the straight up edge to form the seam. Not recommended for metal or corrugated decking seams, use overlap seam instead. b) A joint between two materials where each has a J edge. One J is inverted and slides into the noninverted J of the other material.

Staple gun   A hand held, manual, electric, or pneumatic driven tool used for injecting staples into a material.

Staples   Thin gauge wire bent into a U shape and used to secure one material to another, i.e., shingles to the roof sheathing. Roofing staples may be made from steel, aluminum, or copper.

Starter board   A board nailed along the eave(s), used to raise the front edge of a roofing tile. If not used, the first roofing tile will not align, or seat properly with the succeeding courses.

Starter course   The first application of shingles or roll material to the eaves of a building's roof.

Starter finish course   The second and usually last application of shingles to the eave of a building's roof.

Starter roll   Mineral surfaced roll roofing used along the eave(s) as a first course. The roll is approximately 7 to 8 in. wide and material is as thick as a standard asphalt shingle. This builds up the roof thickness at the eaves to the same thickness the remainder of the roof will have.

Starter strip   (See Starter Roll.) As an alternative to a starter roll one may use cut standard shingles with their tabs pointed toward the ridge.

Steel roof deck   A roof deck consisting of corrugated steel over a system of trusses, rafters, joists, or beams. Usually used as the base for a concrete deck or for the base of an insulated built-up roof. A 1 sq ft area in midspan should not deflect more than 1/240 in. with a 300 lb load placed on it.

Steep   Refers to the slope or pitch of a roof, i.e., the angle of a roof from horizontal.

Steep asphalt   Asphalt formulated with a high melting point. Steep asphalt can be used on a steep roof without the asphalt running off the roof in hot weather.

Steep roof   A roof with a slope above 8:12.

Steepness of roof   The degree of slope from horizontal that a roof varies. Measured in inches of rise for each linear foot of horizontal direction.

Step flashing   Metal shingles or plates used in a stair-step pattern under regular shingles. Step flashing is the recommended flashing whenever a wall or chimney is above the roof line. Also whenever the roof shingles must butt up against the wall or chimney and the shingles transverse from the eaves to the ridge.

Stop bead   A metal flashing that is used as a stopping point when applying stucco.

Storage   Refers to the proper storage of roof materials at the work site. Materials are not to become wet. Maximum site storage for asphalt roofing products is considered 2 weeks.

Storage cover   A cover that protects outside stored roofing materials from the weather. Polyethylene tarps are not to be used since these tarps do not expel trapped moisture. Waterproof canvas is acceptable.

Story pole   a) A full-size layout of the cross section of a wall. b) A full-size layout of the siding on a wall. c) A pole used to find the vertical height of a sloping object anywhere along the slope. The pole is notched at measured intervals and a string is attached via a moving ring. To find the height of a point on the slope, one holds the string at the point and moves the story pole end of the string up the story pole until the string is level. At the point where the string is level, the measurement is taken.

Straight split   Shake shingle that is of constant thickness from end to end.

Straight up method   Method of installing shingles. Shingles are installed in vertical rows from eave to peak rather than in horizontal rows from rake to rake. The method is preferred by some roofers, avoided by others.

Straw nail   Long thin nail used to secure tiles at rakes and ridges.

Stripper   Box worn around the chest or waist that dispenses roofing nails in an organized manner. The box separates the nails and hangs them point down. The use of more modern electric and pneumatics nailers have all but done away with the use of the stripper.

Strip seal   a) Self-sealing asphalt shingle. b) Sealer applied in long narrow strips. c) The sealer strip on the front of an asphalt shingle.

Strip shingle   a) Asphalt or fiberglass shingles. b) Any shingling material sold as short strips.

Strips per square   A shingle is considered a strip. Number of shingles per square.

Structural deficiency   A building structure that does not meet code or does not meet roofing manufacturer's requirements as a roof base.

Struts   The diagonal braces of a roof truss.

Stud   The vertical structural members of a wall. Studs are usually 2×4 or 2×6 lumber, 92 5/8 in. in length.

Substrate   The bottom or base material to which another material is applied. Roof decking can be considered the substrate to the roofing felt and the roofing shingles.

Sun angle   The position from horizontal that the sun is in the sky. Generally considered as 38 1/2 degrees in winter, 85 1/2 degrees in summer for the United States.

Surety maintenance bond   (See Bonded roof.) Type of bond or roof warrantee.

Surface cap sheet   (See Cap sheet.)

Surface color   a) The color of a shingle, tile, or roof. Surface color can be generated with mineral granules, oxides, paints, and stains. b) A color that is applied to the surface of an object rather than throughout the object, i.e., a glaze, a paint, a lamination, or an anodizing.

Surface finish   The texture of the surface of a tile or shingle. The texture can be smooth or rough or made to resemble another object such as shake.

T&G sheathing   Roof sheathing consisting of tongue and groove lumber butted together. The T&G lumber can be 1 by, 2 by, or thicker. Typical T&G sheathing uses 2 × 6 lumber.

Tabs, shingle   (See Shingle tabs)

Tail   The thin portion of a wood or shake shingle. The tail is the end applied toward the roof peak and is covered by the exposure edge, the butt of the next course of shingles. One nails through the shingle tail.

Talc topping   A roofing ply or felt that is covered with a thin layer of talc. The talc acts as a lubricant and allows the sheet to move or be moved easily.

Taper system   Built-up roof that is first covered with varying thicknesses of wedge-shaped insulation panels arranged to provide drainage.

Tapered edge strip   (See Cant Strip)

Tapered split   Shake shingle that is wedge shaped. The thicker butt edge is the weather exposure edge.

Tar bucket   A pail used to transport hot tar or asphalt from the heated container to the roof surface material being installed.

Tar & gravel roof   A near flat, built-up roof covered with a layer of asphalt in which gravel is embedded. The gravel forms the top weather surface.

Tarred felts   See asphalt felts for felt material. Tarred felts use coal tar pitch in place of asphalt.

Tear-offs   a) The act of removing an existing roof covering. b) Term for items that must be or have been removed from a roof's surface.

Temporary roof   Two or more plies of precoated base sheet used to moisture proof a built up roof during temporary work stoppage of the installation of the roof.

Tensile stress   The ability of an object not to break or snap when pulled from opposite directions.

Test cuts   Test cuts are sectional cuts from a roof used to find the reasons for the roof's failure. Usually done to a built up roof to study the ply lamination. Test cuts are considered destructive testing and the roof must be repaired afterward. On new roofs, still being applied, the test cuts should be made before the final roof coating is applied.

Testing agency   An agency or company authorized and qualified to make test on materials and material assembly procedures. May or may not be a profit oriented organization.

Thatch look   (See Dutch Weave)

Thermal barrier   Insulation or any material that aids in the blocking of the flow of heat through an item, wall, floor, ceiling, or roof.

Three-tab shingles   Most common asphalt or fiberglass shingles. Each shingle contains three tabs or weather exposure strips. Three-tab shingles have two full cutouts and two half cutouts.

Tie-in   a) The area of a roof that joins another area of the building's roof. b) The act of joining two separate roof areas. c) The act of shingling the one roof area to another.

Tile clips   Specially designed metal clips used to secure tile to the roof sheathing.

Tile pan   J shaped rain runoff pan installed under the tile at the junction of a tile roof and a vertical wall.

Tile profile   The shape of an individual tile as viewed from its weather surface edge.

Tile roof   A roof that is roofed with metal, slate, concrete or clay tiles.

Tilt up construction   A construction method used primarily for commercial buildings. The walls are precast reinforced concrete and lifted, tilted up into position at the building site.

Toe board   A lumber board nailed to the eave of a roof during roofing installation. The board aids the roofer by giving the roofer, his toes and feet a firm place on which to rest. A toe board also will prevent small items such as tools and nails from rolling from the roof.

Toenail   To nail at an angle. One toenails solid roof decking into place by nailing at an angle and through the edge, rather than the face of the board.

Tongue of square   The narrow extension of a carpenter's square is called the tongue of the square.

Top cord   Truss term given to the two boards that act as rafters.

Top plate   (See Plate.) The topmost board of a wall.

Torch sheet or torch ply   Special roof membrane that, when heated with an open flame, will stick to the roof or another roof ply.

Torched asphalt   a) Asphalt that is flame heated as the roofing plies are being rolled into it. b) Asphalt that can be heated directly with an open flame.

Torches   Hand held or dolly held torches used for applying roofing membranes. Flame is developed with the use of oxygen and acetylene gases or with L.P. (liquefied petroleum or propane). Torch should have a shut-off valve, pressure release trigger, pressure regulator and a stand. Pressures vary with type of torch. Detail torch, 15 to 20 PSI. Field torch, 40 to 45 PSI. Wagon Torch, 40 to 60 PSI.

Trapped water   Water or water vapor in or under the plies of a built-up roof. This condition will result in blisters and eventual roof failure.

Treated wood   a) Wood used in wet or ground contact areas or used in contact with concrete or masonry block. Wood is pressure treated with a chemical preservative, usually CCA, chromated copper arsenate. b) Built-up decking made from oil preserved wood. Must be covered with a resin paper before installing roofing plies.

Trim   Boards or milled strips of lumber used to face off or cover exposed ends of rafters and sheathing. May be plain or fancy in design.

Trimmers   Boards used to trim out or form an opening that is smaller than the original opening. Usually used between two rafters or joists to provide a solid mounting or enclosure for skylights, chimneys, etc.

Truss, king post type   A king post truss has a single vertical post, the king post, from the center of the lower cord to the intersection of the two upper cords.

Truss, scissors type   A scissors truss has both top and bottom cord pairs shaped like an inverted V. It resembles an open pair of scissors.

Truss, W type   A W truss has its top cords shaped like an inverted V and its bottom cord horizontal. The Web member between the lower and upper cords is shaped like a W. This is the most common truss for residential construction.

Tuck pointing   Mason term used for describing the act of placing mortar into a joint with the use of a pointed trowel. Usually done during a repair of an item like a chimney.

Turbine vent   A turbine vent is one of many styles of attic vents. The vent turns when wind blows through its blades and this turning motion sucks the moist, hot air out of the attic.

Two-tab shingle   An asphalt or fiberglass shingle that has two weather exposure tabs or strips. Two-tab shingles have one full cutout and two half cutouts.

Tying-in   (See Tie-in)

U-value   Coefficient of heat transmission. U=BTU/HR/FT^/F. Same as C-value except all components of the roof section, including the air spaces, are included.

UBC code   Uniform Building Code published by the I.C.B.O. The official guide used by contractors and building departments. The UBC gives specifications for materials and their use for all forms of building construction. The UBC is complicated to use and understand by many, thus it has been rewritten and published in simpler form by many publishers.

UL   The abbreviation for Underwriters Laboratories, a highly respected, nonprofit, private testing agency that tests materials and products for a fee.

UL class   a) The fire rating given to roofing materials by the Underwriters Laboratories. b) The wind resistance rating given to roofing materials by the UL Labs.

Umbrella   Cone shape flashing used to direct water off an item. Typically used on legs of roof equipment supports to direct rain away from the mounting bolts.

Undercoating   An approved spray insulation or fireproofing material applied to the bottom side of a steel deck. Class II decks can be made into Class I decks with approved undercoating. Coatings are applied from 1/8 to 1-1/4 in. thick depending on coating material and codes.

Underlayment or underlay   A material, usually roofing felt, that is applied directly to the roofing deck. The underlayment protects the deck from moisture that may enter under the overlying shingles or tile.

Upper cord or top cord   The uppermost lumber in a truss is called the top cord. It replaces a rafter.

Urethane or polyurethane insulation   Approved plastic foam used for insulation sheets on roof decks. Covered by Federal Specification HH-I-1972/2.

Utility knife   A small handled knife with a retractable blade. Uses single edge razor blades and has storage for extra blades in the handle.

Valley   V shaped intersection of two roofs that collects rain water and guides it to the gutter system.

Valley, metal open   A valley where the metal flashing is exposed to the weather.

Valley jack rafter   The rafters that extend from the center of a valley, the valley rafter, to the ridge board.

Valley rafter   The rafter that extends up the center of the valley from the eave to the peak.

Vapor barrier   A material that prevents the passage of water or water vapor through it. Vinyl plastic, aluminum foil, Kraft paper, asphalt felt, asbestos felt, and a laminated combination of these materials are considered vapor barrier materials.

Vapor holes   These are holes used to expel water vapor from a building's walls.

Vapor retarder   (See vapor barrier.) Vapor barrier sheets are required on built-up roofs in areas where the average January temperature is 45 degrees Fahrenheit or less.

Velocity pressure   Pressure exerted by the wind on an object at a set wind speed.

Vent hood   A hood or cap placed over a vent to allow fumes and gases to escape but disallows rain, dirt, bird's, and small animal's entry.

Vent pipe   A pipe that vents fumes or gases into the atmosphere. Usually from 1 to 4 in. in diameter and made from clay, lead, or plastic. The most commonly used plastic is PVC or Poly vinyl chloride. PVC is white or black in color (check local codes for use).

Vent ply   A built-up roofing ply sheet that is ribbed. The channels created by the ribs help to expel hot moist air, trapped under the roof plies, into the atmosphere.

Vent valve stack   a) A pressure release valve that is normally in its off position until the pressure of a container builds up to a dangerous point. Then the valve opens, releasing the pressure. b) A pressure release valve installed in the roof deck to relieve pressure build up under the roofing plies.

Ventilation   The act of venting or expelling vapors from an enclosed area to a non-enclosed area.

Ventilator   Any mechanical device that expels unwanted vapors or fumes from an enclosed area to a nonenclosed area.

Vents   Pipes protruding through the roof surface. Vents expel gases generated by stoves, heaters, and plumbing fixtures into the atmosphere. Attic and room vents expel moisture and hot air into the atmosphere.

Very steep or ht-steep roof   Considered as a built-up roof having an incline of 3 to 6 in. per linear foot.

W-truss   (See Truss, W.)

W-valley   A valley that is made with 'W' flashing. (See Flashing, W.)

Warranty   A short-term agreement to fix something sold, installed, or repaired at little or no cost to the new owner.

Wash down   The act of using a hose or power washer to clean a roof surface or a building's siding. The washing action takes place from the top down so not to force water up and under the siding or roofing.

Waste   Portions of tile or shingle that have been cut from a usable tile or shingle and are no longer usable.

Water absorption   The ability or lack of ability of an item to absorb water or moisture.

Water guard   Part of a valley or wall flashing that has been turned up and forms a dam to, or a channel for, water flow.

Water trough   A valley in which the shingles have been cut back, forming a trough for water flow.

Water vapor transmission   The passing of water in its vapor condition through an object.

Waterproofing   The act of making an item resistant to the entry of, or passing of, water.

Web member   The lumber braces of a roof truss that are located between the lower and upper cords.

Weight, shingle   The weight of a square (100 sq ft), of the shingles applied to a roof. The most common shingle weight is 235 lb.

Welder, roofing   A person qualified to weld metal roofing components.

Wide crown staple   A staple that is wider than 3/4 in. Wide crown staples are sold in widths of 3/4 to 1-1/2 in.

Wind direction   The direction from which the prevailing wind comes, which is important when installing ridge shingles. Ridge shingles should be installed so that their weather edge is facing away or downwind of the prevailing winds.

Wind isotach™   Line on a wind speed map that encircles an area of specific wind velocity. Number on the line is the wind speed or isotach number to be used with a velocity pressure chart to find the wind uplift factor.

Wind resistance   The resistance of the roof or its shingles to being lifted or torn off by the prevailing or expected maximum winds.

Wind speed map   Map of the United States or another country, showing basic wind speeds. Used to figure out the type of roofing and roof installation methods to be used for a given area. Maps may be obtained from the American National Standards Institute, 1430 Broadway, N.Y., N.Y.

Wind uplift   The force exerted by the wind in an upward direction. A roof must withstand being blown off by wind uplift.

Wood fiber deck   A deck that is sheathed with wood.

Woven valley   A full- or half-lace valley.

Zipper tool   A tool used to release or unzip one piece of vinyl or aluminum siding from another.

Index